J. Schmelzer, G. Röpke, R. Mahnke

Aggregation Phenomena in Complex Systems

WILEY-VCH

J. Schmelzer, G. Röpke, R. Mahnke

Aggregation Phenomena in Complex Systems

WILEY-VCH

Weinheim · New York · Chichester · Brisbane · Singapore · Toronto

Dr. habil. Jürn Schmelzer
Prof. Dr. Gerd Röpke
Priv. Doz. Dr. Reinhard Mahnke
Rostock University
Department of Physics
Universitätsplatz 3
D-18051 Rostock
Germany

Library of Congress Card No. applied for.

A catalogue record for this book is available from the British Library.

Die Deutsche Bibliothek – CIP-Einheitsaufnahme
Schmelzer, Jürn:
Aggregation phenomena in complex systems / J. Schmelzer ; G. Röpke ; R. Mahnke. –
1. Aufl. – Weinheim ; New York ; Chichester ; Brisbane ; Singapore ; Toronto :
Wiley-VCH, 1999
ISBN 3-527-29354-X

Printed on acid-free and chlorine-free paper.

Printing: betz druck, D-64291 Darmstadt
Bookbinding: J. Schäffer, D-67269 Grünstadt
Printed in the Federal Republic of Germany

Preface

The present book is devoted to problems of a highly important class of physical processes, the kinetics of phase transformation and aggregation phenomena. We tried to give an overview both of basic fundamental concepts as well as of possible applications. As will be shown, the basic theories are applicable equally well, with certain modifications, to problems of extremely different physical or chemical nature.

The mentioned general attitude – the concentration on fundamental aspects, the broad scope of the discussion ranging from the introduction into basic ideas to the most recent developments – distinguishes, as we hope, the present book from any other of the existing monographs devoted to similar problems.

The realization of such program is, of course, a difficult task. Of enormous help in this respect was the long–standing cooperation with colleagues from different research institutes. In particular, we would like to mention here the common work with colleagues from the Bogoliubov Laboratory of Theoretical Physics of the Joint Institute for Nuclear Research in Dubna, Russia (D. V. Shirkov, V. B. Priezzhev, V. O. Nesterenko), from the Institute of Physical Chemistry of the Bulgarian Academy of Sciences in Sofia, Bulgaria (R. Kaischew, I. Gutzow, A. Milchev, R. Pascova, I. Avramov, D. Kashchiev), the Institute of Physics and Technology in Kharkov, Ukraine (V. V. Slezov), the Department of Theoretical Physics of the University of Riga, Latvia (V. N. Kusovkov). At part, this cooperation is directly reflected in the present monograph in the contributions of A. Milchev, V. N. Kusovkov, V. B. Priezzhev and V. V. Slezov (Slyozov).

Further, we would like to acknowledge the broad discussions during research visits and joint workshops and seminars like the research workshops on Nucleation Theory and Applications in Dubna and other similar meetings. Here we had the possibility and pleasure, to discuss the mentioned topics with a number of colleagues from a variety of institutes beyond the circle mentioned above. In particular, we would like to express our gratitude, among others, to A. P. Grinin, A. K. Shechekin, V. M. Fokin and S. A. Kukushkin (St. Petersburg), Yu. L. Klimontovich, B. M. Smirnov, G. T. Guria and A. Lushnikov (Moscow), V. G. Baidakov (Ekaterinenburg), D. S. Sanditov (Ulan–Ude), Zd. Kozisek and P. Demo (Prague).

In the process of formation of the ideas outlined in the present book we could benefit from further numerous discussions with colleagues at our department, at meetings, colloquia and conferences, from common efforts to find the most appropriate solutions in interpreting and understanding the kinetics of phase transformation and aggregation phenomena. We are indebted to many people who helped us in this way in writing of the book. Here we would like to mention here especially W. Ebeling and L. Schimansky – Geier (Berlin), M. Schlanges (Greifswald), R. Strey (Köln) and H. Baumgartl (Ludwigshafen) who supported our work in many respects.

The authors have been working for years on the problems the book is devoted to. Thus it was natural that many of our own results are reflected in it. Many of the results, outlined in the book, were obtained with the help of colleagues and coworkers. We would like to express our gratitude to all of them. Here we have to mention, in particular, Dr. J. Bartels, Dr. A. Budde, Dr. D. Labudde, F.–P. Ludwig, Dr. J. Möller, Dr. Dr. F. Schweitzer, Dr. H. Tietze and B. Seifert. It is a particular pleasure to acknowledge also the contributions of J. Schmelzer jn., F. Röpke and M. Mahnke.

As already mentioned, the present book is the result of long–standing cooperation between the authors and a number of colleagues working in different research institutes. Such a cooperation could only be realized with the financial support the authors received from different organizations. In particular, we would like to express our gratitude to the Deutsche Forschungsgemeinschaft (Dr. H. Leutner, Dr. D. Schenk), the Bundesministerium für Forschung und Technologie (BMBF), Germany, which supported the common research in the framework of the grants 06RO140 (1) and 06RO745 (0), the Heisenberg – Landau and other specialized programs, as well as to the Russian National Science Foundation, for giving financial support for a number of common research projects including the present monograph. Two of the authors (J.S.; G.R.) would like to express their particular gratitude to the Laboratory of Theoretical Physics of the Joint Institute of Nuclear Research where a major part of the present monograph has been completed.

Last but not least it is a pleasure for us to mention and acknowledge the patience, the understanding and support of our families we could enjoy in the course of writing the present book.

Rostock (Germany) and Dubna (Russia) The Authors

Contents

1 Introduction

First–order phase transformations play a major role in a variety of technological processes and scientific investigations. As some examples here only segregation processes in solid or liquid solutions [187, 308, 309, 310] and solid–to–solid phase transformations [225], bubble formation (cavitation) in liquids [56, 238, 316, 317], condensation processes in the atmosphere [273] or in expanding gases [189, 518], clustering in expanding nuclear matter [32, 105, 378, 482], phase transformations in the early Universe [230], jam formation in vehicular traffic on highways [102, 304, 496, 570] are mentioned. The theoretical interpretation of these processes is therefore of outstanding interest in very many fields of science and technology.

Different aspects of phase transformation processes have been investigated for a long time (an overview over these early attempts is given for example in Volmer's monograph [556]). However, a systematic theoretical treatment of these phenomena was started only in the last century. It is connected, among others, with the names of Young, Laplace, Kelvin, Gibbs and van der Waals. These first attempts were directed mainly to the interpretation of static (thermodynamic) aspects of these processes.

From such historical perspective, the work of Gibbs [170] may be connected with the formulation of the cluster concept in the description of phase transitions. However, in its basic premises Gibbs's method is quite fundamental allowing to give a theoretical interpretation of any kind of heterogenous multi–phase systems.

As an alternative to Gibbs's approach, van der Waals [548, 549] developed a theory of heterogenous systems based on the concept of a continously changing density and composition of the systems under consideration. Latter method turned out in the further development to be the basis for a microscopic statistical – mechanical determination of the properties of inhomogeneous systems [458, 459].

As the next step, different approaches to a kinetic description of phase transitions and aggregation phenomena were anticipated. Important contributions to this field have been developed, again among others, by Einstein, Smoluchowski, Volmer and Weber, Becker and Döring, Stranski and Kaischew, Frenkel and Zeldovich. In their work, the foundations of the classical theory of nucleation, growth and aggregation phenomena have been laid. In its basic premises, the classical theories remain till now, with extensions and generalizations (see, e.g., Refs. [39, 180, 187, 584, 585]), the major tools in the interpretation of experimental results. In contrast to alternative microscopic developments, these approaches have – in application to experimental results – the major advantage that properties of the newly evolving phases are expressed via directly measurable macroscopic properties of the different phases involved.

Presently, a large amount of work is directed to a more precise determination of the properties of small aggregates by applying statistical mechanics, quantum chemical and computer simulation (molecular dynamics, cellular automata and Monte Carlo) methods. One of the most promising new directions in this field is hereby the proper account of the effect of the medium, where clusters are formed, on their properties (in–medium effects). Here, in particular, the methods of quantum statistical physics may be of importance for an accurate description of the properties of clusters. Bound states as quantum systems are well understood. However, in a medium the properties of such states are changed, as demonstrated here for nuclear systems. More difficulties are connected with the proper determination of the non–equilibrium properties including the reaction rates of such and more complex states in a dense medium.

A major impetus, the theoretical description of aggregation phenomena may get generally from the newly developing field of cluster physics [29, 440, 533].

A brief review of milestones in the historical development of the methods of description of cluster growth and aggregation is incomplete without mentioning the work of Lifshitz and Slezov [310, 508] on the kinetic description of the asymptotic stages of the evolution of ensembles of clusters known as coarsening or Ostwald ripening. In addition to its power in the interpretation of the kinetics of phase transitions, this theory gave a further impetus for the search of power laws and self–similarity in the theory of phase transformation processes, in particular, and physics, in general. As mentioned by S. Weinberg [567] "*... physicists have become used to finding the most essential properties are often expressed in terms of laws that relate physical quantities to powers of other quantities ...*"

Simultaneously to the cluster concept in the description of the kinetics of phase transformations, also the ideas of van der Waals found a further development in the kinetic description of phase transitions. Here, in particular, the work of Landau and Ginsburg, Hillert, Cahn, Hilliard and Cook has to be mentioned. In such a picture, the evolution of the new phase is governed by the amplification of fluctuations of the respective order parameters which are small in size but extended to relatively large regions in space compared with the typical sizes of critical clusters in the nucleation – growth model. The kinetics of this process is described in a first approximation by the Cahn – Hilliard – Cook equation [78, 79, 99]. A large amount of work is devoted to the analysis of this approach, possible generalizations and extensions.

Both the nucleation – growth model as well as the model of spinodal decomposition can be considered as limiting models in the description of phase formation and aggregation phenomena. They have the highly valuable property to be analytically tractable. Despite their limitations, they may serve thus as guides in the interpretation of experimental results. Hereby both at a first glance alternative pictures are united more and more. As an example, density functional approaches are employed nowadays widely (Reiss [439], Oxtoby [408]) to determine more accurately the work of critical cluster formation.

Additional tools for an evaluation of the limits of validity of the limiting models and the variety of structure formation processes in aggregation are given by computer simulation methods. Here, by an analysis of simplified model systems, conditions for the validity of limiting models are established and more general ideas concerning the course of first–order phase transitions, like the generalized cluster model by Binder [40], may be tested.

As it turns out from this brief and necessarily incomplete overview, the proper description of phase formation and aggregation is a highly complex task. This statement is true already if the boundary conditions for the systems under consideration are fixed, i.e., if the external conditions do not change with time. However, in a number of cases, like in clustering in expanding gases, the thermodynamic parameters of the ambient phase vary with time already in the absence of clustering processes. Aggregation phenomena are affected, on one hand, by such variations. On the other hand, they determine widely the further evolution of the state of the system. This way, aggregation processes and the evolution of the system as a whole have to be described adequately in a self–consistent way.

It is the aim of the present book, to give an overview over some of the basic methods of analysis of phase transformation and aggregation phenomena and the variety of possible applications.

In the first part (I. General concepts), the monograph introduces the reader into the physics of such highly non–linear, complex phenomena and the basic methods of their description. Hereby the outline of different analytical theories (nucleation and growth, spinodal decomposition etc.), their merits and limitations, is supplemented by an overview on computer simulation methods (stochastic approaches, Monte Carlo methods, cellular automata models) of aggregation phenomena. Due to their particular importance, separate chapters are devoted to the concepts of self–similarity and self–organized criticality and to a critical analysis of the methods of chemical reaction kinetics employed widely in nucleation – growth and aggregation models.

In a second part (II. Selected applications), the basic theories are extended and applied to different specific processes (determination of the number of clusters formed in nucleation – growth processes, nucleation and growth in expanding matter, multifragmentation in nuclear collisions, evolution of the element size distribution in the early Universe, segregation in porous materials, spinodal decomposition in adiabatically isolated systems, aggregation in traffic flow). Hereby the interplay between aggregation and response of the system, where such processes take place, increases even more the complexity of the problems under investigation and the variety of spatio–temporal patterns which may evolve.

This way, the monograph may serve as an introduction to the theory of nucleation, growth and aggregation processes as well as a source of new results in the highly exciting and rapidly developing field of the analysis of complex systems.

Part I

General Concepts

2 Classical Theories of Phase Transformation Processes: Nucleation, Cluster Growth, Coarsening, and Spinodal Decomposition

J. Schmelzer

2.1 Classical Nucleation Theory

2.1.1 Introduction

Classical nucleation theory was developed in the 1920s and 1930s by a number of scientists. First of all the names Farkas [139], Volmer and Weber [556], Volmer [557], Kaischew and Stranski [228], Becker and Döring [26], Frenkel [158] and Zeldovich [579] have to be mentioned. As noted by Farkas, the basic kinetic model, underlying classical nucleation theory, was proposed by L. Szilard.

The analysis was initially directed mainly to the case of droplet formation in a one–component vapor (Farkas [139]). Kaischew and Stranski [228] investigated, as early as 1934, the formation of crystals from vapors. The first derivation of nucleation kinetics for the case of crystallization of an undercooled melt was also developed relatively early by Volmer and Weber [556], while Reiss [438] was the first to apply the outlined ideas to nucleation in multi–component systems.

A summary and thorough discussion of earlier attempts at determining nucleation rates for different systems may be found in Volmer's and Frenkel's monographs [158, 557]. Further developments are summarized in the monographs by Hirth and Pound [209], Zettlemoyer [584, 585], the recent review article by Mutaftschiev [379] and with particular emphasis to phase formation processes in glasses and glass–forming melts by Gutzow, Schmelzer [187].

The classical theory of nucleation is, with modifications, still the most widely applied tool for the interpretation of nucleation processes in many fields. Therefore we will start the discussion of nucleation with an outline of its basic ideas.

Here we will develop the theory taking as an example precipitation in a quasibinary solid or liquid solution at constant values of pressure p and temperature T, when only one of the components segregates to form clusters of the newly evolving phase. Generalizations can be carried out straightforwardly.

In classical nucleation theory, a spatially homogeneous system is considered (*assumption 1*), where, in the simplest case, particles of one of the components (atoms, molecules) aggregate to form clusters of the new phase. Moreover, it is supposed that clusters B_j consisting of j ambient phase particles (also denoted as monomers, molecules, building units) grow and decay by addition or evaporation of monomers B_1 only according to the scheme (*assumption 2*)

$$B_1 + B_j \rightleftharpoons B_{j+1} \tag{2.1}$$

as in a binary chemical reaction. Such an assumption is quite reasonable for a wide range of applications, since the number of monomers exceeds, by many orders of magnitude, the concentration of clusters with $j > 1$. Monomers have, moreover, the highest mobility. Reactions of other types are excluded in this scheme.

At the advanced stages of cluster formation and for highly mobile clusters, the probability of collisions between clusters of equal or different sizes is, in general, not equal to zero and assumption 2 has to be replaced by a more general scheme involving in addition reactions of the type $B_i + B_j \rightleftharpoons B_{i+j}$. In these cases, relations similar to Smoluchowski's coagulation equations describe the kinetics of the transformation (von Smoluchowski [519, 520]; cf. also [39, 40, 237]). In phase transformation processes proceeding in solids, in general, and glass–forming melts, in particular, the low mobility, respectively, high viscosity of the system excludes such reaction paths to a large extent.

With assumption 2 the change of the number of clusters per unit volume $N(j,t)$, consisting of j monomers, with time t is connected with two possible reaction channels, with two processes of the form as given with Eq.(2.1) involving (B_{j-1}, B_j, B_1) and (B_j, B_{j+1}, B_1), respectively. The basic equations for the kinetic description of these processes are given by (*assumption 3*)

$$\frac{dN(j,t)}{dt} = J(j-1,t) - J(j,t) \quad \text{for} \quad j \geq 2 \tag{2.2}$$

with

$$J(j,t) = w^{(+)}(j,t)N(j,t) - w^{(-)}(j+1,t)N(j+1,t) \, . \tag{2.3}$$

Here $w^{(+)}(j,t)$ is the average number of monomers which is incorporated into a cluster of size j per unit time, while the coefficients $w^{(-)}(j,t)$ describe similarly the rate of decay processes.

Equation (2.2) is implicitely subject to the assumption that the state of the cluster is determined solely by the number of monomers contained in it (*assumption 4*: equilibrium shape of clusters). This assumption is motivated by the argument that a cluster rapidly goes over into the equilibrium shape and structure corresponding to the respective monomer number contained in it, representing thus the only configuration which has to be taken into account.

An investigation of the kinetics of nucleation, considering non–equilibrium configurations of the clusters and the resulting additional reaction paths, may be found, e.g., in the following references (Ziabicki [588]; Kaischew, Stoyanov [229], Binder [42]; for more recent generalizations see, e.g., Oxtoby et al. [408] and Reiss et al. [439]).

The kinetic coefficients $w^{(+)}$ and $w^{(-)}$ have to be determined based on the analysis of the growth and decay kinetics of the clusters which may differ in dependence on the particular system considered, while the general equations (2.2) – (2.3) remain the same so far as the assumptions 1 – 4 are fulfilled.

Once the kinetic coefficents $w^{(+)}$ and $w^{(-)}$ are known, the evolution of the cluster size distribution and related quantities can be determined numerically (see, e.g. Bartels [21]; Bartels et al. [22]; Mirold and Binder [349]; Olson and Hamill [398], Schmelzer et al. [482, 483]; Wehner and Wolfer [565]). However, an analytic approach is also possible and this was the way the

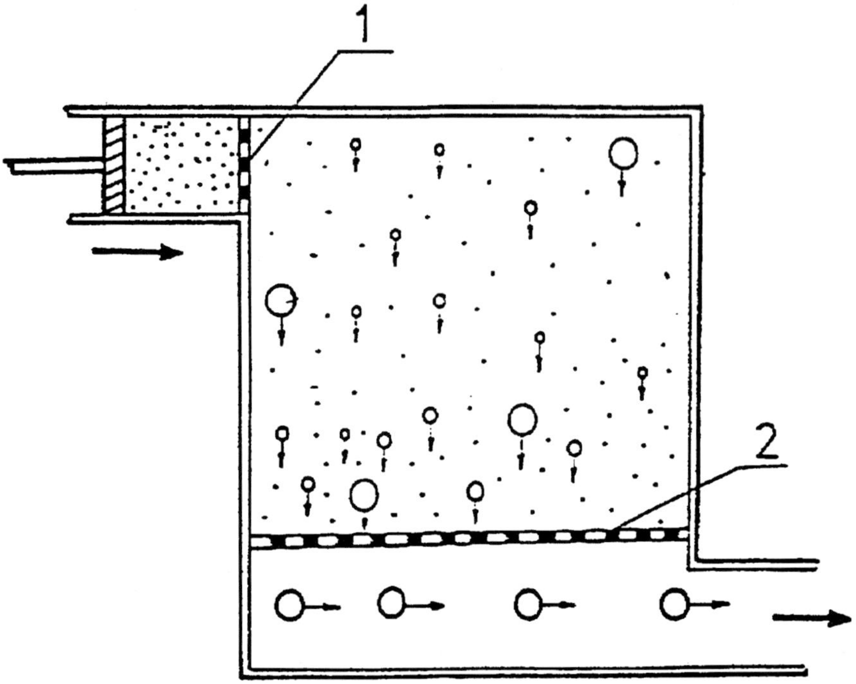

Fig. 2.1 Schematic representation of Szilard's model used in the derivation of the classical expression for the steady–state nucleation rate here shown for the process of vapor condensation. In an isothermal chamber B, connected with a reservoir of ambient phase particles A, the process of condensation takes place and a population of subcritical and supercritical liquid clusters is formed. The clusters with particles numbers $j \geq g \gg j_c$ are removed from the chamber instantaneously via the semipermeable grate *(2)*. It is inpenetrable for clusters with sizes $j < g$. Simultaneously an equivalent number of monomers is assumed to enter the system through the membrane *(1)*, which is inpenetrable for clusters with monomer numbers $j > 1$. In such a way, a constant supersaturation is sustained in the system and a time–independent nucleation rate may be established.

theory of nucleation was originally developed. The analytical results are obtained for the case that the state of the system is not changed in the course of nucleation, or in other words, if the supersaturation remains constant during the process (*assumption 5*).

In classical nucleation theory this situation is realized by using the following model proposed by Szilard (see Fig. 2.1; Becker, Döring [26]; Becker [28]). It is assumed that in a certain volume of the ambient phase clusters are formed by processes of the type as expressed through equations (2.1) – (2.3). Once a cluster reaches an upper limiting size $j = g \gg j_c$ it is removed from the system and g monomers are added to it (*assumption 6*).

Consequently, the condition

$$N(j,t) = 0 \qquad \text{for} \qquad j \geq g \gg j_c \tag{2.4}$$

is always fulfilled in the model system.

Moreover, since the number of monomers is conserved, in addition, the relation

$$N(1,t) + \sum_{j=2}^{g-1} jN(j,t) \equiv N = \text{ constant} \tag{2.5}$$

holds.

Starting with a distribution of monomers only, after a certain time interval τ_1 a time–independent steady–state distribution with respect to cluster sizes is established in the system. In the classical theory the steady–state is assumed to be established immediately (*assumption 7: steady–state approximation*).

As an intermediate step in the development of the classical theory also the so–called equilibrium distribution of clusters $N^{(e)}(j)$ is determined. This distribution is calculated based on the following additional assumptions (*assumptions 8–10*, see Frenkel [158]) that

(i) the ensemble of clusters in the matrix can be considered as a perfect solution (similar to a perfect mixture of gases if vapor condensation is discussed),

(ii) the equilibrium distribution $N^{(e)}(j)$ corresponds to a restricted minimum of the Gibbs free energy G, for which the constraints (2.1), (2.4) and (2.5) have to be fulfilled,

(iii) the number of monomers aggregated in clusters with sizes $j \geq 2$ is small compared with their total number N.

Based on these assumptions this distribution may be obtained in the form (Frenkel [158]; Demo, Kosizek [111]; Mutaftschiev [379])

$$N^{(e)}(j) = N\left[\exp\left(-\frac{\Delta G(j)}{k_B T}\right)\right] . \tag{2.6}$$

This expression is used as a reference state for the determination of the kinetic coefficients $w^{(-)}$.

$\Delta G(j)$ is the change of the Gibbs free energy if a cluster consisting of j monomeric units is formed in the system. In classical theory, $\Delta G(j)$ is commonly expressed in the form

$$\Delta G(j) = -j\Delta\mu(p,T) + \sigma A\,, \quad \Delta\mu(p,T) = \mu_\beta(p,T) - \mu_\alpha(p,T)\,. \tag{2.7}$$

Here $\Delta\mu$ is the difference of the chemical potential of the segregating particles in the ambient phase (specified by the subscript β), respectively, the newly evolving phase (specified by α) both taken at a pressure p and temperature T. σ is the specific interfacial energy (surface tension), taken in classical approaches as equal to the respective value for the case of an equilibrium coexistence of both phases at a planar interface (*assumption 11: capillarity approximation*). A is the surface area of the cluster.

The critical cluster radius R_c corresponding to the maximum of ΔG in dependence on the cluster radius R is given by

$$\frac{\partial \Delta G(j)}{\partial R} = 0\,, \qquad R_c = \frac{2\sigma}{c_\alpha \Delta\mu} . \tag{2.8}$$

In Eq.(2.8), c_α is the concentration of the segregating particles in the newly evolving phase.

A detailed discussion of thermodynamic aspects of nucleation including a possible account of a curvature dependence of surface tension in the calculation of the work of formation of clusters of the newly evolving phase can be found in the monographs by Ulbricht et al. [544], Gutzow and Schmelzer [187] and in a recent paper by Schmelzer et al. [481].

In the outline of the basic ideas and results of classical nucleation theory we will use here an alternative approach as developed by Slezov, Schmelzer [510, 516, 517] and Slezov, Schmelzer and Tkatch [511]. This approach allows to avoid some of the controversial points of the classical approach, in particular, the derivation of the so-called equilibrium distribution (cf. also Schmelzer et al. [482, 483]) and the application of the principle of detailed balancing in the determination of the emission rates $w^{(-)}$. Moreover, a number of results are obtained beyond the scope traditionally followed in the classical theory, its modifications and generalizations.

2.1.2 Determination of the Kinetic Coefficients

While the attachment rates $w^{(+)}$ may be determined more or less easily based on a macroscopic approach (see e.g. Bartels et al. [22]; Slezov, Schmelzer [515, 510, 517]; Gutzow, Schmelzer [187], Schmelzer et al. [483] and the subsequent discussion of growth phenomena) the calculation of the rate of detachment of monomers from the cluster requires, in principle, microscopic considerations.

Though first attempts for a microscopic determination of the detachment rates have been formulated in recent years (Narsimhan, Ruckenstein [388]) the most common approach till now remains the application of the principle of detailed balancing (*assumption 12*, see, e.g., Yourgreau et al. [577]). To demonstrate how this principle can be applied, we consider first cluster formation in a thermodynamic equilibrium state. For thermodynamic equilibrium states (when $\Delta\mu < 0$ holds), Eq.(2.7) is equivalent to a Boltzmann or Gibbs type cluster size distribution evolving by thermal fluctuations in an equilibrium state.

According to the principle of detailed balancing in an equilibrium state the fluxes in each of the reaction channels are in balance (cf. Eqs.(2.1)). This condition yields

$$\begin{aligned}
&w^{(+)}(1)N^{(e)}(1) - w^{(-)}(2)N^{(e)}(2) = 0\\
&w^{(+)}(2)N^{(e)}(2) - w^{(-)}(3)N^{(e)}(3) = 0\\
&w^{(+)}(3)N^{(e)}(3) - w^{(-)}(4)N^{(e)}(4) = 0\\
&\qquad\ldots\\
&w^{(+)}(j)N^{(e)}(j) - w^{(-)}(j+1)N^{(e)}(j+1) = 0\,.
\end{aligned} \tag{2.9}$$

Here the reference distribution $N^{(e)}(j)$ was applied being well-defined for thermodynamic equilibrium states (cf. Eqs.(2.2) and (2.3)).

As a consequence from these equations we obtain with Eq.(2.6)

$$\frac{w^{(+)}(j)}{w^{(-)}(j+1)} = \frac{N^{(e)}(j+1)}{N^{(e)}(j)} = \exp\left\{-\frac{\Delta G(j+1) - \Delta G(j)}{k_B T}\right\} . \tag{2.10}$$

As pointed out by Katz et al. [240, 241, 242, 243, 244], Slezov and Schmelzer [510], in application to cluster formation in thermodynamically unstable states Eqs.(2.10) both the introduction of the so–called equilibrium distribution and the determination of the emission rates from such distributions by using the principle of detailed balancing for clusters of supercritical sizes become highly questionable (see also Schmelzer et al. [482, 483], chapter 10).

However, as shown by Slezov and Schmelzer [510, 517], a relation similar to Eqs.(2.10) can be obtained without the application of the principle of detailed balancing and the introduction of the so–called equilibrium distribution with respect to cluster sizes. This relation reads

$$\frac{w^{(+)}(j)}{w^{(-)}(j+1)} = \exp\left[\frac{\mu_\beta - \mu_\alpha(j+1)}{k_B T}\right] . \tag{2.11}$$

μ_β is the chemical potential of a primary building unit in the ambient phase, while $\mu_\alpha(j+1)$ is its value in a cluster of size $(j+1)$.

This relation is valid not only for one–component but also for multi–component systems, when clusters with a given stoichiometric composition are formed (see, for example, [510, 517] and also Slezov [512]). Denoting by ν_i the stoichiometric coefficients describing the composition of the building units of the cluster, we have in this general case

$$\mu_\beta = \sum_i \mu_{i\beta}\nu_i . \tag{2.12}$$

By $\mu_{i\beta}$ the chemical potentials of the different components in the solution are denoted.

The difference of the Gibbs free energies $\Delta G(j)$ between the heterogeneous (one cluster consisting of j primary building units in the ambient phase) and the homogeneous initial state can be expressed then also as

$$\Delta G(j) = j\left[\mu_\alpha(j) - \mu_\beta\right] . \tag{2.13}$$

We obtain further

$$\begin{aligned} \Delta G(j+1) - \Delta G(j) &= \\ &= (j+1)\left[\mu_\alpha(j+1) - \mu_\beta\right] - j\left[\mu_\alpha(j) - \mu_\beta\right] , \end{aligned} \tag{2.14}$$

respectively,

$$\Delta G(j+1) - \Delta G(j) \cong \mu_\alpha(j+1) - \mu_\beta . \tag{2.15}$$

In this way, we arrive at

$$\mu_\beta - \mu_\alpha(j+1) = -\left[\Delta G(j+1) - \Delta G(j)\right] . \tag{2.16}$$

By a substitution of this expression into Eq.(2.11) we get, finally,

$$\frac{w^{(+)}(j)}{w^{(-)}(j+1)} = \exp\left\{-\frac{\Delta G(j+1) - \Delta G(j)}{k_B T}\right\} . \tag{2.17}$$

Now, if one introduces an auxiliary function $N^{(0)}(j)$ as

$$N^{(0)}(j) = \left[\exp\left(-\frac{\Delta G(j)}{k_B T}\right)\right] \tag{2.18}$$

we may write also for thermodynamically unstable states independent of cluster size

$$\frac{w^{(+)}(j)}{w^{(-)}(j+1)} = \frac{N^{(0)}(j+1)}{N^{(0)}(j)} = \exp\left\{-\frac{\Delta G(j+1) - \Delta G(j)}{k_B T}\right\} . \tag{2.19}$$

In this way, it turns out that $N^{(0)}$ is widely identical to $N^{(e)}$, however, without assigning the meaning of a cluster size distribution in a real or artificial model system to this quantity.

The approach sketched above can be generalized straightforwardly to nucleation processes in multi–component systems and to arbitrary boundary conditions. The details of such generalization can be traced in Ref. [517].

2.1.3 Continuums Approximation: The Frenkel–Zeldovich Equation

For an analytical description of the basic characteristics of the process of phase formation we go over from the set of kinetic equations (2.2) and (2.3) to a continuous description in form of a Fokker–Planck equation (see e.g. Slezov, Schmelzer [510, 517]; Gutzow, Schmelzer [187]). In doing so we make the following replacement

$$N(j,t) \Rightarrow f(j,t) , \qquad N(1,t) \Rightarrow c . \tag{2.20}$$

Here and in the subsequent derivations, by c the concentration of the segregating particles in the ambient phase is denoted.

In a continuums approximation the evolution of an ensemble of clusters formed by nucleation and growth processes can be described by an equation of the form (see, e.g., [235, 510, 517, 511])

$$\frac{\partial f(j,t)}{\partial t} = \frac{\partial}{\partial j}\left\{w^{(+)}(j,t)\left[\frac{\partial f(j,t)}{\partial j} + \frac{f(j,t)}{k_B T}\frac{\partial \Delta G(j)}{\partial j}\right]\right\} \tag{2.21}$$

denoted commonly as Frenkel–Zeldovich equation. $f(j,t)$ is the cluster size distribution function obeying the condition

$$\int_0^\infty f(j,t)dj = N^{(tot)} , \tag{2.22}$$

where $N^{(tot)}$ is the total number of clusters per unit volume in the system with sizes $j > 1$.

Equation (2.21) retains the same form also for more complex cases of phase formation like bubble formation, crystallization and nucleation–growth in multi–component solutions of a new phase with a given stoichiometric composition (see, e.g., Gutzow, Schmelzer [187]; Slezov, Schmelzer [510, 517]; Slezov [512, 513, 515]).

The change of the cluster size distribution function $f(j,t)$ with time is connected with the flux $J(j,t)$ in cluster size space by

$$\frac{\partial f(j,t)}{\partial t} = -\frac{\partial J(j,t)}{\partial j} , \tag{2.23}$$

$$J(j,t) = -w^{(+)}(j,t)\left\{\frac{\partial f(j,t)}{\partial j} + \frac{f(j,t)}{k_BT}\frac{\partial \Delta G(j)}{\partial j}\right\} . \tag{2.24}$$

Equation (2.21) has the same structure as the relation describing the macroscopic deterministic flow as well as diffusion processes of particles characterized by a volume concentration c [305], i.e.,

$$\frac{\partial c}{\partial t} = -\nabla\left\{[c(\vec{r},t)\vec{v}] - D\nabla c(\vec{r},t)\right\} . \tag{2.25}$$

Here $\vec{v}$ is the macroscopic (hydrodynamic) velocity of the particles while D is the diffusion coefficient, connected with the diffusive motion of the considered component in real space.

It follows that the deterministic (macroscopic) velocity of motion in cluster size space $v(j,t)$ is given by

$$v(j,t) = -w^{(+)}(j,t)\left\{\frac{1}{k_BT}\frac{\partial \Delta G(j)}{\partial j}\right\} \tag{2.26}$$

while the diffusion coefficient in cluster size space equals $w^{(+)}(j,t)$.

2.1.4 Basic Thermodynamic Relationships

The surface area A of the cluster, assumed to be of spherical shape with a radius R, can be expressed through the number j of particles in the cluster by

$$A = 4\pi R^2 = 4\pi j^{2/3} \left(\frac{3\omega_s}{4\pi}\right)^{2/3} . \tag{2.27}$$

Here the notation

$$\omega_s = \frac{4\pi}{3} a_s^3 \tag{2.28}$$

is introduced. ω_s is the volume of a monomeric building unit of the segregating particles in the newly evolving phase, a_s their linear size. Moreover, the relation

$$\left(\frac{4\pi}{3\omega_s}\right) R^3 = j \tag{2.29}$$

is used.

In terms of the number of monomers in the cluster, Eq.(2.7) can be reformulated thus to give

$$\Delta G(j) = -j\Delta\mu + \alpha_2 j^{2/3} , \qquad \alpha_2 = 4\pi\sigma \left(\frac{3\omega_s}{4\pi}\right)^{2/3} . \tag{2.30}$$

With these notations, the critical cluster size j_c, corresponding to a maximum of the Gibbs free energy, is given by

$$\left.\frac{\partial \Delta G(j)}{\partial j}\right|_{j=j_c} = -\Delta\mu + \frac{2}{3}\alpha_2 j_c^{-1/3} = 0 , \qquad j_c^{1/3} = \frac{2\alpha_2}{3\Delta\mu} . \tag{2.31}$$

Moreover, for the second derivative of ΔG we obtain

$$\left.\frac{\partial^2 \Delta G(j)}{\partial j^2}\right|_{j=j_c} = -\frac{2}{9}\alpha_2 j_c^{-4/3} = -\left(\frac{9}{8}\right)\left[\frac{(\Delta\mu)^4}{\alpha_2^3}\right] . \tag{2.32}$$

These expressions are needed for the subsequent derivations.

In particular, the value of G at $j = j_c$ gets the form

$$\Delta G(j_c) = \frac{1}{3}\alpha_2 j_c^{2/3} = \left(\frac{4}{27}\right)\frac{\alpha_2^3}{(\Delta\mu)^2} \,. \tag{2.33}$$

Moreover, the range of j–values δj in the vicinity of the extremum of ΔG, where the partial derivative $(\partial \Delta G(j)/\partial j)$ is nearly equal to zero and thermal fluctuations determine the motion of the clusters in cluster size space, is determined by the inequality

$$\Delta G(j_c) - \Delta G(j) \leq k_B T \,. \tag{2.34}$$

The respective range of j–values is given by

$$|j - j_c| \leq \delta j = \frac{1}{\sqrt{-\dfrac{1}{2k_BT}\left(\left.\dfrac{\partial^2 \Delta G}{\partial j^2}\right|_{j=j_c}\right)}} \,. \tag{2.35}$$

For simplicity of the notations we consider segregation in a perfect solution (the analysis may be carried out also for the more general case of real solutions which leads, however, only to quantitative modifications of the parameters [514, 516]).

For perfect solutions, the chemical potential of the segregating particles in the ambient phase may be expressed in the form

$$\mu_\beta(p,T,c) = \mu_\beta(p,T,c_\infty) + k_BT\ln\left(\frac{c}{c_\infty}\right) \,, \tag{2.36}$$

resulting in (cf. Eq.(2.7))

$$\Delta\mu = k_BT\ln\left(\frac{c}{c_\infty}\right) \,. \tag{2.37}$$

Here c_∞ is the concentration of the segregating particles in the ambient phase in equilibrium with the newly evolving phase at a planar interface at the given values of pressure p and temperature T. By definition of c_∞, the relation $\mu_\alpha(p,T) = \mu_\beta(p,T,c_\infty)$ holds. This relation is used, in addition to Eq.(2.7), in the derivation of Eq.(2.37).

2.1.5 Steady–State Nucleation Rate and Steady–State Cluster Size Distribution

When a system is brought into a metastable state by a sudden quench, the initial cluster size distribution is determined by the spectrum of heterophase fluctuations existing in the initial equilibrium state. This initial cluster size distribution is a rapidly decreasing function of j, i.e., the relation $(\partial f(j,t)/\partial j) < 0$ holds.

The negative gradient $(\partial f(j,t)/\partial j)$ in cluster size space results in the metastable state in a flux into the positive direction of the j–axis compensated at part by the deterministic flow term proportional to $(\partial \Delta G/\partial j)$ (cf. Eq.(2.21)).

Starting with the considered initial state it takes some time τ_1, the so–called time–lag in nucleation, to establish a time–independent flux in the interval $0 < j \leq g$. The length of this time–lag depends, in general, on the chosen value of g.

We demand that g fulfils the condition $g \geq (j_c + \delta j)$ due to the following considerations: Once a cluster has reached this size, the further evolution in cluster size space is dominated by the deterministic growth term in Eq.(2.21) and not by diffusion–like processes as it is the case in the interval $(j_c - \delta j, j_c + \delta j)$. Thus the clusters with sizes $j \geq g$, defined in above described way, may serve as centers for the evolution of the newly evolving phase. A more precise specification of the value of g will be given somewhat later.

Assuming a steady–state has been established in the system in the range $0 < j < g$, $g \geq j_c + \delta j$, i.e., the flux in this region of cluster size space has become independent of j $(J(j,t) = J = \text{const.})$, Eq.(2.24) yields

$$J(j,t) = J = -w^{(+)}(j,t)\left\{\frac{\partial f(j,t)}{\partial j} + \frac{f(j,t)}{k_B T}\frac{\partial \Delta G(j)}{\partial j}\right\}, \qquad \text{for} \quad j \leq g\,. \tag{2.38}$$

Once J is a constant, the cluster size distribution in the considered range does not depend on time (cf. Eq.(2.23)).

For the determination of the steady–state cluster size distribution function f_j in the considered range of n–values we introduce, in addition, another so far unknown function Ψ_j via

$$f_j = \Psi_j \exp\left(-\frac{\Delta G(j)}{k_B T}\right) , \qquad 0 < j \leq g . \tag{2.39}$$

Taking into account the relation $\Delta G(j \to 0) = 0$ we have

$$f_j\big|_{j\to 0} = \Psi_j\big|_{j\to 0} = c . \tag{2.40}$$

A substitution of the ansatz Eq.(2.39) into Eq.(2.38) yields

$$J = -w^{(+)}(j,t)\left(\frac{\partial \Psi_j}{\partial j}\right) \exp\left(-\frac{\Delta G(j)}{k_B T}\right) . \tag{2.41}$$

An integration of Eq.(2.41) with respect to j in the range from zero to j leads further to

$$\Psi_j = c - J \int_0^j \frac{\exp\left(\frac{\Delta G(j')}{k_B T}\right)}{w^{(+)}(j')} dj' . \tag{2.42}$$

By determining g, moreover, as equal to the smallest value of j for which

$$\Psi_j\big|_{j=g} \cong 0 \tag{2.43}$$

holds we arrive at

$$J = \frac{c}{\int_0^g \frac{\exp\left(\frac{\Delta G(j')}{k_B T}\right)}{w^{(+)}(j') dn'}} . \tag{2.44}$$

Taking into consideration that $\Delta G(j)$ has a sharp maximum in the vicinity of $j = j_c$, one obtains in a good approximation the following expression for the steady–state nucleation rate

$$J = cw^{(+)}(j_c)\Gamma_{(z)} \exp\left(-\frac{\Delta G(j_c)}{k_B T}\right) , \tag{2.45}$$

$$\Gamma_{(z)} = \sqrt{-\frac{1}{2\pi k_B T}\left(\left.\frac{\partial^2 \Delta G}{\partial j^2}\right|_{j=j_c}\right)} . \tag{2.46}$$

In the derivation of above given equations, $\Delta G(j)$ was expanded into a Taylor series including second order terms. Moreover, $w^{(+)}(j)$ was set equal to the respective value at $j = j_c$ and the integration was extended to infinity. By rewriting Eq.(2.42) in the form

$$\Psi_j = c - J\left\{\int_0^g \frac{\exp\left(\frac{\Delta G(j')}{k_B T}\right)}{w^{(+)}(j')} dj' + \int_g^j \frac{\exp\left(\frac{\Delta G(j')}{k_B T}\right)}{w^{(+)}(j')} dj'\right\} \tag{2.47}$$

and employing Eq.(2.44) we have

$$\Psi_j = J\int_j^g \frac{\exp\left(\frac{\Delta G(j')}{k_B T}\right)}{w^{(+)}(j')} dj' . \tag{2.48}$$

With the same approximations as in the derivation of Eq.(2.45) (Taylor expansion of ΔG, $w^{(+)}(j) \cong w^{(+)}(j_c)$, $g \to \infty$) one gets, finally,

$$\Psi_j = \left(\frac{c}{2}\right) \operatorname{erfc}\left\{\Gamma_{(z)}\sqrt{\pi}(j - j_c)\right\} , \tag{2.49}$$

$$\operatorname{erfc}(\xi) = \frac{2}{\sqrt{\pi}} \int_\xi^\infty dz \exp(-z^2) \tag{2.50}$$

or (with Eq.(2.39))

$$f_j = \left(\frac{c}{2}\right) \exp\left(-\frac{\Delta G(j)}{k_B T}\right) \operatorname{erfc}\left\{\Gamma_{(z)}\sqrt{\pi}(j - j_c)\right\} . \tag{2.51}$$

Taking into account Eqs.(2.35) and (2.46), Eq.(2.51) may be rewritten in the form (cf. also ([39, 543]))

$$f_j = \left(\frac{c}{2}\right) \exp\left(-\frac{\Delta G(j)}{k_B T}\right) \operatorname{erfc}\left\{\frac{(j - j_c)}{\delta j}\right\} . \tag{2.52}$$

Since erfc$(\xi = 1) \cong 0$ holds, the value of g, for which the steady–state distribution is nearly equal to zero, can be estimated as (cf. Eqs.(2.35), (2.46))

$$(g - j_c) \cong \frac{1}{\sqrt{\pi}\Gamma_{(z)}} = \delta j . \tag{2.53}$$

In this way, the specification of g is completed.

2.1.6 Estimate of Characteristic Time Scales

For the subsequent analysis, in addition to Eq.(2.21), a similar partial differential equation determining the change of the flux $J(j,t)$ with time and number of particles in a cluster j is required. A derivation of Eq.(2.24) with respect to time results with Eq.(2.23) in

$$\begin{aligned}\frac{\partial J(j,t)}{\partial t} = w^{(+)}(j,t) &\left\{\frac{\partial^2 J(j,t)}{\partial j^2} + \left[\frac{1}{k_B T}\frac{\partial \Delta G(j)}{\partial j}\right]\left(\frac{\partial J(j,t)}{\partial j}\right)\right\} \\ + &\left(\frac{\dot{w}^{(+)}(j,t)}{w^{(+)}(j,t)}\right) J(j,t) .\end{aligned} \tag{2.54}$$

According to Eqs.(2.33) and (2.45) the nucleation rate decreases rapidly with a decrease of the supersaturation. Therefore, in the transient stage to a steady state (for $0 < j \leq g$) and in the stage, where intensive nucleation processes occur, the supersaturation, respectively, concentration can be considered as nearly time–independent. This result implies on the other hand that for the description of nucleation $w^{(+)}(j,t)$ in Eq.(2.24) can be taken as a constant. The term

$$\left(\frac{\dot{w}^{(+)}(j,t)}{w^{(+)}(j,t)}\right) J(j,t) \cong 0 \tag{2.55}$$

will be omitted, therefore, in the further discussion of nucleation due to the assumed approximate constancy of the supersaturation in the nucleation stage.

The time required for the establishment of a steady state for $j \leq g$ is determined mainly by the time interval of diffusive motion in the vicinity of the extremum of ΔG. In this range $(j_c - \delta j,\ j_c + \delta j)$, the driving force for the deterministic flow ($\propto (\partial \Delta G/\partial j)$) is small, and, instead of the general equation (2.54), the motion in cluster size space can be described by a diffusion equation of the form

$$\frac{\partial J(j,t)}{\partial t} = w^{(+)}(j_c) \left\{ \frac{\partial^2 J(j,t)}{\partial j^2} \right\} . \tag{2.56}$$

The time interval τ_1 after which a steady state is established in the considered range of j–values can be written thus as

$$\tau_1 \cong \frac{(\delta j)^2}{w^{(+)}(j_c)} . \tag{2.57}$$

On the other hand, a second characteristic time scale τ_2 may be introduced via

$$\tau_2 = \frac{\delta j}{\dot{j}_c} . \tag{2.58}$$

It is the time interval the critical cluster size requires to pass the distance δj in cluster size space.

In the initial stages of the transformation, the state of the system is only insignificantly changed and the relation $\tau_1 < \tau_2$ holds. This inequality gives us at the same time the condition when the expressions for the steady–state nucleation rate can be applied.

Vice versa, if τ_2 becomes equal or greater than τ_1 a steady state and a steady–state nucleation rate cannot be sustained any more and the evolution of the system is dominated by a deterministic growth of the already formed supercritical clusters.

Consequently, the moment of time, when nucleation processes in the system become insignificant, can be determined by

$$\tau_1 \cong \tau_2 \ . \tag{2.59}$$

Above formulated dependencies give the basis for an estimate of the moment of time when dominating nucleation is replaced by a stage of dominating independent growth of the already formed supercritical clusters. Moreover, as will be shown in chapter 9, these dependencies allow also a determination of the number of clusters formed in the first stages of the transformation and their average size. However, the derivations imply the knowledge of the expressions for the kinetic coefficients $w^{(+)}(j,t)$.

As it is evident from Eqs.(2.44), (2.57) and (2.59), the knowledge of these coefficients is also required if the general expressions for the steady–state nucleation rate, the estimates for the time–lag in nucleation and the length of the interval of steady–state nucleation are applied to special cases. Therefore, in the next section we discuss some particular mechanisms of growth widely applied in a number of applications.

2.2 Description of Cluster Growth Processes

2.2.1 Diffusion–Limited Segregation and Related Growth Mechanisms

Once a cluster of supercritical size is formed, its further growth is determined by diffusional fluxes of the condensing particles to it from relatively distant parts of the matrix. The determination of the growth rate for such mechanism of growth is usually based on the following simplified model (see also Crank [101]).

It is assumed that at some distance $r = R$ from a center of spherical symmetry and at large distances $(r \to \infty)$ the concentrations of the segregating particles are fixed, i.e. that the relations

$$c(r = R) = c_R \,, \qquad c(r \to \infty) = c \,. \tag{2.60}$$

hold.

The density of fluxes of particles j_i of the different components are determined by the diffusion equation

$$\vec{j}_i = -\left(\frac{D_i c_i}{k_B T}\right) \text{grad } \mu_i \,. \tag{2.61}$$

Here D_i is the partial diffusion coefficient of the i–th component and c_i its volume density in the system.

We consider for simplicity of the notations a quasi–binary system, where only one of the components segregates to form a cluster of the new phase. In this case the expression (2.61) is simplified to

$$\vec{j} = -\left(\frac{Dc}{k_B T}\right) \text{grad } \mu \,. \tag{2.62}$$

For a perfect solution, the relation

$$\mu(p, T, x) = \mu_0(p, T) + k_B T \ln x \tag{2.63}$$

holds and Eq.(2.62) can be written in the well–known form of the first Fick's law

$$\vec{j} = -D \text{grad } c \,. \tag{2.64}$$

In this expression, D is the partial (bulk) diffusion coefficient of the segregating particles and c (respectively x) their volume concentration (molar fraction).

With the continuity equation (2.65)

$$\frac{\partial c}{\partial t} + \text{div } \vec{j} = 0 \,, \tag{2.65}$$

which reflects the conservation of the number of segregating particles, the diffusion equation is obtained (for the spherically symmetric case) as

$$\frac{\partial c}{\partial t} = \frac{\partial}{\partial r}\left(r^2 D \frac{\partial c}{\partial r}\right) . \tag{2.66}$$

By applying the boundary conditions (2.60), for large times ($t \to \infty$) a time–independent concentration field is established. This concentration field is found, from the stationary solution of Eq.(2.66), to be

$$c(r) = -\left(\frac{c - c_R}{r}\right) R + c . \tag{2.67}$$

The density of fluxes through the surface with the radius R is given in this case by

$$j_R = -D\left(\frac{\partial c}{\partial r}\right)_{r=R} \tag{2.68}$$

or

$$j_R = -D\left(\frac{c - c_R}{R}\right) . \tag{2.69}$$

If one assumes that in the vicinity of a growing cluster at each moment of time a practically stationary concentration profile is established as given by Eq.(2.67) [steady–state approximation for cluster growth (compare Zener [583])], the change of the cluster radius with time may be described by

$$\frac{dR}{dt} = -\frac{1}{c_\alpha} j_R \tag{2.70}$$

or by

$$\frac{dR}{dt} = \frac{1}{c_\alpha}\left(\frac{c - c_R}{R}\right) = \frac{c}{c_\alpha R}\left(1 - \frac{c_R}{c}\right) . \tag{2.71}$$

Here $c_\alpha = 1/\omega_s$ is the concentration of the segregating particles in the newly evolving phase.

It is assumed, in addition that in the immediate vicinity of the clusters in the matrix a local equilibrium concentration is established. Thus c_R may be expressed as (cf. [187])

$$c_R = c_\infty \left[\exp\left(\frac{2\sigma\omega_s}{k_B T} \frac{1}{R} \right) \right] . \tag{2.72}$$

As already noticed, c_∞ is the equilibrium concentration of the segregating particles for an equilibrium coexistence of both phases at a planar interface.

The concentration c in the undisturbed matrix corresponds, on the other hand, to a critical cluster size R_c, determined by

$$c = c_\infty \left[\exp\left(\frac{2\sigma\omega_s}{k_B T} \frac{1}{R_c} \right) \right] . \tag{2.73}$$

A substitution of Eqs.(2.72) and (2.73) into Eq.(2.71) and a subsequent Taylor–expansion of the exponential functions yields

$$\frac{dR}{dt} = \frac{2\sigma D c}{c_\alpha^2 k_B T} \left[\frac{1}{R} \left(\frac{1}{R_c} - \frac{1}{R} \right) \right] . \tag{2.74}$$

Equation (2.74) is the basic relation for the description of diffusion–limited segregation processes. It can be generalized to describe other mechanisms of growth as well.

The result of such an extension can be written in the general form (see Slezov, Sagalovich [508]) as

$$\frac{dR}{dt} = \frac{2\sigma D^{(m)} c}{c_\alpha^2 k_B T} \left[\frac{a^{(m-1)}}{R^m} \left(\frac{1}{R_c} - \frac{1}{R} \right) \right] , \tag{2.75}$$

where a is a length parameter reflecting specific properties of the considered growth mechanism, $D^{(m)}$ is the diffusion coefficient for the respective mechanism of growth.

Different growth mechanisms are described by this equation for different values of m ($m = 0$: ballistic or interface kinetic limited growth; $m = 1$: diffusion–limited growth; $m = 2$: diffusion along grain boundaries; $m = 3$: diffusion in a dislocation network; (Slezov, Sagalovich [508])).

Equation (2.74) serves, moreover, as one of the basic equations for the development of the theoretical description of Ostwald ripening, the competitive growth of an ensemble of clusters in the late stages of the phase transformation, which we are going to discuss in section 2.3.

2.2.2 Application to the Description of Nucleation

Equation (2.75) may be reformulated in terms of the number of particles j in the cluster. Such a reformulation is required in order to apply Eq.(2.26) for a determination of the kinetic coefficients $w^{(+)}(j,t)$. With Eqs.(2.8), (2.29) and (2.31) we obtain

$$v(j,t) = \frac{dj}{dt} = 4\pi D^{(m)} c \left(\frac{\Delta\mu}{k_B T}\right) \left[\frac{a^{(m-1)}}{R^{(m-2)}}\left(1 - \frac{R_c}{R}\right)\right] \tag{2.76}$$

or

$$\frac{dj}{dt} = 4\pi D^{(m)} c \left(\frac{\Delta\mu}{k_B T}\right) \left\{ \frac{a^{(m-1)}}{\left[\left(\frac{3\omega_s}{4\pi}\right) j\right]^{(m-2)/3}} \left[1 - \left(\frac{j_c}{j}\right)^{1/3}\right] \right\} . \tag{2.77}$$

According to Eqs.(2.26) and (2.31) we get further

$$\begin{aligned} w^{(+)}(j,t) &= -\frac{v(j,t)}{\dfrac{1}{k_B T}\dfrac{\partial \Delta G(j)}{\partial j}} \\ &= 4\pi D^{(m)} c \left\{ \frac{a^{(m-1)}}{\left[\left(\frac{3\omega_s}{4\pi}\right) j\right]^{(m-2)/3}} \right\} . \end{aligned} \tag{2.78}$$

In particular, for bulk diffusion limited growth ($m = 1$), we find

$$w^{(+)}(j,t) = 4\pi D c \left(\frac{3\omega_s}{4\pi}\right)^{1/3} j^{1/3} \tag{2.79}$$

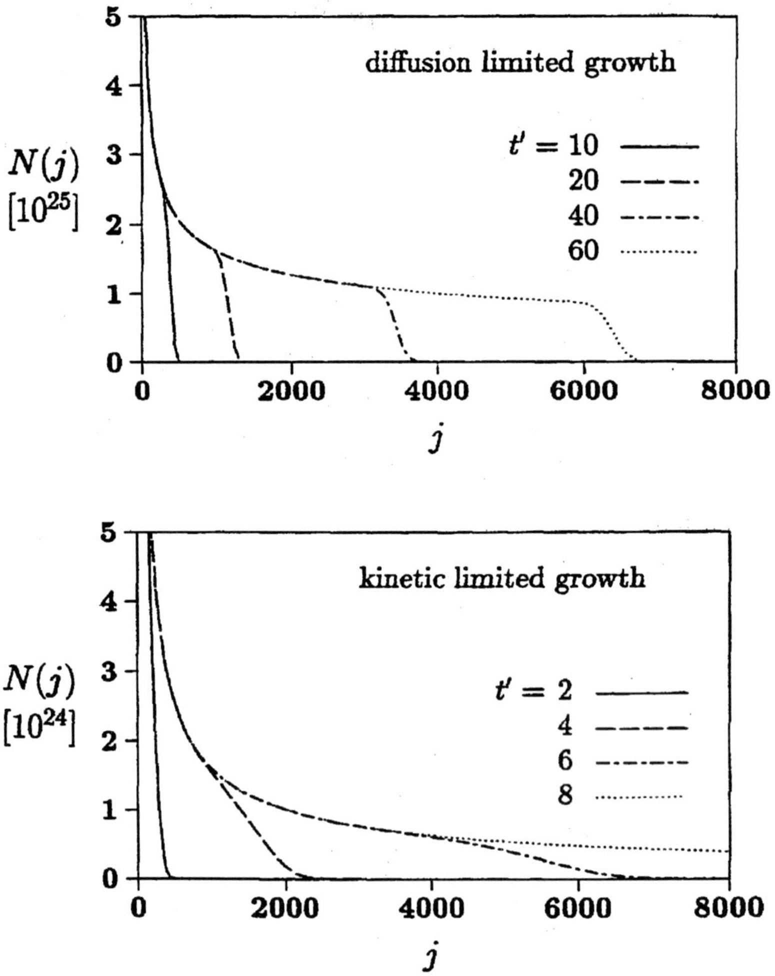

Fig. 2.2 Different stages of the time evolution of the cluster size distribution $N(j,t)$ as obtained by the numerical solution of the set of equations (2.2) and (2.3) with the kinetic coefficients (2.79) and (2.80). In the calculations, it is supposed that constancy of the supersaturation (concentration of monomeric building units) is sustained in the system by some appropriate mechanism. In the course of time, a steady-state is approached as it has to be expected according to Eq.(2.51). For further details of the calculations, the values of the parameters and the definition of the reduced time scale t', see, e.g., [322, 482, 483].

while for kinetic limited growth ($m = 0$)

$$w^{(+)}(j,t) = 4\pi D^{(*)} c a_m \left(\frac{\omega_s}{\omega_m}\right)^{2/3} j^{2/3} \tag{2.80}$$

holds.

In Eq.(2.80) the parameter a (cf. Eq.(2.76)) is identified with the lattice constant a_m of the considered solution. Moreover, similar to Eq.(2.28) the volume of a segregating particle ω_m occupied in the ambient phase is introduced as

$$\omega_m = \frac{4\pi}{3} a_m^3 \ . \tag{2.81}$$

For the case of condensation of a vapor the kinetic coefficients $w^{(+)}(j,t)$ are given by [187]

$$w^{(+)}(j,t) = c \left(\frac{3v_\alpha}{4\pi}\right)^{2/3} \left(\frac{8\pi k_B T}{M}\right)^{1/2} j^{2/3} \tag{2.82}$$

while the respective quantity for melt crystallization may be written as

$$w^{(+)}(j,t) \cong \varsigma \left(\frac{k_B T}{d_0^3 \eta}\right) j^{2/3} \ . \tag{2.83}$$

In Eq.(2.82), v_α is the volume of a particle in the liquid phase, M the mass of a particle of the gas while c refers to the particle concentration in the gas. In Eq.(2.83), η is the viscosity of the melt, d_0 a size parameter specifying the size of the ambient phase particles in the melt, while ς is the so-called sticking coefficient (cf. [187]).

In all discussed so far and other cases (cf., again, [187]), the knowledge of the kinetic coefficients $w^{(+)}$ allows, together with the thermodynamic results, a detailed analysis of the process of nucleation and growth based on the formalism outlined above. In particular, for the considered here case, with Eqs.(2.30), (2.32), (2.33), (2.45), (2.46), (2.79) and (2.80) we obtain the follwing expressions for the steady–state nucleation rate

$$J = \left(\frac{4\pi}{3}\right)\left(\frac{Dc^2 a_s}{j_c^{1/3}}\right)\sqrt{\frac{\alpha_2}{\pi k_B T}}\exp\left\{\frac{-B(T)}{\left[\ln\left(\frac{c}{c_\infty}\right)\right]^2}\right\} \tag{2.84}$$

for diffusion limited growth and

$$J = \left(\frac{4\pi}{3}\right)\left[Dc^2 a_m\left(\frac{\omega_s}{\omega_m}\right)^{2/3}\right]\sqrt{\frac{\alpha_2}{\pi k_B T}}\exp\left\{\frac{-B(T)}{\left[\ln\left(\frac{c}{c_\infty}\right)\right]^2}\right\} \tag{2.85}$$

for kinetic limited growth. Moreover, the notation

$$B(T) = \left(\frac{4}{27}\right)\left(\frac{\alpha_2}{k_B T}\right)^3 \tag{2.86}$$

is introduced.

Numerical solutions of the set of kinetic equations with above given kinetic coefficients are shown in Figs. 2.2.

Further details concerning the course of nucleation and growth processes in the first stages of first–order phase transitions are discussed in detail by Slezov and Schmelzer [516]. Some of them will be analyzed also in connection with special applications in chapters 9 and 10.

2.3 Coarsening of Ensembles of Clusters

2.3.1 The Lifshitz–Slezov Theory

If the state of the system is not changed significantly by nucleation and growth processes of the already formed supercritical clusters, the critical cluster size is nearly constant, the supercritical clusters grow at the expense of primary building units in the ambient phase and (for $R \gg R_c$) from

Eq.(2.74) a time dependence of the average cluster radius as $\langle R \rangle^2 \sim t$ is to be expected for diffusion–limited growth.

The situation becomes different if due to the growth of the ensembles of clusters the number of segregating particles is decreased to a value near the equilibrium concentration in the matrix ($c = c_\infty$) and the critical cluster radius is increased to a size comparable with the dimensions of the majority of clusters formed.

In this stage, according to Eq.(2.74) and the results of the thermodynamic analysis, clusters with radii $R < R_c$ shrink and are dissolved supplying in this way the larger clusters with monomers for a further growth. The critical cluster radius cannot be considered any more as a constant but has to be determined in a self–consistent way.

From a kinetic point of view, this mechanism is determined by the cluster size dependence of the equilibrium solubility c_R (compare Eq.(2.72)). Small clusters are characterized by higher values of c_R, thus, for a given value c of the average concentration of the segregating particles in the melt clusters with equilibrium solubilities larger this value ($R < R_c$) shrink, while clusters with a lower value of the solubility ($R > R_c$) are capable to a further growth.

The thermodynamic driving force of dissolution, respectively, growth is hereby proportional to the differences $(c - c_R)$ or $[\mu(c) - \mu(c_R)]$. The dissolution of the smaller cluster provides the monomeric building units for the growth of the supercritical ones.

From a macroscopic thermodynamic point of view the evolution of the ensemble of clusters is governed by the general thermodynamic evolution criteria, the decrease in the respective thermodynamic potential due to the decrease in the interfacial contributions to it.

The process is completed, in general, only after one superlarge cluster has won the competition process (cf. also chapter 3). However, the situation may be different, if additional factors – like elastic strains – influence the coarsening process (see, for example, chapter 13).

Though the process of competitive growth considered here has been a well–known experimental fact since the first detailed investigation by W. Ostwald (see Ostwald (1901) [404]), a satisfactory theory was developed first in the late 1950s by Lifshitz, Slezov [308]–[310] and repeated for kinetically limited growth by Wagner [560].

The Lifshitz–Slezov theory was formulated initially for diffusion–limited growth, described by Eq.(2.74). Here the main assumptions and results are summarized briefly.

If we denote by R_{co} the critical cluster radius for the homogeneous metastable initial state, as a first step a dimensionless time scale may be introduced by

$$\tilde{t} = \left(\frac{2\sigma D c_{\infty}}{c_{\alpha}^2 k_B T R_{co}^3} \right) t \tag{2.87}$$

and Eq.(2.74) can be rewritten in the simpler form

$$\frac{dR}{d\tilde{t}} = \frac{R_{co}^3}{R} \left(\frac{1}{R_c} - \frac{1}{R} \right) . \tag{2.88}$$

As a next step, the dimensionless variables u, x and τ are introduced via

$$u = \frac{R}{R_c} , \qquad x = \frac{R_c}{R_{co}} , \qquad \tau = 3 \ln[x(\tilde{t})] . \tag{2.89}$$

In these reduced variables, the growth equation (2.88) reads

$$\frac{du^3}{d\tau} = \gamma(\tau)[u-1] - u^3 \qquad \text{with} \qquad \gamma(\tau) = \frac{1}{\left(x^2 \dfrac{dx}{d\tilde{t}} \right)} . \tag{2.90}$$

Possible curves for the growth rate v_u in reduced variables

$$\frac{du}{d\tau} = v_u = \frac{\left[\gamma(\tau)(u-1) - u^3 \right]}{(3u^2)} \tag{2.91}$$

are shown in Fig. 2.3.

From a physical point of view for large times only the situation corresponding to a value of γ equal to $\gamma = (27/4)$ in Fig. 2.3 is of relevance (only one cluster size exists for which the growth rate is equal to zero). This particular

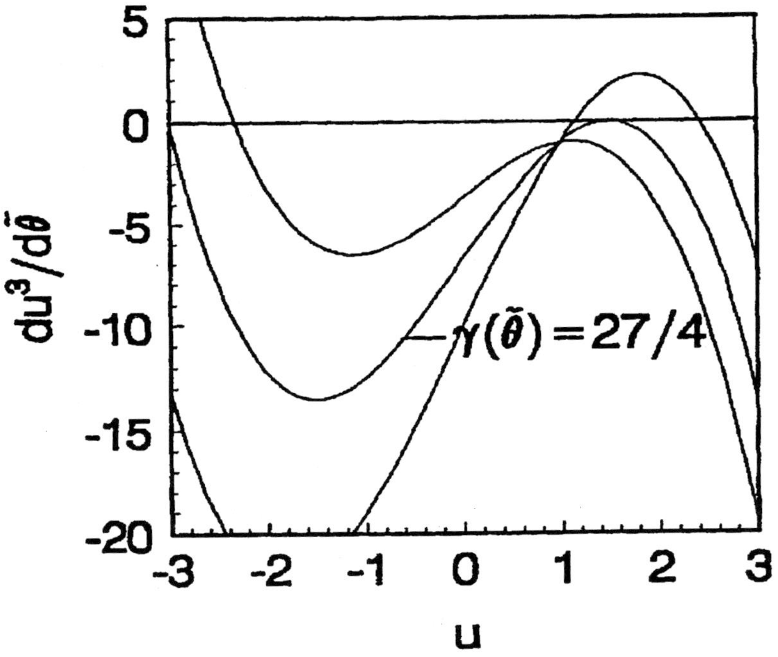

Fig. 2.3 Possible types of curves (v_u vs u). For the late stage of coarsening only the situation corresponding to $\gamma = 27/4$ is relevant.

value of γ is found taking into account that in the asymptotic region $\gamma(\tau)$ is determined by

$$v_u = \frac{\partial v_u}{\partial u} = 0 \,. \tag{2.92}$$

With these conditions Eqs.(2.91) – (2.92) yield

$$\gamma = \frac{27}{4} \,, \qquad u = \frac{3}{2} \,. \tag{2.93}$$

In this way with Eq.(2.90)

$$\frac{dx^3}{d\tilde{t}} = \frac{4}{9} \qquad \text{or} \qquad R_c^3(\tilde{t}) = R_{co}^3(1+\tilde{t}) \tag{2.94}$$

is obtained. Thus, for diffusion–limited growth the critical cluster radius grows as

$$R_c(t) \sim t^{1/3} . \tag{2.95}$$

In the Lifshitz–Slezov theory in addition to the cluster size distribution $f(R,t)$, describing the cluster ensemble in absolute variables, also the cluster size distribution function $\varphi(u,\tau)$ in reduced variables is introduced. It is connected with $f(R,t)$ by

$$f(R,t)dR = f(R,t)R_c(t)d\left(\frac{R}{R_c(t)}\right) = \varphi(u,\tau)du \tag{2.96}$$

or

$$\varphi(u,\tau) = f(R,t)R_c(t) . \tag{2.97}$$

For the determination of $\varphi(u,\tau)$, only the condition of conservation of the number of monomers is used.

The derivation is somewhat lengthy and cannot be given here. The main results are the following (see [308] – [310]):

(i.) The cluster size distribution function $\varphi(u,\tau)$ can be expressed in the form

$$\varphi(u,\tau) = A\exp(-\tau)P(u) , \qquad A - \text{constant} . \tag{2.98}$$

(ii.) The time–independent part of the cluster–size distribution function $P(u)$ obeys the normalization condition

$$\int_0^\infty P(u)\, du = 1 . \tag{2.99}$$

For diffusion–limited growth in addition the relations

$$P(u) = \frac{3^4 e u^2 \exp\left\{-\dfrac{3}{2[(3/2)-u]}\right\}}{2^{5/3}(u+3)^{7/3}[(3/2)-u]^{11/3}} \tag{2.100}$$

$$\int_0^\infty uP(u)du = 1 \tag{2.101}$$

hold, resulting in

$$\langle R \rangle = R_c(t) \ . \tag{2.102}$$

(iii.) According to Eqs.(2.98) and (2.99) the evolution of the number of clusters is determined by the prefactor $A\exp(-\tau)$ only, which gives

$$\frac{N(t)}{N(0)} \sim \exp(-\tau) = \left(\frac{R_{co}}{R_c}\right)^3 \sim \frac{1}{t} \ . \tag{2.103}$$

In this way, the whole process of structural reorganization considered is characterized by two equations giving the time–dependence of the average cluster size and the number of clusters in the system. Both the average cluster size and the critical Gibbs–Thomson cluster radius behave for diffusion–limited growth as $\sim t^{1/3}$.

The approach of a time–independent form of the cluster size distribution function in appropriately chosen coordinates (R/R_c) can be considered as a particular form of dynamically self–organized criticality. As will be shown in chapter 14, such behavior is also found in spinodal decomposition in adiabatically isolated systems. A general discussion of this concept including a variety of different applications can be found in chapter 7.

2.3.2 Modifications and Generalizations

Equations (2.95) – (2.103) form the basic content of the theoretical description of Ostwald ripening as developed first by Lifshitz and Slezov [308] – [310]. This theory was modified and extended subsequently, the main results remained, however, unchanged (see also Voorhees [559]; Slezov, Schmelzer, Möller [509]).

The series of extensions of the theory to different growth mechanisms, e.g., of the type given by Eq.(2.75), was opened by Wagner [560], who applied the outlined ideas to the case of kinetically limited growth (for other results in this direction see the overview given by Slezov and Sagalovich [508]).

A second line of generalizations of the theory is connected with the incorporation of direct diffusional interactions of the clusters into the kinetic description of Ostwald ripening (Voorhees, Glicksman [558]; Marqusee, Ross [355]; Tokuyama, Kawasaki [537]; Marder [353]). The analysis shows, however that the asymptotic power laws remain unchanged, only some additional prefactors in the expressions for R_c and $\langle R \rangle$ occur, which are functions of the volume fraction of the segregating phase.

In the case of diffusional interactions, a time–independent cluster size distribution is also established in the course of time, however, the curves are broader and more symmetric than those of the Lifshitz–Slezov distribution.

As another factor, which may diminish the gap between experiment and theory the influence of thermal fluctuations or externally generated noise has recently found to be of importance at intermediate stages of this process (Ludwig, Schmelzer, Bartels [320]; Möller [374]; Slezov, Schmelzer, Tkatch [514]). An extended discussion of further factors which may account for a deviation from the Lifshitz–Slezov distribution is given by Jayanth and Nash (1989) [224].

Other generalizations of the Lifshitz–Slezov theory deal with an alternative approach to the theory of Ostwald ripening, allowing the description of the first non–asymptotic stage of coarsening (Schmelzer [468]; Schmelzer, Ulbricht [471]) and a thermodynamic interpretation of this process.

A further problem of intensive research is connected with the influence of elastic strains on the process of Ostwald ripening, in particular, its asymptotic behavior (Schmelzer [468]; Schmelzer, Gutzow [472]; Kawasaki, Enomoto [245]; Pascova et al. [415, 416]; Schmelzer et al. [474, 475]). For the case that only elastic matrix–cluster interactions have to be taken into account it can be shown that qualitatively the growth kinetics is modified as compared with the results obtained first by Lifshitz and Slezov, if the total energy of elastic deformations due to the formation and growth of one cluster increases more rapidly than linear with the volume of the cluster.

A particularly interesting example in this respect consists in coarsening in a porous solid. A more detailed discussion of this topic is given in chapter 13.

2.4 Spinodal Decomposition

2.4.1 Introduction

Spinodal decomposition is another important mechanism first–order phase transformations may proceed (cf. [39, 54, 180, 397, 507]). In contrast to nucleation, discussed in detail in section 2.1, it is characterized by a continuous diffusional amplification of initially small variations of density or concentration in the system (*up–hill diffusion*).

A theoretical description of this process was developed first by Hillert [206], Cahn and Hilliard [78] based on the van der Waals theory of interfacial effects ([549], see also Rowlinson [458]). The theory was extended, among others, by Filipovich [148] and Cook [99] to include the influence of stochastic fluctuations into the theory and by Langer, Bar–on and Miller [300] to account for non–linear terms in spinodal decomposition.

In the present introductory chapter, we will restrict ourselves to the consideration of the linear Cahn–Hilliard–Cook theory of spinodal decomposition. But already in such a simplified approach, non–linearities may enter the picture by a particular choice of the boundary conditions, i.e., the assumption of an adiabatic closure of the system. This additional assumption will lead to a very complex behavior and is believed to give a key for the understanding of the scenario of spinodal decomposition, in general (see chapter 14).

2.4.2 The Cahn–Hilliard–Cook Equation

Following van der Waals ([549], see also Rowlinson [458], Rowlinson, Widom [459], Cahn, Hilliard [78]) the free enthalpy G of a binary inhomogeneous solution can be written in a first approximation in the form

$$G = \int \left[g(c) + \kappa (\nabla c)^2 \right] dV \ . \tag{2.104}$$

Here c is the volume concentration of one of the components in the solution, again, $g(c)$ is the volume density of the free enthalpy and κ ($\kappa > 0$) a coefficient describing the contributions to the thermodynamic potential due to the inhomogeneities in the system.

If the deviations from the initial homogeneous concentration c_0 are relatively small, then a Taylor expansion of Eq.(2.104) results in the following

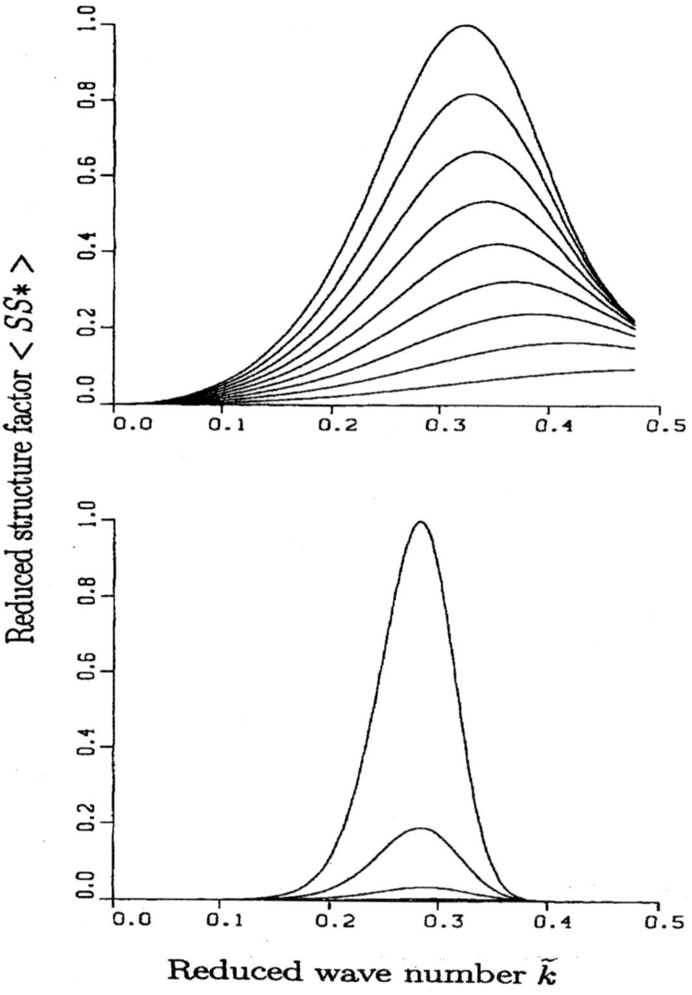

Fig. 2.4 Evolution of the structure function $\langle S^2(\tilde{k},\tilde{t})\rangle$ for the first stages of the decomposition process. For the curves shown in the upper part the evolution is determined by stochastic fluctuations. The different curves in the upper part correspond (with increasing values of S^2) to the following values of time (in reduced units): $\tilde{t} = 30$, 60, 90, 120, 150, 180, 210, 240, 270. In the lower part similar curves are shown for times $\tilde{t} = 300$, 600, 900, 1200, 1500. The curves are normalized in such a way that the maximum of S^2 equals one.

expression for the change of the free enthalpy connected with the evolution of the concentration field $c(\vec{r},t)$

$$\Delta G = \int \left[\frac{1}{2} \left(\frac{\partial^2 g}{\partial c^2} \right)_{c_0,T} (c - c_0)^2 + \kappa (\nabla c)^2 \right] dV \, . \tag{2.105}$$

In agreement with the thermodynamic stability conditions (see, e.g., Kubo [269]), a spontaneous growth of the density fluctuations takes place only for $g''(c_o, T) < 0$, since only in this case may the amplification of the density profile be accompanied by a decrease of the free enthalpy of the system.

In the framework of the Cahn–Hilliard–Cook theory, the kinetics of the decomposition process is described by a generalized diffusion equation connecting the variations in the thermodynamic potential G with the kinetics of the decomposition process. It follows from the set of equations

$$\frac{\partial c}{\partial t} + \mathrm{div}\vec{j} = 0 \, , \qquad \vec{j} = \vec{j}_D + \vec{j}_A \, , \tag{2.106}$$

$$\vec{j}_D = -M\nabla \frac{\delta G}{\delta c} \, , \qquad \vec{j}_A = -\nabla A(\vec{r}, t) \tag{2.107}$$

and has the form

$$\frac{\partial c(\vec{r}, t}{\partial t} = M g''(c_0, T)\nabla^2 c(\vec{r}, t) - 2M\kappa\nabla^4 c(\vec{r}, t) + \nabla^2 A(\vec{r}, t) \, . \tag{2.108}$$

$\vec{j}_D$ is the vector describing the deterministically determined density of fluxes of particles, while $\vec{j}_A$ represents the flow connected with the fluctuating scalar field $A(\vec{r}, t)$ superimposed on the deterministic flow. M is a mobility coefficient.

For a solution of this equation, the $c(\vec{r}, t)$ and $A(\vec{r}, t)$ fields are expressed through Fourier expansions via

$$c(\vec{r}, t) = c_0 + \sum_{-\infty}^{\infty} S(\vec{k}_n, t) \exp\left(i\vec{k}_n\vec{r}\right) d\vec{r} \, , \tag{2.109}$$

$$S(\vec{k}_n, t) = \frac{1}{V}\int [c(\vec{r}, t) - c_0] \exp\left(-i\vec{k}_n\vec{r}\right) d\vec{r} \, , \tag{2.110}$$

$$A(\vec{r}, t) = \sum_{-\infty}^{\infty} B(\vec{k}_n, t) \exp\left(i\vec{k}_n\vec{r}\right) d\vec{r} \, , \tag{2.111}$$

$$B(\vec{k}_n, t) = \frac{1}{V} \int A(\vec{r}, t) \exp\left(-i\vec{k}_n \vec{r}\right) d\vec{r} \,, \tag{2.112}$$

V is the volume of the system under consideration.

Eqs.(2.106) – (2.112) result in the following differential equation for the spectral function $S(\vec{k}_n, t)$

$$\frac{\partial S(\vec{k}, t)}{\partial t} = R(\vec{k}, t) S(\vec{k}, t) - \kappa^2 B(\vec{k}, t) \,, \tag{2.113}$$

where the amplification factor $R(\vec{k}, t)$ is determined by

$$R(\vec{k}, t) = -M\vec{k}^2 \left[g''(c_0, T) + 2\kappa \vec{k}^2 \right] . \tag{2.114}$$

The subscript n in $\vec{k}_n$ is omitted here and subsequently for simplicity of the notations.

Based on Eq.(2.114), in analogy to the critical cluster size in nucleation (cf. Eq.(2.8)) a critical wave number k_c may be introduced. It is defined by the condition that the deterministic amplification rate $R(k_c, t)$ is equal to zero. This condition yields

$$\vec{k}_c^2 = -\frac{1}{2\kappa} g''(c_0, T) \,. \tag{2.115}$$

Concentration waves in the Fourier spectrum with wave numbers $k > k_c$ decay while the modes with $k < k_c$ grow. The value of the wave number corresponding to the highest amplification rate is given by

$$k_{max} = \frac{1}{\sqrt{2}} k_c \,. \tag{2.116}$$

Moreover, in experimental studies of phase transformation processes not the spectral function itself but a quantity proportional to the average of the square of the spectral function $\langle S^2 \rangle = \langle S S^* \rangle$ is of relevance. Assuming that in the average the concentration fluctuations are equal to zero and uncorrelated

$$\langle A(t) \rangle = 0 \,, \qquad \langle A(\xi) A(\chi) \rangle = Q(\vec{k}) \delta(\xi - \chi) \tag{2.117}$$

we get (see, e.g., [321])

$$\frac{\partial \langle S^2(\vec{k},t)\rangle}{\partial t} = 2R(\vec{k},t)\langle S^2(\vec{k},t)\rangle + \vec{k}^4 Q(\vec{k}) \ , \tag{2.118}$$

$$Q(\vec{k}) = \frac{2Mk_BT}{V}\frac{1}{k^2} \ . \tag{2.119}$$

Finally, by introducing dimensionless wave numbers $\tilde{k}$ and a dimensionless time scale $\tilde{t}$ via

$$\tilde{k} = ak \ , \qquad \tilde{t} = \frac{4\kappa M}{a^4}t \tag{2.120}$$

we obtain with Eq.(2.115)

$$\frac{\partial \langle S^2(\tilde{k},\tilde{t})\rangle}{\partial \tilde{t}} = \tilde{k}^2 \left\{ \left[\tilde{k}_c^2 - \tilde{k}^2\right] \langle S^2(\tilde{k},\tilde{t})\rangle + \frac{k_BTa^2}{2\kappa V} \right\} \ . \tag{2.121}$$

The parameter a is a measure of the intermolecular distance in the solution.

2.4.3 The Initial Stages of Spinodal Decomposition

Numerical solutions of Eq.(2.121) are shown in Fig. 2.4 (for the details of the calculations see [321]). After a relatively short initial period, determined by the stochastic generation of the initial distribution (upper curve), the typical features of the Cahn–Hilliard scenario of spinodal decomposition – constant values of the critical wave number and the wave number of highest amplification rates, intersection of the different curves at the time–independent value of k_c – are found.

However, in real systems and also in Monte–Carlo simulations of spinodal decomposition, the situation is different. Both the critical wave number as well as the wave number of highest amplification rate are shifted in the course of time to lower values of k (see also chaper 5). Such effects may be accounted for by the introduction of non–linear terms in the respective kinetic equation (see, e.g., [39, 54, 180, 300]).

As already mentioned, in chapter 14 we will consider a special form of non–linearity connected with changes of the temperature in the system due to an assumed adiabatic closure. The resulting variations in the temperature result for such constraints in a very interesting and complex behavior exhibiting already in the linear Cahn–Hilliard–Cook description many features which are characteristic for decomposition in real systems.

2.5 Discussion

In the present chapter, only a brief introduction into the basic ideas of the theory of nucleation, cluster growth and spinodal decomposition could be given as far as they are required for the understanding of nucleation and aggregation phenomena analyzed in the subsequent chapters. For a more detailed discussion of topics not covered here the reader is referred to the already cited review articles or monographs (e.g., [1, 39, 40, 54, 67, 180, 184, 187, 254, 379, 544, 584, 585]) as well as to the subsequent chapters of the present book.

3 A Stochastic Approach to Nucleation

D. Labudde, R. Mahnke

3.1 Introduction

In the present chapter, we propose a stochastic description of the process of formation and growth of clusters. Hereby we consider as an example condensation in supersaturated vapors.

The method of description discussed is restricted to a mesoscopic time scale, i.e., a scale where the microscopic processes are not considered in detail but reflected by small changes of the macroscopic parameters of the system (e.g. pressure, temperature). The clusters are described again by the classical droplet model. This statement implies that they are characterized by macroscopic values of density and surface tension.

Once the vapor becomes supersaturated, a phase transition by homogeneous nucleation may occur in the system. The basic quantity describing this process is the cluster size distribution $\mathbf{N}$ as a function of time t. We may write

$$\mathbf{N}(t) = (N_1(t), N_2(t), \ldots, N_j(t), \ldots, N_N(t)) \ . \tag{3.1}$$

$\mathbf{N}$ represents the number of clusters N_j of size j [128, 330]. Single particles (atoms, molecules) are called monomers of size $j = 1$. Bound states are clusters of size $j \geq 2$.

We investigate a finite system with a fixed total number of particles N_0 [246, 259, 297]. The boundary condition

$$N_0 = N_1(t) + \sum_{j=2}^{N} j N_j(t) = \text{const.} \tag{3.2}$$

is fulfilled since the particles are either free or bounded in clusters. In addition, we assume that $N^{(0)}$ non–reactive bath particles are present in the system.

Concerning other thermodynamic boundary conditions we can choose different constraints corresponding to different conditions of phase formation. The respective conditions are reflected in the theory by the adequate choice of the thermodynamic potential.

3.2 The Model

We start the derivation of the basic equations at a microscopic level of description. The reactive particles $i = 1, 2, \ldots, N_0$ are characterized each by their positions $s_i = (s_{1i}, s_{2i}, s_{3i})$ and momenta $q_i = (q_{1i}, q_{2i}, q_{3i})$. This first kind of reactive particles are the monomers. We can identify these particles with the basic units (atoms or molecules) of the system. Out of these particles, dimers, trimers etc. are formed. We denote the bound particles as clusters of size j.

Clusters of size j are characterized in addition to their positions and momenta also by the number of monomeric particles contained in such a cluster (or its mass $m_j = jm$). The cluster size coordinate we consider as a continuous variable, too.

On the microscopic level we can describe the motion of the reactive and non–reactive particles by a set of coordinates of the form

- Cluster: $x_i(t) = (r_i(t), p_i(t), n_i(t))$,
- Bath particles: $y_i(t) = (s_i(t), q_i(t))$.

However, the analytical solution of the set of microscopic equations of motion cannot be determined, in general.

As a next step, we introduce therefore the microscopic density ϕ in phase space via

$$\begin{aligned} \phi\ (x, y, t) &= \sum_i \delta(x - x_i(t))\delta(y - y_i(t)) \qquad (3.3) \\ &= \sum_i \delta(r - r_i(t))\delta(p - p_i(t))\delta(j - j_i(t))\delta(s - s_i(t))\delta(q - q_i(t)) \, . \end{aligned}$$

The quantity $\phi(r, p, j, s, q, t)\, dr\, dp\, dj\, ds\, dq$ gives the number of particles of the background gas at the position $(s, s+ds)$ with the momentum $(q, q+dq)$ and the number of clusters of the size $(j, j+dj)$ at the position $(r, r+dr)$ with the momentum $(p, p+dp)$ at time t.

By integrating the microscopic density ϕ over the total phase space, we get

$$\int dx \int dy\ \phi(x, y, t) = N^{(0)} + N(t)\ . \tag{3.4}$$

The total number of particles of the system is not constant. It consists of a constant number of nonreactive particles ($N^{(0)}$ = const) and a time dependent number of reactive particles $N(t)$. The particle number changes permanently by the reactive collisions between the clusters and monomers.

The mass of an inert gas particle is assumed to be much smaller than the mass m of a reactive particle. Therefore an adiabatic decoupling of the cluster gas from the background gas can take place [297]. We may integrate thus over the coordinates of the bath particles. The inert gas acts then as a fluctuating background and has no direct effect on the evolution of the cluster size distribution [330].

According to the general approaches of the physics of equilibrium– and nonequilibrium systems [246, 259], we introduce further the partition function with respect to cluster sizes as the average value over the microscopic density. We assume hereby that we can assign well–defined values of the thermodynamic parameters (temperature and mobility) to the bound states of the particles. We define thus single–particle distributions by

$$f_1(r, p, j, t) = \langle \phi \rangle\ . \tag{3.5}$$

The term $f_1(r, p, j, t)\, dr\, dp\, dj$ gives the number of particles with the size $(j, j+dj)$ at the position $(r, r+dr)$ with the momentum $(p, p+dp)$ at time t. The other particles may occupy any available positions with any arbitrary momenta. After the integration over all spatial coordinates and momenta, we get

$$\int f_1(r, p, j, t) dr dp = N_j(t)\ . \tag{3.6}$$

$N_j(t)$ is the number of particles of size j with any arbitrary momenta and spatial positions.

The state of the cluster distribution is characterized in the reduced description by the number of monomers $N_1(t)$ and the number of aggregates of different sizes $N_j(t)$, $j = 2, 3, \ldots, N_0$. This way, we arrive at a description as specified by Eq.(3.1) where the constraints Eq.(3.2) have to be fulfilled.

3.3 Thermodynamic Description of the Cluster Ensemble

3.3.1 The Helmholtz Free Energy

In order to develop a kinetic description of cluster formation, we determine first the thermodynamic properties of the cluster ensemble. Hereby we consider the canonical ensemble, where volume and temperature of the system are fixed.

We start with the relation

$$F(T, V, \mathbf{N}) = -k_B T \ln Z(T, V, \mathbf{N}) \tag{3.7}$$

well–known from statistical physics. As the first step, the canonical partition function $Z = Z(T, V, \mathbf{N})$ is determined. The partition function is given by

$$Z(T, V, \mathbf{N}) = \prod_{j=1}^{N} \frac{1}{N_j!\, h^{3N_j}} \int d^{3N_j}p\, d^{3N_j}q\ \exp(-\beta H_j)\ , \tag{3.8}$$

where H_j is the Hamilton function of a cluster of size j and $\beta = 1/k_B T$. The Hamilton function H_j for a cluster of size j we write as the sum of the kinetic energy $[\vec{p}_j^{\,2}/2m_j]$ and potential energy term φ_j

$$H_j = \frac{\vec{p}_j^{\,2}}{2m_j} + \varphi_j\ . \tag{3.9}$$

The function φ_j results from the interactions of the j particles in one single cluster. Cluster–cluster interactions are not considered so far.

The evaluation of the partition function can be carried out then by standard methods of statistical mechanics. The integral splits into two parts, the ideal part $Q_{N_j}^{ideal}$ and the binding energy part $Q_{N_j}^{bin}$. Assuming that all reactive monomers have the same mass, $m_j = jm$, and that the potential energy depends only on the size j of clusters and on temperature T ($\varphi_j = \varphi_j(T)$; possible cluster–cluster interactions are excluded, as mentioned, here from the consideration), we have

$$Z(T,V,\mathbf{N}) = \prod_{j=1}^{N_0} \frac{1}{N_j!} \left\{ \frac{V}{\lambda_j^3} \exp(-\beta\varphi_j) \right\}^{N_j} , \tag{3.10}$$

$$\lambda_j(T) = \frac{h}{\sqrt{2\pi m_j k_B T}} = \frac{\lambda_1(T)}{\sqrt{j}} , \qquad \lambda_1(T) = \frac{h}{\sqrt{2\pi m k_B T}} . \tag{3.11}$$

Here by λ_j the de–Broglie wave length of a cluster of size j is denoted, h is Planck's constant.

Using Stirling's formula $x! \sim x^x e^{-x}$ in the ideal part of the partition function (Eq.(3.10)), the following expression for the free energy of a mixture of clusters is obtained

$$F(T,V,\mathbf{N}) = k_B T \sum_j N_j \left[\ln\left(\frac{\lambda_j^3 N_j}{V} \right) - 1 \right] + \sum_j N_j \varphi_j(T) . \tag{3.12}$$

The two terms in the free energy expression can be interpreted as an ideal part, resulting from the kinetic energy and describing the mixture of the clusters; the binding energy part, resulting from the potential energy of the clusters itself, i.e., describing the clusters as bound states.

The thermal equation of state of the system can be obtained from

$$p = -\left(\frac{\partial F}{\partial V} \right)_{T,\mathbf{N}} . \tag{3.13}$$

We get

$$p = \frac{k_B T}{V} N_{tot}^{(cl)} , \qquad N_{tot}^{(cl)} = \sum_{j=1}^{N_0} N_j(t) , \tag{3.14}$$

i.e., the equation of state of a mixture of perfect gases.

Finally, for the chemical potential of the gas we find (in the absence of clustering)

$$\mu\left(T, \frac{N_1}{V}\right) = \left(\frac{\partial F}{\partial N_1}\right)_{V,T} = k_B T \ln\left(\frac{N_1}{V}\lambda_1^3\right) . \tag{3.15}$$

By taking the derivative of the state function with respect to temperature, we can determine further thermodynamic quantities [269, 336]. However, before doing so we have to specify the expression for the binding energy $\varphi_j(T)$.

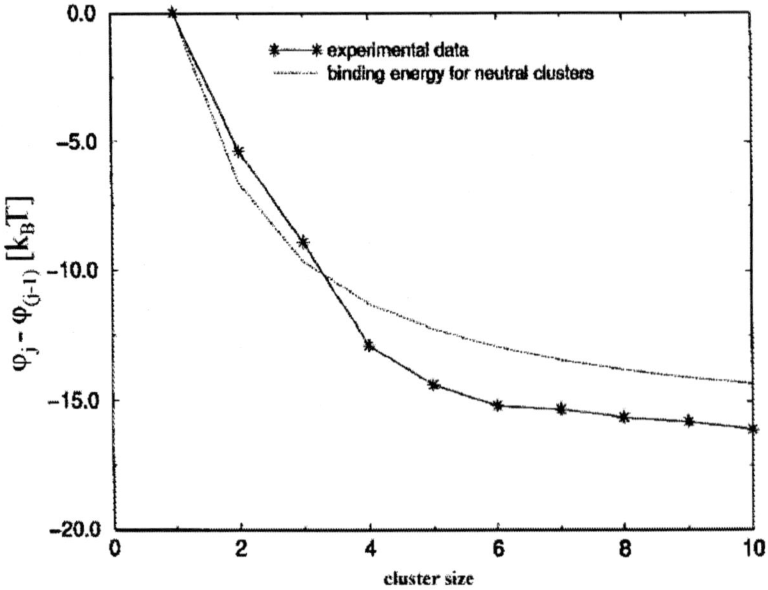

Fig. 3.1 Binding energy of neutral methanol clusters. The experimental data are due to Buck, 1997

3.3.2 Thermodynamic Potentials at Different Boundary Conditions

For different thermodynamic boundary conditions and for an analysis of the results, the expressions for further appropriate thermodynamic potentials are required. Based on the results outlined above, the respective derivations can be carried out straightforwardly.

The entropy is determined generally via

$$S(T, V, \mathbf{N}) = -\left(\frac{\partial F}{\partial T}\right)_{V,\mathbf{N}} \tag{3.16}$$

resulting initially in

$$S(T, p, \mathbf{N}) = k_B \sum_{j=1}^{N_0} N_j(t) \left[\frac{5}{2} - \ln\left(\frac{\lambda_j^3(T) N_j(t) p}{N_{tot}^{(cl)} k_B T}\right)\right] - \\ - \sum_{j=2}^{N_0} N_j(t) \frac{\partial \varphi_j(T)}{\partial T} \; . \tag{3.17}$$

For the internal energy ($U = F + TS$) of the considered state we have, in addition,

$$U = \sum_{j=1}^{N_0} N_j(t) \left(\frac{3}{2} k_B T + \varphi_j(T) - T \frac{\partial \varphi_j}{\partial T}\right) \; . \tag{3.18}$$

In the calculations for the case of the absence of clustering we set $N_1(t) = N_0$, $N_j(j \geq 2, t) = 0$ and the respective relations may be evaluated immediately. In general, we have to find an expression for $\varphi_j(T)$ and the partial derivative $[\partial \varphi_j(T)/\partial T]$.

In order to determine $\varphi_j(T)$ adequately, we write down first the expression for the Gibbs free energy $G = F + pV$. It reads

$$G(T, V, \mathbf{N}) = \sum_{j=1}^{N_0} N_j(t) \left\{\varphi_j(T) + k_B T \ln\left(\frac{N_j(t)}{V} \lambda_j^3\right)\right\} \; . \tag{3.19}$$

The binding energy contributions to the Gibbs free energy of one cluster of size j are given for the case of droplet formation in the vapor by

$$G(j) = j \mu_{eq}^{(\infty)}(T) + \alpha_2 j^{2/3} + k_B T \tau \ln j \; , \tag{3.20}$$

$$\alpha_2 = 4\pi\sigma \left(\frac{3v_\alpha}{4\pi}\right)^{2/3} , \qquad v_\alpha = \frac{1}{c_\alpha} . \tag{3.21}$$

Here c_α denotes the concentration of the liquid phase and σ is the surface tension.

We will have, in general,

$$\begin{aligned} \varphi_j(T) = 0 \quad & \text{for} \quad j = 1 \\ G(j) \quad & \text{for} \quad j \geq 2 . \end{aligned} \tag{3.22}$$

If we assume φ_j as independent of cluster size, we may write, thus, approximately [502],

$$\varphi_j(T) \cong 0 \quad \text{for} \quad j = 1 \tag{3.23}$$

$$\varphi_j \cong \mu_{eq}^{(\infty)}(T) \quad \text{for} \quad j \geq 2 . \tag{3.24}$$

For the case of an equilibrium coexistence of liquid and gas phases at a planar interface, the chemical potential $\mu_{eq}^{(\infty)}$ of a monomer in the gas is given by (cf. Eq.(3.15))

$$\mu_{eq}^{(\infty)}(T) = k_B T \ln \left[\lambda_1(T)^3 c_{eq}^{(\infty)}(T)\right] . \tag{3.25}$$

Here $c_{eq}^{(\infty)}$ is the particle density of the vapor in such an equilibrium state.

The equilibrium vapor density $c_{eq}^{(\infty)}(T)$ is related to the equilibrium vapor pressure $p_{eq}(T)$ by the perfect gas law

$$c_{eq}^{(\infty)}(T) = \frac{p_{eq}(T)}{k_B T} . \tag{3.26}$$

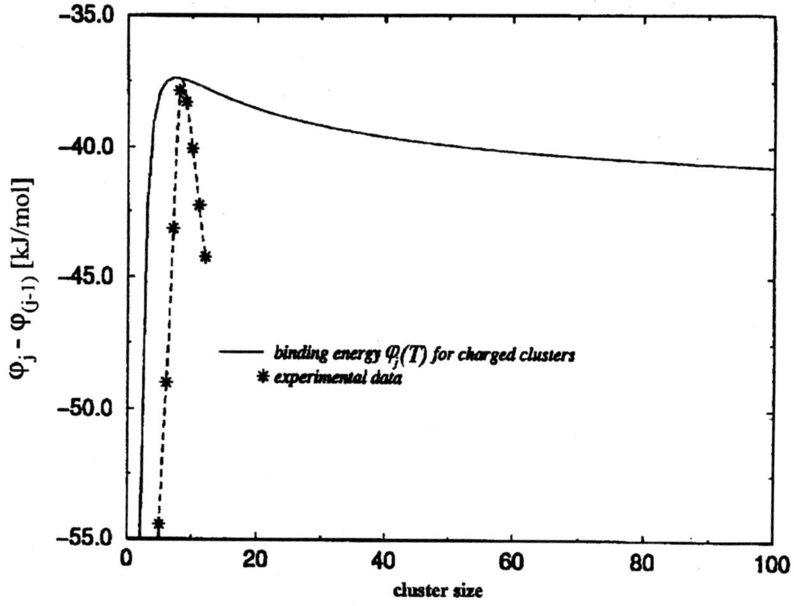

Fig. 3.2 Binding energy differences of a charged water droplet [466].

For the temperature dependence of the equilibrium vapor pressure $p_{eq}(T)$ we may use the following expression

$$p_{eq}(T) = p_{eq}(T_0) \exp\left[\frac{\hat{q}}{k_B}\left(\frac{1}{T_0} - \frac{1}{T}\right)\right] , \tag{3.27}$$

where $p_{eq}(T_0)$ is the equilibrium vapor pressure at temperature T_0, and $\hat{q}$ is the heat of vaporization per molecule.

With these relations, we get the simple result

$$\left(\frac{\partial \varphi_j(T)}{\partial T}\right) \cong k_B \left(\frac{\mu_{eq}^{(\infty)}(T) + \hat{q}}{k_B T} - \frac{5}{2}\right) j . \tag{3.28}$$

Thus we obtain, finally,

$$S(T,p,\mathbf{N}) = k_B \left\{ \sum_{j=1}^{N_0} N_j(t) \left[\frac{5}{2} - \ln \left(\frac{\lambda_j^3(T) N_j(t) p}{N_{tot}^{(cl)} k_B T} \right) \right] - \right.$$
$$\left. - \sum_{j=1}^{N_0} (1-\delta_{1,j}) j N_j(t) \left(\frac{\mu_{eq}^{(\infty)}(T) + \hat{q}}{k_B T} - \frac{5}{2} \right) \right\} , \qquad (3.29)$$

$$U(T,p,\mathbf{N}) = \sum_{j=1}^{N_0} N_j(t) \left\{ \frac{3}{2} k_B T + (1-\delta_{1,j}) \left[\frac{5}{2} k_B T - \hat{q} j \right] \right\} . \qquad (3.30)$$

Here by $\delta_{1,j}$ the Kronecker symbol is denoted ($\delta_{1,j} = 1$ for $j = 1$, $\delta_{1,j} = 0$ otherwise).

In the absence of clustering Eq.(3.29) is reduced to the well–known Tetrode–Sackur formula (cf. [258]) and Eq.(3.30) represents the energy of a one – component perfect gas (with $f = 3$).

The expressions for the thermodynamic functions may be used to express them in dependence on the appropriate thermodynamic variables, e.g, the internal energy as a function of $U = U(S, V, \mathbf{N})$ or the entropy as $S = S(U, V, \mathbf{N})$. The respective transformations are straightforward and will not be given here.

3.3.3 Comments on the Determination of the Binding Energy $\varphi_j(T)$

The negative binding energy $\varphi_j(T)$ in dependence on cluster size has to be determined theoretically from microscopic considerations or introduced empirically. We will use here not Eqs.(3.22) and (3.29) but a modification of the well known Bethe–Weizsäcker ansatz [329, 328, 494]. For neutral clusters we write

$$\varphi_j(T) = \mu_{eq}^{(\infty)}(T)\, j(1 - j^{-\alpha}) + \sigma A_1 j^{2/3}(1 - j^{-\beta}) . \qquad (3.31)$$

$$\frac{\partial \varphi_j}{\partial T} = k \left[\frac{\mu_{eq}^{(\infty)} + \hat{q}}{k_B T} - \frac{5}{2} \right] n(1 - n^{-\alpha}) . \qquad (3.32)$$

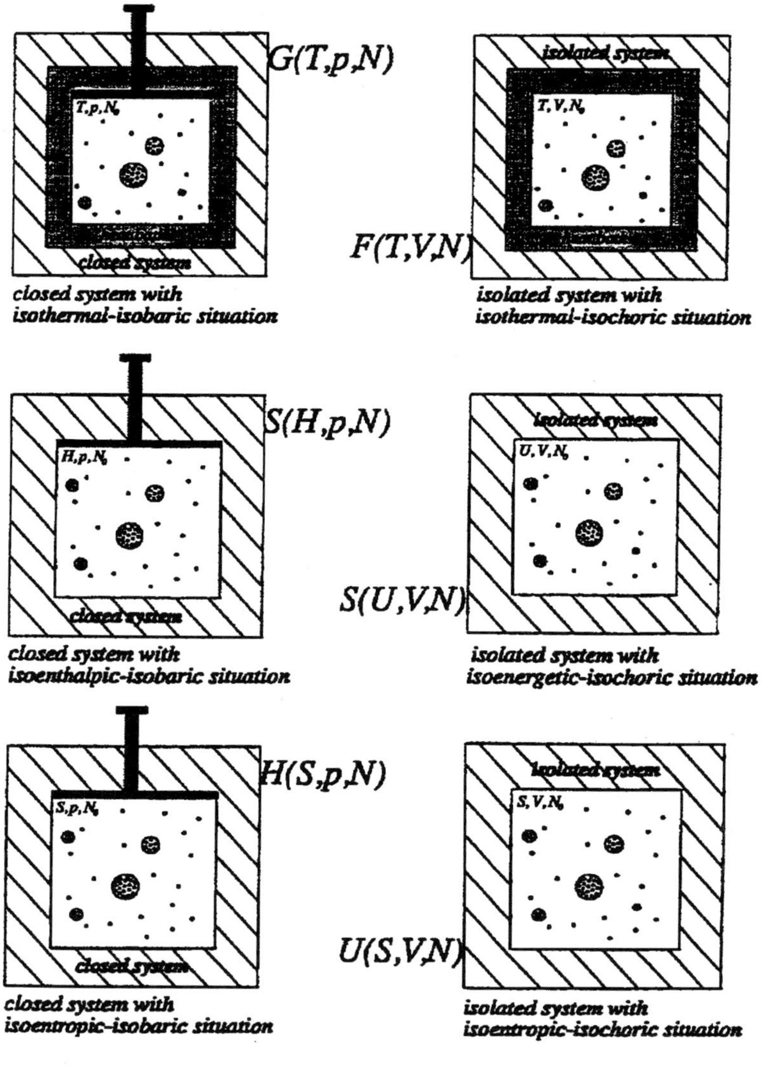

Fig. 3.3 Thermodynamic systems under different boundary conditions in diagram form.

Here it is taken into account that the Bethe–Weizsäcker ansatz is not valid for clusters of small sizes. Therefore a Padé approximation is used [329, 328, 336] to correct the results in this range.

The values of the parameters are taken from experiment. They are found as $\alpha = 1.1$ and $\beta = 2.4$ for water. These values will be applied here also for the description of cluster formation in methanol vapors (cf. Fig. 3.1).

In Fig. 3.1 binding energy differences for methanol are shown. It is evident that the experimental results are in good agreement with the theoretical curve as obtained from Eq.(3.31).

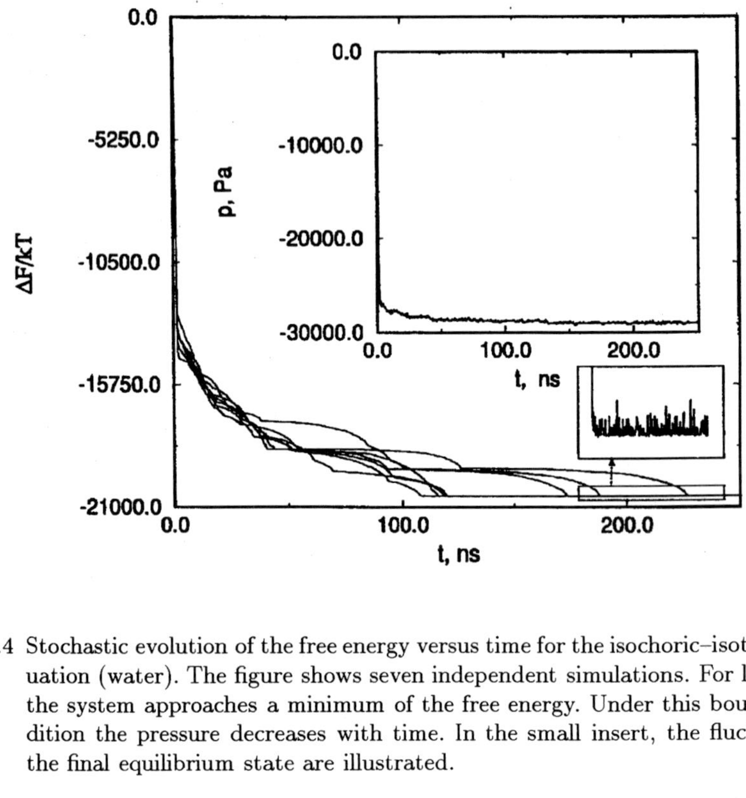

Fig. 3.4 Stochastic evolution of the free energy versus time for the isochoric–isothermal situation (water). The figure shows seven independent simulations. For large times, the system approaches a minimum of the free energy. Under this boundary condition the pressure decreases with time. In the small insert, the fluctuations in the final equilibrium state are illustrated.

We will also consider cluster formation on charged particles. For charged clusters we have to add an additional term reflecting the influence of a charge on cluster formation.

Applying classical electrodynamics, we get for the Coulomb energy of a charge in vacuum the following relation

$$U_{Coulomb} = \frac{3}{5} \frac{Q^2}{4\pi\varepsilon_0} \frac{1}{R_j} \tag{3.33}$$

or

$$U_{Coulomb} = \frac{3}{5}\frac{Q^2}{4\pi\varepsilon_0}\left(c_\alpha \frac{4\pi}{3}\right)^{1/3} j^{-1/3} . \tag{3.34}$$

The equation for the binding energy of charged clusters can be written then as

$$\varphi_j(T) = \mu_{eq}^{(\infty)}(T)j + \sigma A_j + \frac{3}{5}\frac{Q^2}{4\pi\varepsilon_0}\left(c_\alpha \frac{4\pi}{3}\right)^{1/3} j^{-1/3} . \tag{3.35}$$

The total potential energy of a single cluster contains thus three terms (cf. Fig. 3.2), a volume term $\sim j$, determined by the macroscopic value of the chemical potential $\mu_{eq}^{(\infty)}$ of the monomers, a surface term $\sim j^{2/3}$, determined by the macroscopic value of the surface tension σ surface tension, a charge term $\sim j^{-1/3}$ proportional to the charge of the cluster squared.

A comparison with the experimental data (cf. Fig. 3.2) shows, again, a quite reasonable agreement.

3.4 Kinetics of Cluster Formation

3.4.1 Kinetic Equation and Transition Probabilities

Similarly to classical nucleation theory, we assume that the cluster size distribution $\mathbf{N}(\mathbf{t})$ is changed by monomer – cluster reactions of the form

$$N_1 + N_j \rightleftharpoons N_{j+1} \tag{3.36}$$

(formation and dissolation of clusters by one monomer), only. The probability $P(\mathbf{N}, t)$ that the system has the cluster distribution $\mathbf{N}$ at time t, is given by the master equation [144, 211]

$$\frac{\partial}{\partial t}P(\mathbf{N}, t) = \sum_{\mathbf{N}'}\{W(\mathbf{N}|\mathbf{N}')P(\mathbf{N}', t) - W(\mathbf{N}'|\mathbf{N})P(\mathbf{N}, t)\} . \tag{3.37}$$

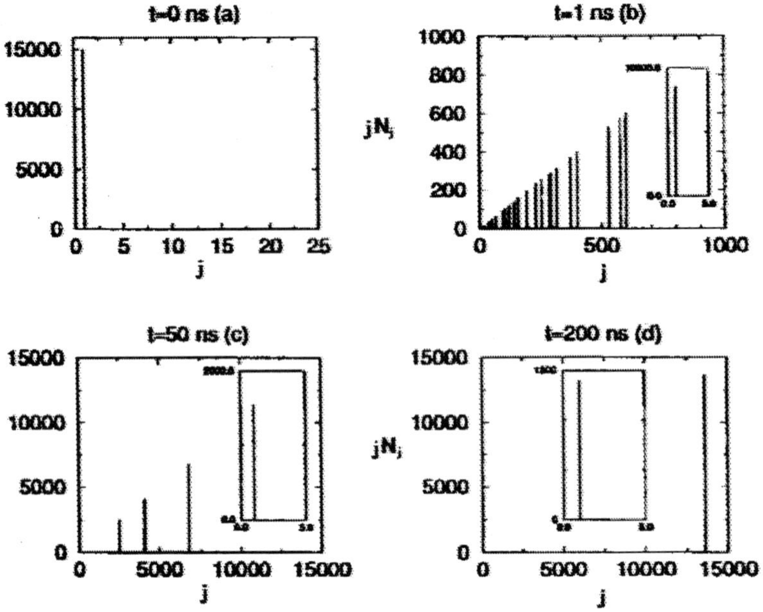

Fig. 3.5 Stochastic evolution of the cluster size distribution for condensation of water vapor. The quantity jN_j vs cluster size j is presented for different times. The final state is a two–phase state. The equilibrium cluster size distribution for water is $N_1 = 1388$ and $N_{13612} = 1$.

The equilibrium states in this system correspond to the extrema of the thermodynamic potential. The appropriate thermodynamic potentials for the different boundary conditions are shown in Fig 3.3.

On the other side, the equilibrium cluster distribution $\mathbf{N}^{(eq)}$ is determined by the stationary solution of the master equation $(\partial P_{st}/\partial t = 0)$. For any thermodynamic constraints, reflected by the characteristic thermodynamic potential Φ, we may write

$$P_{st}(\mathbf{N}) = P_{norm}^{-1} \exp(-\Phi/k_B T) \tag{3.38}$$

with a normalizing factor P_{norm}. The states $\mathbf{N}^{(eq)}$ have the highest possible equilibrium probabilities $P_{st}(\mathbf{N}^{(eq)}) \to \max$ [126].

In order to solve the master equation, we need the transition probabilities $W(\mathbf{N}|\mathbf{N}')$ from the cluster distribution $\mathbf{N}$ to the allowed by Eq.(3.36) distributions $\mathbf{N}'$ and vice versa. Following the approach as developed by Slezov and Schmelzer [510, 517], we may write

$$W(\mathbf{N}'|\mathbf{N}) = W(\mathbf{N}|\mathbf{N}') \exp\left(-\frac{\Delta\Phi}{k_B T}\right) \tag{3.39}$$

or

$$w_j^{(-)}(N_j) = w_{j+1}^{(+)}(N_{n-1}) \exp\left(-\frac{\Delta\Phi}{k_B T}\right) . \tag{3.40}$$

Here $w_j^{(-)}(N_j)$ is the probability of evaporation and $w_{j+1}(N_{j-1})$ is the probability of attachment of single particles.

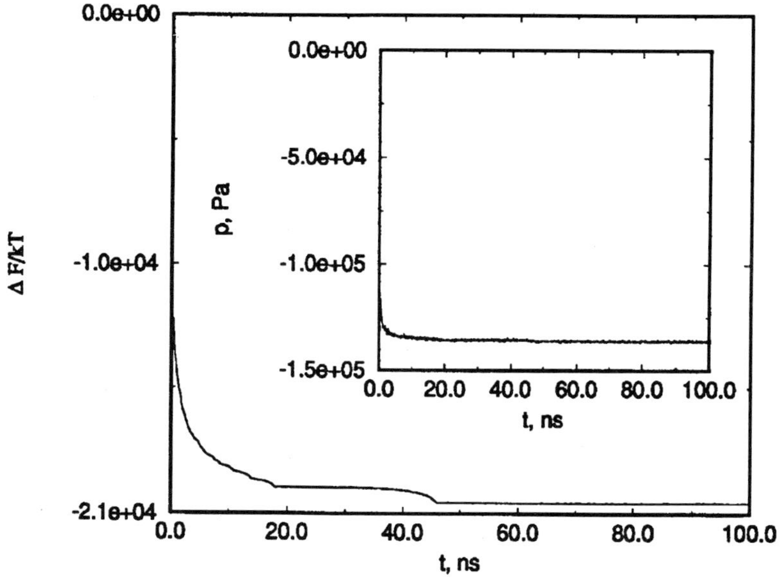

Fig. 3.6 Stochastic evolution of the free energy of the cluster ensemble for methanol vs time.

As a plausible expression for the rate of attachment of a monomer to a cluster of size j ($j \geq 2$), we use the coefficient of restitution ansatz. The rate

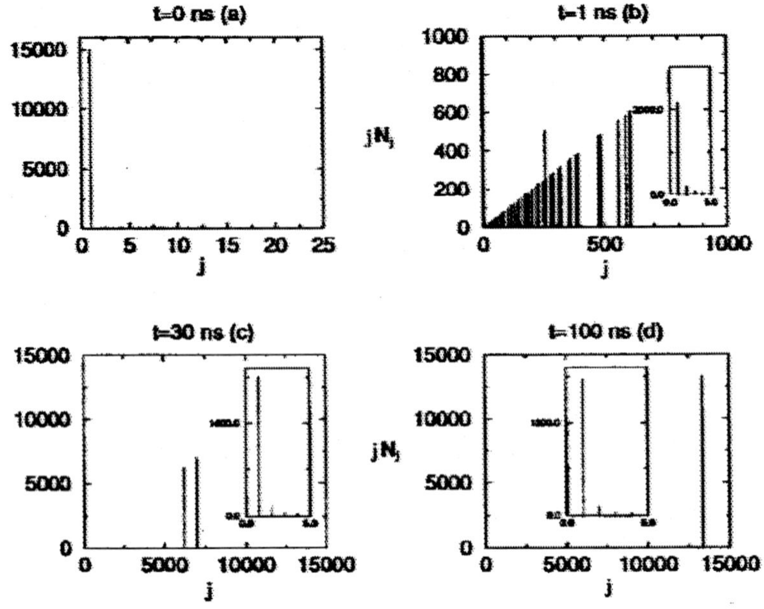

Fig. 3.7 Stochastic evolution of the cluster size distribution for methanol. The quantity jN_j vs cluster size j is presented for different times, again. The equilibrium state consists of one large cluster ($j = 13611$) surrounded by vapor particles ($N_1 = 1389$).

of attachment must be proportional to the number of clusters N_j of size j, the density of monomers N_1/V and the cluster surface A_j. We have

$$w_j^{(+)}(N_j) = \alpha(T) A_j N_j \frac{N_1}{V} \qquad \text{for} \qquad j \geq 2\,. \tag{3.41}$$

For the formation of a dimer from two monomers we have the similar expression

$$w_1^{(+)}(N_1) = \alpha(T) A_1 \frac{N_1(N_1 - 1)}{V}\,. \tag{3.42}$$

The coefficient $\alpha(T)$ is given by $\alpha(T) = D l_0^{-1}$, where D is the diffusion coefficient, $l_0(T) = 2\sigma / c_\alpha kT$ is the capillary length, σ is the surface tension and c_α is the concentration of particles in a liquid phase [329].

3.4.2 Solution of the Kinetic Equations

The time evolution of the system into the equilibrium state can be described as the solution of the master equation (3.37). This solution is obtained here by applying the Monte–Carlo simulation method (cf. [52] and chapter 5). Hereby a special procedure has been designed to convert MC–steps into real time scale.

Here the Metropolis method [365] for a constant temperature Monte–Carlo simulation is applied. In contrast to the microcanonical ensemble where all states have the same weight, in the canonical ensemble to different states different weights are assigned to.

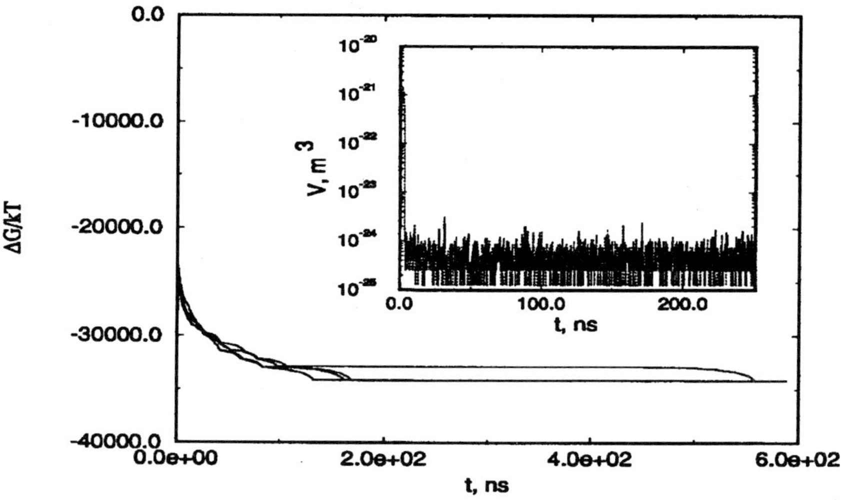

Fig. 3.8 Evolution of the free enthalpy and the volume of the system vs time under isobaric–isothermal boundary conditions.

A simple random walk through phase space is not applicable for the evaluation of observables in the (V, T, N) ensemble. To generate such a path that the different possible states occur with the correct probability, a Markov process has to be constructed, yielding a limit distribution corresponding to the equilibrium distribution of the canonical ensemble.

In the canonical ensemble the particle number N_0, the volume V and the temperature T are fixed. In such a situation an observable A is computed as

$$\langle A \rangle = \frac{1}{Z} \int\limits_{\Omega} A(x) \exp\left(-\frac{H(x)}{k_B T}\right) dx \ , \tag{3.43}$$

$$Z = \int\limits_{\Omega} \exp\left(-\frac{H(x)}{k_B T}\right) dx \ . \tag{3.44}$$

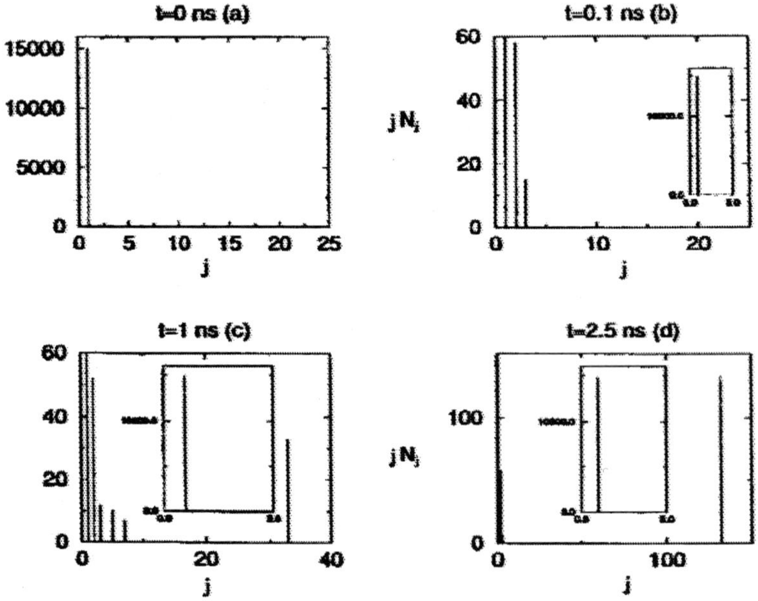

Fig. 3.9 Stochastic evolution of the cluster distribution for the isoenergetic–isochoric situation. The quantity jN_j vs cluster size j is presented for different times. The equilibrium cluster size distribution for water is $N_1 = 14860$ and $N_{140} = 1$.

To develop the appropriate Monte–Carlo method we note that in equilibrium the distribution of states is given by (cf. Eq.(3.38))

$$P(x) = Z^{-1} \exp\left[-H(x)/k_B T\right] dx \ . \tag{3.45}$$

The ratio of the transion probabilities depends only on the change in energy $\Delta\Phi$ in passing from one state to another, i.e.,

$$\frac{W(\mathbf{N}'|\mathbf{N})}{W(\mathbf{N}|\mathbf{N}')} = \exp(-\Delta\Phi/k_B T) \ . \tag{3.46}$$

The further calculations are carried out following the usual prescription:

1. Specify an initial configuration $\mathbf{N} = \mathbf{N}(t = 0)$.

2. Generate a new configuration $\mathbf{N}'$.

3. Compute the energy change $\Delta\Phi$.

4. If $\Delta\Phi < 0$, accept the new configuration and return to step 2.

5. Compute the quantity $\exp[-\Delta\Phi/k_B T]$.

6. Generate a random number $\epsilon \in [0, 1]$.

7. If ϵ is less than $\exp[-\Delta\Phi/k_B T]$, accept the new configuration and return to step 2.

8. Otherwise, retain the old configuration as the new one and return to step 2.

At this point we see more clearly the meaning of the used here choice of the transition probabilities. The system is driven towards the minimum of the free energy for given values of the parameters (N_0, T, V). Step 4 accounts that we always accept a new configuration having a smaller free energy than the previous one. Configurations which increase the free energy are only accepted with a Boltzmann factor less than one.

3.5 Results of the Simulations for Different Boundary Conditions

3.5.1 Isochoric–Isothermal Boundary Conditions

In the considered case we get the transition probabilities in the form

$$
\begin{aligned}
w_j^{(-)}(N_j) &= w_{j-1}^{(+)}(N_{j-1}) \exp\left[\frac{(F(T,V,\mathbf{N}') - F(T,V,\mathbf{N}))}{k_B T}\right] , \\
w_j^{(+)}(N_j) &= \alpha(T) A_j N_j \frac{N_1}{V} \qquad \text{for} \qquad j \geq 2 , \\
w_1^{(+)}(N_1) &= \alpha(T) A_1 \frac{(N_1 - 1) N_1}{V} \qquad \text{for} \qquad j = 1 .
\end{aligned}
\tag{3.47}
$$

As the initial configuration, we assume always that all particles exist in the system as monomers, only. Moreover, in order to specify the parameter values, we use the data for water and methanol (cf. Table 3.1).

property/substance	water	methanol
liquid density c_α	$3.35 \cdot 10^{28}$ m^{-3}	$1.49 \cdot 10^{28}$ m^{-3}
surface tension σ	$7.3 \cdot 10^{-2}$Jm^{-2}	$2.25 \cdot 10^{-2}$Jm^{-2}
diffusion coefficient D	10^{-5}m^{2}s^{-1}	10^{-5}m^{2}s^{-1}
molar mass M	18 g mol^{-1}	32 g mol^{-1}
heat of boiling $\hat{q}$	$7.35 \cdot 10^{-20}$ J	$5.88 \cdot 10^{-20}$ J
reference temperature T_0	293 K	293 K
equilibrium concentration of vapor c_0 at $T = T_0$	$5.78 \cdot 10^{23}$m^{-3}	$3.25 \cdot 10^{23}$m^{-3}
total number of particles N_0	15000	15000
initial volume V	$1.9 \cdot 10^{-21}$m^3	$4 \cdot 10^{-22}$m^3
initial temperature T	293 K	293 K

Tab. 3.1 Data for water and methanol [156, 298].

We demonstrate our numerical results in several figures (Figs. 3.4 and 3.5 for water and Figs. 3.6 and 3.7 for methanol). Qualitatively, the results for water and methanol are identical. The equilibrium cluster distribution for water is $N_1 = 1388; N_{13612} = 1$ and for methanol $N_1 = 1389; N_{13611} = 1$. However, the time scales are different due to different values of the parameters affecting the nucleation–growth process.

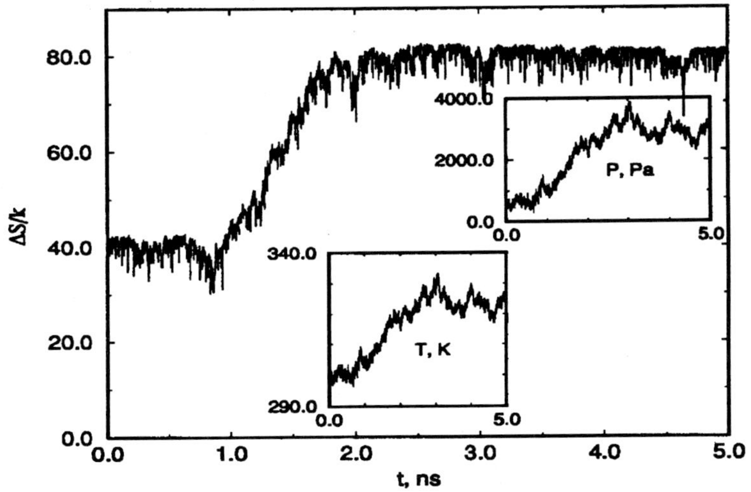

Fig. 3.10 Stochastic evolution of the entropy for water vapor under isoenergetic–isochoric boundary conditions. In the small inserts we show the time evolution of temperature and pressure.

3.5.2 Isothermal–Isobaric Boundary Conditions

The transition probabilities are given by Eqs.(3.47), again. Figure 3.8 illustrates the evolution of the free enthalpy difference from the initial to the final states. Here the final state consists of a compact liquid phase and practically no monomers. For the assumed, now, boundary condition of constancy of the external pressure, a stable two–phase state cannot be formed.

Again, the same equilibrium state is formed for both substances (water and methanol). The cluster distribution in the equilibrium is shifted to one large cluster (liquid phase).

3.5.3 Isoenergetic–Isochoric Boundary Conditions

Under isoenergetic–isochoric conditions the probabilities of attachment are expressed via the entropy $S(U, V, \mathrm{N})$ of the different allowed states. We obtain

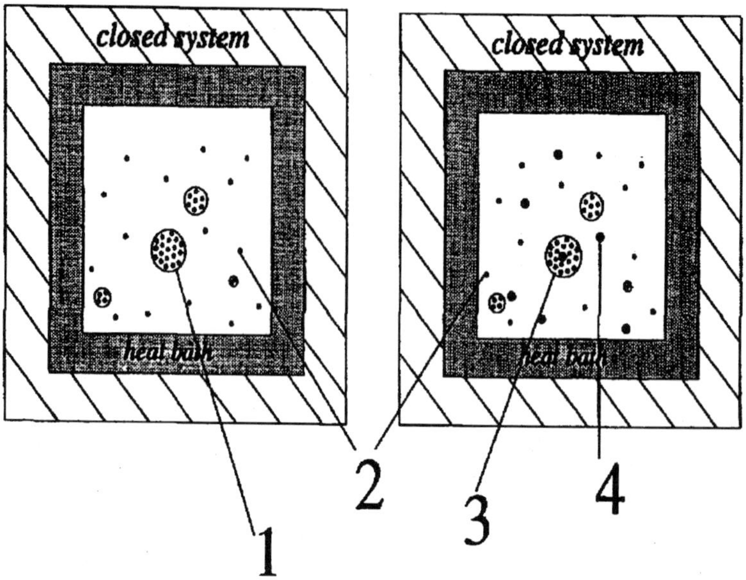

Fig. 3.11 Model applied for the comparison of the course of homogenous and heterogenous cluster formation in a thermodynamic system under isothermal–isochoric boundary conditions (1: neutral cluster, 2: neutral monomers, 3: charged cluster, 4: charged monomer).

$$\begin{aligned} w_j^{(-)}(N_j) &= w_{j-1}^{(+)}(N_{j-1}) \exp\left[\frac{(S(U,V,\mathbf{N}') - S(U,V,\mathbf{N}))}{k_B}\right] , \\ w_j^{(+)}(N_j) &= \alpha(T) A_j N_j \frac{N_1}{V} \qquad \text{for} \qquad j \geq 2 , \\ w_1^{(+)}(N_1) &= \alpha(T) A_1 \frac{N_1(N_1-1)}{V} \qquad \text{for} \qquad j = 1 . \end{aligned} \tag{3.48}$$

The final cluster distribution in the equilibrium differs compared with the isothermal–isochoric case. We obtain the equilibrium distribution for water as $N_1 = 14870$, $N_{130} = 1$ (Fig. 3.9).

The reason for the small final cluster (liquid phase) is the emission of the latent heat of condensation. The system starts to heat up and the process

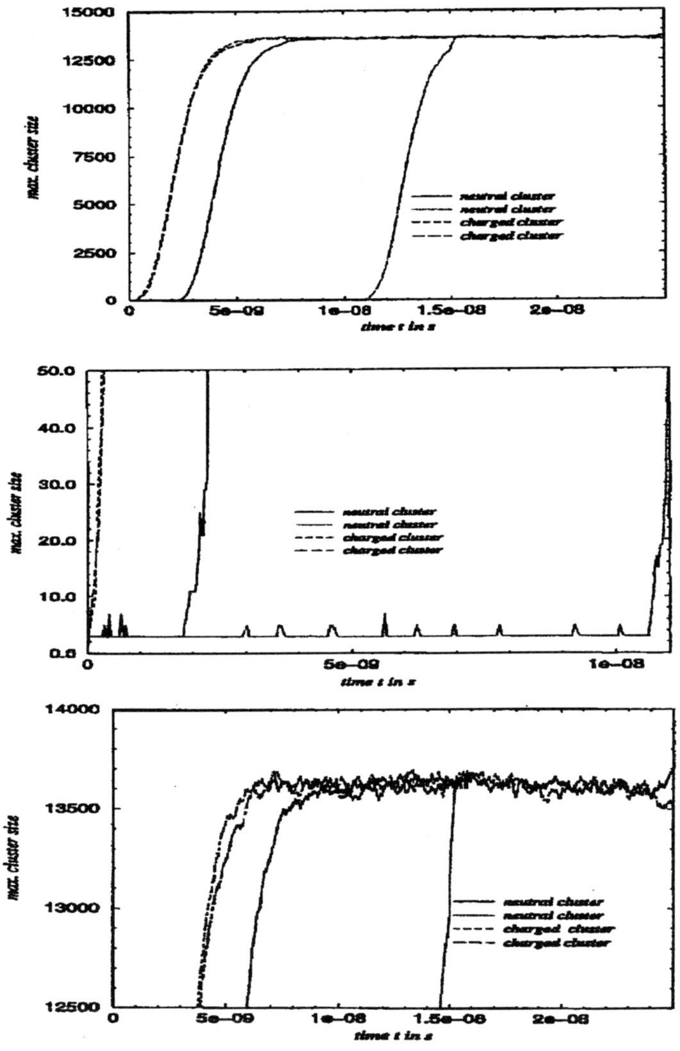

Fig. 3.12 Stochastic evolution of the largest cluster in both considered systems for homogenous and heterogenous nucleation. For each of the cases, two independent simulations of the process are shown. Top: overall course of the evolution; center: initial stage; bottom: final stage.

of cluster growth terminates. The equilibrium state of the system will be reached faster in comparison with the isothermic–isochoric situation.

3.5.4 Heterogeneous Cluster Formation under Isothermal–Isochoric Boundary Conditions

As a further application we consider, now, cluster formation under isothermal – isochoric conditions but assume the possible existence of charged particles as additional nucleation cores. In the simulations, we compare the time evolution of only one cluster in the system in presence and absence of charged particles, respectively, for the both considered situations.

The transition probabilities Eqs.(3.47) remain the same as for the case of homogeneous nucleation. However, the expressions for the free energies change. For the binding energy we use here Eq.(3.35).

Fig. 3.12 shows the time evolution of the largest cluster into the equilibrium state (two-phase-state). We suppose here in the calculations that only one nucleus is formed.

It is evident that in the case of heterogeneous nucleation the system quickly goes over into the equilibrium state. The existence of the charge accelerates the growth of the cluster. The size of the critical cluster is smaller then in the case of homogeneous nucleation.

The final equilibrium state remains, however, nearly the same independent of the existence of the charge in the cluster (for more details on heterogeneous nucleation mechanisms see, e.g., Gutzow, Schmelzer [187]).

3.6 Conclusion

In the present chapter, we have shown how the stochastic theory of cluster kinetics in isolated and closed systems may be developed and applied to different situations. It is shown that the choice of the boundary conditions may affect significantly the course and the final stages of the nucleation–growth process. In the average over different simulations, the results coincide with the kinetic description as outlined in chapter 2. However, for the consideration in systems containing a small number of particles the stochastic approach to nucleation has serious advantages due to the account of the possible effect of fluctuations.

4 Bound State – Cluster – Correlations in Macroscopic Matter: A Quantum–Statistical Approach

G. Röpke

4.1 Isolated Clusters

Phase transformation processes which proceed by nucleation and growth involve the formation of relatively small aggregates or clusters of the newly evolving phase. Their adequate description may be of decisive importance for the understanding of transformation phenomena. On the other hand, the properties of clusters may be used directly in various technological applications.

The physics of clusters exhibits new and interesting aspects characterized by the transition between the limiting cases of few particle bound states and to a condensed phase. Being a system of finite size, surface effects are of importance. On the other hand, collective effects appear if the clusters are sufficiently large. The local density approach can be used to interpolate between both limiting cases.

As examples, nucleonic, atomic and molecular clusters are considered here. Only isolated clusters are studied first. A quantum statistical treatment of the interacting many particle system is needed for a rigorous approach to the clustered state of matter. The basic ideas of such treatment are given in the subsequent sections.

4.1.1 Clusters – a Specific State of Matter

Standard disciplines of physics are atomic and molecular physics on one hand (with a discrete spectrum of well–separated levels of energy), and the physics of the solid and the liquid states on the other hand (characterized by collective excitations with a quasi–continuous spectrum of energy levels). The new subject of cluster physics gives a connection between these different disciplines. Lacking for a sharp definition, clusters can be characterized as a complex or aggregate consisting of A particles, a considerable part of which is located at the surface. Further characteristics of clusters are the complexity of the energy level spectra, the variability of their geometric structures, the sensitivity in respect to the interaction with their surroundings.

We may consider isolated clusters where the interaction with the surrounding matter can be neglected, but also clusters imbedded in matter of finite density and temperature, or clusters interacting with other clusters. Whereas the treatment of isolated clusters leads to the solution of an A particle problem, the description of the physical properties of a system of clusters can be given within the many–particle theory. If the density is low, the interaction between the cluster and the medium or between different clusters, respectively, can be considered to be weak, so that perturbation theories can be applied. At high densities, the properties of the clusters can be strongly modified so that a perturbation approach starting from isolated clusters may become inappropriate.

Clusters can be found in very different situations. Examples for clusters to be described by quantum statistics are atomic clusters formed in adiabatic expansion, atomic nuclei, many exciton clusters in excited semiconductors, clusters in dense alkaline plasmas, clusters in alloys, small crystallites in crystal growth processes. Examples for clusters to be described by classical statistics are clusters of stars and galaxies, clusters in biology, solvatation, etc.

The most important questions to be answered in the analysis of clusters are: how the macroscopic properties are formed from the properties of atoms and molecules such as the work function from the ionization potential or the electron affinity, the phonon dispersion from the vibration frequencies of a molecule, collective plasmon excitations, conductivity, magnetic susceptibility, superconductivity and other electric properties, vapor pressure, temperature of melting, and other characteristics of phase transitions. The technological aspects of the physics of clusters are of high interest (photography, nuclear energetics, catalysis, microelectronics, aerosols, etc.).

The main problem in the theoretical analysis consists in the systematic treatment of the A – particle system and the search for the origin of collective modes in finite size systems. With increasing A, one can derive bulk properties (proportional to A) and surface properties (proportional to $A^{2/3}$). While for small cluster we can apply the concept of bound states with properties strongly varying with the particle number A, for large clusters other concepts like the liquid drop model are applicable. Systems in the transition region where macroscopic behavior establishes are also denoted as mesoscopic systems. Presently there is a large research activity in that field. One of the typical examples are clusters.

4.1.2 Bound States

A few–body system with attractive interaction, consisting of A particles ($A = 2, 3, \ldots$), can form bound states, where the elementary particles stay together for a long period of time. Compared with the constituents, bound states are characterized by new properties. For example, the excitation spectrum of the clusters is of specific nature and depends on the cluster size. Also the response to an external field for quantum clusters is different, depending on the particle number and the state of excitation. The method of theoretical analysis is different whether classical or quantum systems are considered.

Classical Systems

In classical mechanics, the equation of motion for the two–body problem can be solved introducing the center of mass and relative coordinates. For attractive interaction, bound states can be formed where the motion is restricted to finite regions in space. If the energy is higher than a threshold value, scattering states appear where the motion is unbound.

Classical systems with more than two particles are, in general, not integrable. Under certain conditions, perturbation theory can be applied where the unperturbed system can be solved by separation. It is the subject of nonlinear mechanics to investigate whether the trajectories remain stable under the influence of the perturbers. As an example, we refer to the motions within the solar system. Different concepts such as regular and chaotic motion have been worked out to characterize the time evolution of classical A – particle systems. In particular, the distinction between bound states and unbound motion is not sharp. Also for energies above the threshold of unbound motion, clusters can be formed which stay together for a relatively long time interval. As an example we can consider stellar clusters in astronomy. However, particles can always escape not coming in conflict with the conservation laws of energy and momentum.

The treatment of an A – particle system of interacting particles in classical mechanics can be performed by molecular dynamics simulation. The classical equations of motion are solved numerically, and the evolution of the system, the formation of clusters and their behavior under external perturbations can be calculated.

Star clusters are examples for the formation of structures and their evolution in classical mechanics.

Quantum systems

On a microscopic level, bound states have to be described within a quantum mechanical approach. The Schrödinger equation of the A – particle system reads

$$H^{(A)}\psi_{A\alpha p}(1,\ldots,A) = E_{A\alpha p}\psi_{A\alpha p}(1,\ldots,A)\ . \tag{4.1}$$

The Hamiltonian

$$H^{(A)} = \sum_{i=1}^{A} \frac{p_i^2}{2m_i} + \frac{1}{2}\sum_{i\neq j=1}^{A} V(ij,i'j') \tag{4.2}$$

contains two, the kinetic energy and the potential energy, terms. The variables $\{i\} = \{1,\ldots,A\}$ are single–particle observables such as momentum p_i, spin σ_i, isospin τ_i, etc. We prefer this set of state variables compared with others such as position etc.

The solutions of the Schrödinger equation should be normalizable, i.e.,

$$\int |\psi_{A\alpha p}(1,\ldots,A)|^2 = 1\ . \tag{4.3}$$

They are characterized by the quantum numbers $\{A\alpha p\}$, where besides the particle number A, the total momentum $\vec{p}$, also further "internal" quantum numbers α are taken into account. If the system is included in a normalization volume Ω_0, the dependence of the wave function on the normalization volume can be different. If the amplitude of the wave function of internal motion remains finite in the limit $\Omega_0 \to \infty$, the corresponding eigenstate is a bound state describing a finite motion. If the amplitude goes to zero

($\propto \Omega_0^{-1/2}$), the corresponding state is a scattering state describing an unbound motion. The spectrum of energy eigenvalues $E_{A\alpha p}$ can be discrete or continuous, respectively.

A simple example is the hydrogen atom, consisting of two particles (electron, proton) and interacting via the Coulomb potential

$$V(12,1'2') = \frac{e_1 e_2}{4\pi\epsilon_0} \frac{1}{(\vec{p}_1 - \vec{p}_1{}')^2 \Omega_0} \delta_{p_1+p_2, p_1'+p_2'} \delta_{\sigma_1,\sigma_1'} \delta_{\sigma_2,\sigma_2'} . \tag{4.4}$$

The solutions of the Schrödinger equation can be found explicitly in this case, bound states occur if the energy eigenvalues are negative.

A new aspect occurs if identical particles are considered. Then, the wave function has to fulfil symmetry relations. In the case of bosons, it has to be symmetric, and in the case of fermions, it has to be antisymmetric with respect to transpositions of two particles. Solutions of the Schrödinger equation, which do not fulfil the corresponding symmetry relation, are not admitted.

As a simple example, the electrons of a H_2 – molecule can be found in a symmetric orbital if the spins form a singlet state (binding state) or in a antisymmetric orbital if the spins form a triplet state (antibinding state). In atoms with many electrons, the condition of antisymmetry of the electronic wave function leads to the Pauli exclusion principle. A well–known consequence is the formation of shells in atoms. Another example for a quantum system of identical particles is the atomic nucleus, where the protons as well as the neutrons obey the Pauli exclusion principle.

Bound states show new properties compared with the elementary particles. They may be considered as new species (chemical picture), each having a specific chemical potential $\mu_{A\alpha}$. From a more fundamental point of view, however, these bound states are composites and have an internal structure. As long as no internal degrees of freedom are excited, these bound states may be considered as 'elementary'. If chemical reactions are taken into account, the different chemical potentials $\mu_{A\alpha}$ are related within each other, see the following section. It is an interesting aspect of statistical physics that clusters of different size, molecules, atoms, nuclei, nucleons, etc., may be considered as elementary, depending on the energies of excitation in the process under consideration.

4.1.3 Local Density Approximation

The solution of the A–particle Schrödinger equation to calculate isolated clusters is a difficult task if the number A of particles is large. Compared with infinite systems, the surface is essential in determining the properties of clusters. More general, properties such as the particle density are space dependent what makes the treatment more complex compared with the homogeneous case.

Different methods have been worked out to solve the many–particle Schrödinger equation, including perturbation expansions, variational approaches and numerical simulations. A standard approach to the many particle problem is the Hartree–Fock approximation, where correlations between the particles are neglected. The wave function of the A – particle system is represented by a (anti–)symmetrized product of single–particle wave functions. The optimum form of the single–particle wave functions is found from the solution of the Hartree–Fock equations.

A very efficient approach to finite, inhomogeneous quantum systems is the density functional theory [226, 120], see also [136]. This method is able to account for correlations in the considered systems. The starting point in this method is the energy as a functional of the density. For charged particle systems the respective expression is given by

$$
\begin{aligned}
E[n(\vec{r},t),\tau(\vec{r},t)] = & \frac{\hbar^2}{2m}\int \tau(\vec{r},t)\,d\vec{r} + \int V(\vec{r})n(\vec{r})\,d\vec{r} \\
& + \frac{1}{2}\int\int \frac{\rho(\vec{r},t)\rho(\vec{r}',t)}{|\vec{r}-\vec{r}'|}d\vec{r}d\vec{r}' + \int E_{\mathrm{xc}}[\vec{r},t]d\vec{r}\,.
\end{aligned} \tag{4.5}
$$

This expression includes the kinetic energy, the potential energy, the Coulomb energy and the exchange–correlation energy term, respectively. With the single–particle wave function $\phi_\alpha(\vec{r},t)$ the density is given as $n(\vec{r},t) = \sum_\alpha |\phi_\alpha(\vec{r},t)|^2$ and the kinetic–energy density as $\tau(\vec{r},t) = \sum_\alpha |\nabla\phi_\alpha(\vec{r},t)|^2$.

For the exchange–correlation term $E_{\mathrm{xc}}[\vec{r},t]$ a gradient expansion can be performed. In the local density approximation, appropriate expressions for the exchange energy and the correlation energy of the homogeneous quantum system can be used.

The single–particle wave functions $\phi_\alpha(\vec{r},t)$ follow from the solution of the Kohn–Sham [260] equations within a variational procedure to find the minimum of the energy functional, where the time–dependent single–particle Hamiltonian is obtained as

$$H(\vec{r},t)\phi_\alpha(\vec{r},t) = \frac{\delta E}{\delta\phi_\alpha^*(\vec{r},t)} \, . \tag{4.6}$$

For a more exhaustive presentation of the local–density approximation we refer to the topical literature (see e. g. [69, 136]). We only like to mention that also extensions such as to include inhomogeneous systems [445], spin density, quantum condensates and finite temperatures are possible.

4.1.4 Atomic Clusters

Hartree–Fock calculations have been performed for Na_A clusters [61] and further alkaline clusters [68]. Within that quantum chemical ab initio approach, the electronic degrees of freedom are treated fully quantum–mechanically, whereas the positions of the nuclei are varied adiabatically as described by the Born–Oppenheimer approximation [62]. A remarkable result is that different geometries for the ion distribution give nearly the same value of the binding energy so that structure isomers should be considered for such clusters in order to evaluate thermodynamic and relaxation properties. The motion of the ions can be included within the molecular dynamics method [81] by solving their classical Newton equations, coupled to the quantum–mechanical Kohn–Sham equations for the electrons [165, 193].

Within the local density functional approximation for simple metal clusters, magic numbers have been obtained corresponding to the closure of shells in correspondence to the shell structure of atoms. These magic numbers have also been found in experiments with metallic clusters (for a comprehensive review of shells and supershells in metallic clusters see [196, 69]).

A well–investigated property of metallic clusters is the photoionization potential [196], which follows within a simple electrostatic approach as

$$W_R = W_{\text{inf}} + \frac{3}{8}\frac{e^2}{R} \, . \tag{4.7}$$

Here R is the radius of the metallic droplet, W_R the work function for a cluster of size R while W_{inf} corresponds to the respective value for a macroscopic system [61].

An interesting phenomenon is found in Hg_A clusters [434]. For small clusters, $A < 20$, a van–der–Waals cluster of dielectric atoms was observed, whereas the metallic properties were present for $A > 70$ with the correct value $W_{\text{inf}} = 4.49$ eV of the work function. This behavior is consistent with a metal–non-metal transition, which gradually occurs if the number of atoms in the cluster is smaller then $A \cong 70$.

A very rapid progress can be observed in the investigation of optical properties of simple metal clusters. Within the theoretical questions to be investigated are the decay of the electronic excited states, the coupling of the collective states with single–particle states, the coupling of these states with vibrational modes, and the effect of temperature of both the position and the width of the resonances. Recently, multipole oscillations have been considered in different alkaline clusters [390, 575, 536, 132].

Another interesting effect is the observation of multiply charged clusters. A minimum size A of a metallic cluster is necessary to stabilize it against Coulomb explosion. Minimum values A^* have been found for double, triple, and higher charged clusters of different species.

There exists a broad literature on atomic clusters which will not be reviewed here (for further details see [196, 69]). Instead we will discuss as examples the properties of nucleonic clusters.

It should be mentioned that both disciplines – analysis of atomic and nucleonic clusters – strongly interact at present. Many methods developed in, e.g., nuclear physics have been applied also to atomic systems (see [9, 489]) and vice versa.

4.1.5 Nucleonic Clusters

Atomic nuclei occur preferably in the mass number region $A = 1, \ldots, 300$. Extended experimental data such as bound state energies and reaction cross sections are well investigated for these nucleonic clusters, consisting of protons (number Z) and neutrons (number $N = A - Z$).

The bound state energy $E_b(A, Z)$ of the most stable isotope with mass number A is shown in Fig. 4.1, see Ref. [60]. For comparison, the curve resulting from the Bethe–Weizsäcker formula

$$E_b(A, Z) = b_{\text{vol}} A - b_{\text{surf}} A^{2/3} - \frac{1}{2} b_{\text{symm}} \frac{(N - Z)^2}{A} - \frac{3}{5} \frac{Z^2 e^2}{R_c} \tag{4.8}$$

is also shown, with the parameter values $b_{\rm vol} = 15.56$ MeV, $b_{\rm surf} = 17.23$ MeV, $b_{\rm symm} = 46.57$ MeV, and $R_c = 1.24A^{1/3}$ fm. It is clearly seen that this four–parameter fit reproduces the binding energies over a very large region of A fairly well. The deviations can be explained by pairing energies and shell structure effects near the magic numbers, they are also shown in Fig. 4.1.

The Bethe–Weizsäcker formula can be understood within a droplet model of nuclear matter, and may be derived from a microscopic description within a local density approximation.

The evaluation of the bound state properties starting from the Schrödinger equation demands the knowledge of the nucleon–nucleon interaction potential $V(12, 1'2')$. Such a potential does not have a simple form like the Coulomb interaction in condensed matter physics and is reconstructed from empirical data such as nucleon–nucleon scattering phase shifts. It depends on internal quantum numbers such as spin and isospin, and it seems to be possible only in a rather restricted region of energy to parametrize these data by a potential. The reason is that nucleons are bound states consisting of quarks and gluons, and the description of the interaction between nucleons by a potential is similar non–fundamental like the use of potentials such as the Lennard–Jones potential to describe the interaction between atoms.

A possible parametrization of the nucleon–nucleon interaction can be given in terms of separable potentials. For nuclear matter, separable representations of realistic nucleon–nucleon potentials such as the Paris potential [423] are available. The rank N interaction in the two–particle channel α has the form

$$V_\alpha^{LL'}(k, k') = \sum_{i,j=1}^{N} w_{\alpha i}^{L}(k)\ \lambda_{\alpha ij}\ w_{\alpha j}^{L'}(k') \ , \tag{4.9}$$

where k denotes the relative momentum, L is the angular momentum. The separable potential has the advantage that the two–particle Schrödinger equation can immediately be solved. It is not a restriction compared to other, local forms of the interaction because it can be shown that an arbitrary local potential can be represented by a sum of separable interactions. For simplicity one can start with a rank 1 potential neglecting coupling between different channels.

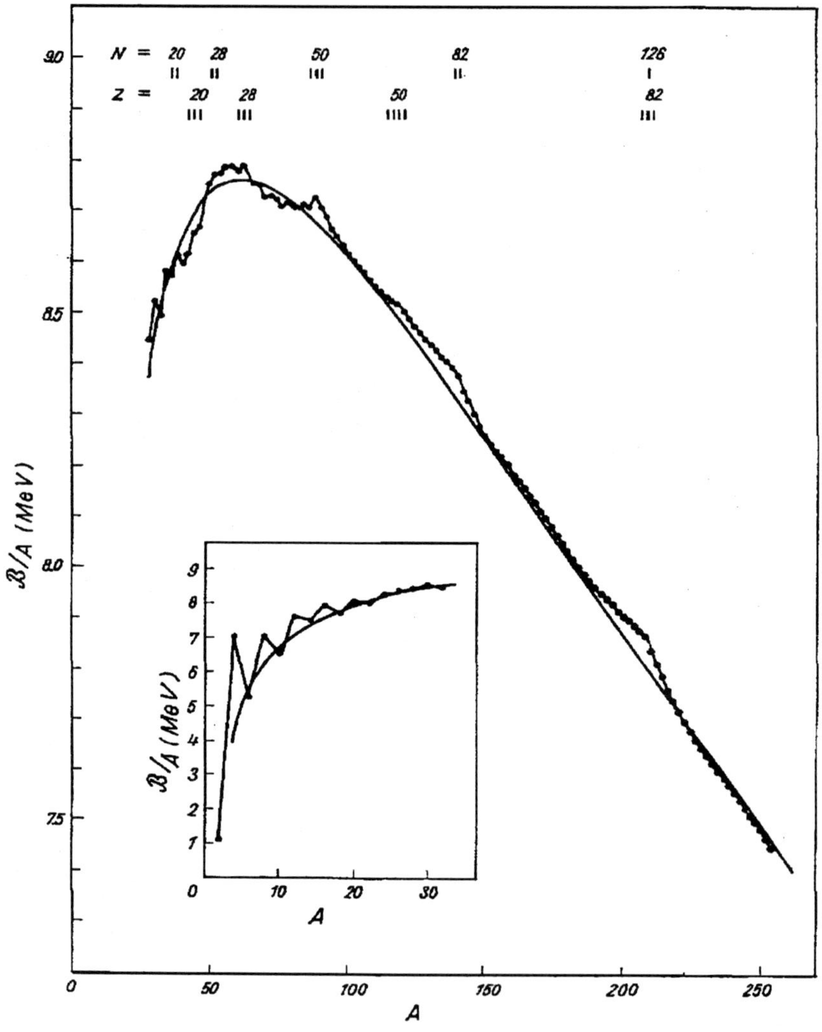

Fig. 4.1 Binding energy per nucleon E_b/A of the most stable isobars. Measured values are compared with the Bethe–Weizsäcker formula Eq. (4.8) with $b_{\mathrm{vol}} = 15.56$ MeV, $b_{\mathrm{surf}} = 17.23$ MeV, $b_{\mathrm{symm}} = 46.57$ MeV, $R_c = 1.24A^{1/3}$ fm.

Obviously the two–particle properties are reproduced from the two–nucleon interaction potential at not too high excitation energies. However, the precise reproduction of three–nucleon properties etc. as well as nuclear matter properties from that potential is presently not possible. One of the reasons may be that nucleons are composite particles so that three–particle and higher order interaction terms may occur.

Despite the different forms of the interaction potential, nucleonic clusters have many properties similar to simple metal clusters. Shell structure, magic numbers, excitations, giant resonances as collective modes in analogy to plasmon excitations are examples for many–particle effects, which occur in both systems, see also [9, 489] for the strong interrelations between the fields of atomic and nuclear clusters.

4.2 Clusters in a Medium

Clusters of elementary particles are described on a quantum mechanical level as bound states by solving the A – particle Schrödinger equation. Bound states may be considered as new species, with an effective interaction, and which are able to perform reactions. In a dense medium, the properties of clusters are modified. A quantum statistical approach is outlined here which allows to describe clusters in a dense medium in a self–consistent way.

4.2.1 Ideal Mixtures

The fact that bound states are composites becomes evident when internal degrees of freedom are excited. For instance, a reaction describes the change of the internal structure of the composites, for instance, as a consequence of collisions.

We will consider a many–particle system where bound states can be formed, confined in a volume Ω_0. In the thermodynamic limit, the volume Ω_0 as well as the total particle number N tend towards infinity, but the density N/Ω_0 remains finite. The simplest possibility to describe the considered system is as a mixture of different non–interacting clusters, i. e. elementary particles ($A = 1$), as well as different bound states. The interaction between the clusters is neglected in such approach, but accidentally reactions can occur if the clusters collide. Such an ideal mixture of reacting species is familiar from a chemical description.

The reaction is described by an equation of the form

$$\sum_{A\alpha} \nu_{A\alpha} B_{A\alpha} = 0 \,, \tag{4.10}$$

where $\nu_{A\alpha}$ are the stochiometric coefficients, and $B_{A\alpha}$ are the different components. Note that the total momenta $\vec{p}$ of the clusters are integrated over.

In the chemical approach, it is assumed that within each of the components thermal equilibrium is achieved due to frequent elastic collisions. In contrast, chemical equilibrium between the different components can be established due to relatively rare inelastic, reactive collision only after some time.

For each of the components, a chemical potential $\mu_{A\alpha}$ can be introduced, which is determined by the cluster density and the temperature. Since the number of a particular cluster is not a conserved quantity in reacting systems, a precise definition of $\mu_{A\alpha}$ can only be given within a nonequilibrium approach as presented in the following section. The quantity

$$\sum_{A\alpha} \nu_{A\alpha} \mu_{A\alpha} \tag{4.11}$$

is denoted as the affinity of the reaction, and chemical equilibrium is achieved if the affinity is equal to zero. Considering in an ideal mixture the chemical potential $\mu_{A\alpha}(T, n_{A\alpha})$ as a function of the temperature T and the cluster density $n_{A\alpha}$, the equilibrium concentrations of different clusters at given temperature are determined by the law of mass action.

A special version of this equation is the Saha equation, which describes the thermal equilibrium between atoms, ions and electrons in a plasma.

In non–equilibrium states, reaction rates can be introduced. The derivation of these reaction rate equations can be performed from the appropriate master equations, as shown below.

The chemical picture where each of the bound states can be considered as a new species, is applicable only in the low–density limit, where the interaction between the clusters is negligible (ideal mixture). If the density becomes higher, interaction effects between the different components are of importance, and the concept of isolated clusters is not applicable any more. At high densities clusters strongly overlap and the chemical picture breaks down totally. In such cases, the adequate description of the many–particle systems has to be given within a quantum statistical approach (see, e.g., [227, 267, 592]).

Another situation where the chemical picture appears to be inapplicable as compared with the physical picture is strong nonequilibrium. If the number of clusters $\{A\alpha\}$ is not nearly conserved so that a quasiequilibrium can be assumed, instead of $\mu_{A\alpha}$ only the chemical potential of the conserved elementary particle $\mu_{1\alpha}$ can be introduced. The index α denotes the internal state such as spin and isospin (proton or neutron) in nuclear systems. If β reactions transforming neutrons into protons and leptons are taken into account, only the baryon number is a conserved quantity and can be used to define a baryonic chemical potential.

4.2.2 Second Quantization

It should be pointed out that bound states are obtained from the solution of the Schrödinger equation. Therefore, the formation of clusters has to be described within a quantum statistical approach to a many–particle system. A further aspect is that some kind of interaction between the particles is required to form clusters. At the same time it yields an interaction of the clusters with its surroundings, what demands a many–particle treatment. An elegant approach can be given within second quantization.

Starting from a microscopic picture the many–particle system will be described by a Hamiltonian (cf. Eq. (4.2))

$$H = \sum_{1} E(1)\, a_1^\dagger a_1 + \frac{1}{2} \sum_{121'2'} V(121'2')\, a_1^\dagger a_2^\dagger a_{2'} a_{1'} \tag{4.12}$$

with $1 = \{\alpha_1, \vec{p}_1\}$; $\vec{p}_1$ denotes the momentum, α_1 spin and further internal quantum numbers (the generalization to arbitrary single–particle states is straightforward). The creation operator $a^\dagger$ and the annihilation operator a for the elementary particles obey fermion or boson commutation relations (if the spin is odd or even, respectively); $E(1) = p_1^2/(2m_1)$ is the kinetic energy.

For the pair interaction potential we adopt the form

$$V(\alpha p, \beta k; \alpha' p', \beta' k') = V_{\alpha\beta}(pk, p'k')\; \delta_{\alpha\alpha'}\, \delta_{\beta\beta'}\, \delta_{p+k,p'+k'}\ . \tag{4.13}$$

The generalization to interactions which modify also the internal quantum numbers is straightforward.

As a consequence of the interaction, correlations will occur in equilibrium as well as in non–equilibrium states of the system. If the interaction is weak, we can use the concept of a nearly free system of quasi–particles where the energies are shifted by mean field terms (Hartree–Fock approximation). Essentially new properties arise if the interaction is strong.

In strongly coupled fermion systems with attractive interaction, according to the Schrödinger equation (4.1), for an A – particle system the formation of **bound states** may be possible. This is an important quantum effect which modifies macroscopic properties such as the thermodynamic equations of state or transport properties. In the low–density limit, we can apply the chemical picture where the system is treated as a nearly ideal mixture of different components, and reactions due to collisions may occur.

In a dense medium, however, the single–particle and bound state properties are modified. For instance, a two–particle bound state is not an exact boson. Let us introduce the creation operator

$$a^\dagger_{2\alpha p} = \sum_{12} \psi^*_{2\alpha p}(12) a^\dagger_2 a^\dagger_1 \,, \tag{4.14}$$

which may be considered as creation operators of two–particle clusters in a state $\{\alpha, \vec{p}\}$. With the corresponding annihilation operator $a_{2\alpha p}$, the commutation relations

$$[a_{2\alpha p}, a^\dagger_{2\alpha' p'}] = \delta_{\alpha\alpha'}\delta_{pp'} - \sum_{121'2'} \psi_{2\alpha p}(12)\psi^*_{2\alpha' p'}(1'2') \times (\delta_{11'} a^\dagger_{2'} a_2 + \delta_{22'} a^\dagger_{1'} a_1 - \delta_{12'} a^\dagger_{1'} a_2 - \delta_{21'} a^\dagger_{2'} a_1) \tag{4.15}$$

are not purely bosonic, but contain additional contributions which are in the average proportional to the single–particle density. These additional density corrections arise from the fact that the constituents of the composite are fermions and have to obey the Pauli exclusion principle also with respect to the medium (Pauli blocking). As a consequence to be detailed below, bound states in a dense medium are dissolved (**Mott effect**) because the phase space occupied by fermions is no more available to form a bound state wave function.

A rigorous treatment of clusters in a dense medium can be performed only within certain approximations. Powerful methods have been developed such as path integration and the method of thermodynamic Green functions which will be used here, see [227, 267, 147]. Recently also Monte–Carlo or Quantum Molecular Dynamics simulations have been developed to describe the properties of these strongly coupled many–particle systems.

To describe clusters in a dense system, we start with the quantum – statistical many – particle approach. Macroscopic properties $\langle \mathcal{O} \rangle^t = \mathrm{Tr}[\rho(t)\mathcal{O}]$ are evaluated by averaging with the statistical operator $\rho(t)$. Particularly we are interested in stationary solutions where the determination of the statistical operator $\rho = \exp(-S)$ is more simple.

We will discuss first the thermal equilibrium which will be described by the grand canonical ensemble, $S = \beta(H - \sum_\alpha \mu_\alpha N_\alpha) - \ln \mathrm{Tr} \exp[\beta(H - \sum_\alpha \mu_\alpha N_\alpha)]$, where H is the hamiltonian, and $N_\alpha = \sum_p a^\dagger_{1\alpha p} a_{1\alpha p}$ the particle number in the internal quantum state α. For thermal equilibrium states, we will discuss the composition of the system as given by A – particle correlations, and we are able to derive thermodynamic relations like the equations of state.

Our main task is the evaluation of correlation functions such as $\langle a^\dagger_1 a_{1'} \rangle^t$, $\langle a^\dagger_1 a^+_{1'} \rangle^t$, $\langle a^\dagger_1 a^\dagger_2 a_{2'} a_{1'} \rangle^t$, $\langle a^\dagger_1 a^\dagger_2 a^\dagger_{2'} a^\dagger_{1'} \rangle^t$, etc, using the method of thermodynamic Green functions. In equilibrium, these reduced density matrices are not depending on time t.

The more general case of non–equilibrium states will be considered in the following section. There, the non–equilibrium statistical operator $\rho(t)$ will be constructed according to the method of Zubarev [592].

4.2.3 Green's Function Approach

Let us start with the analysis of states of a many–particle system in equilibrium (cf. also Ref. [227, 267]). The averaging has to be performed with respect to the grand canonical ensemble depending on the temperature $T = \beta^{-1}$ and the chemical potential μ (or μ_1, if different elementary particles are considered). The single–particle Green function G_1 is defined according to [147]

$$G_1(\tau_1, \tau_1') = \frac{1}{i\hbar} \langle \mathrm{T}[a_1(\tau_1) a^\dagger_{1'}(\tau_1')] \rangle \tag{4.16}$$

and can be interpreted as the propagation of a single–particle excitation within the many–particle system. The T operation means ordering with respect to the variable τ, which enters the description via a Heisenberg–like dependence

$$\mathcal{O}(\tau) = \exp(\tau S)\mathcal{O}\exp(-\tau S) \ . \tag{4.17}$$

An additional sign arises in the case of reordering fermionic operators.

Due to the periodicity of boundary conditions (Kubo–Martin–Schwinger relation), after a Fourier transform a presentation in terms of discrete so–called Matsubara frequencies $z_\nu = (\pi\nu)/(i\hbar\beta)$ (ν – odd for Fermions; even for bosons) can be given. On one hand, the Matsubara Green function allows for a systematic perturbative treatment in terms of Feynman diagrams. On the other hand, the relation to physical properties is obtained in the following way: We take the analytical continuation of $G_1(1, z_\nu)$ from the discrete Matsubara frequencies z_ν into the whole complex z plane (we take the representation where $G_1(1, 1', z_\nu) = G_1(1, z_\nu)\delta_{11'}$ is diagonal). The gap at the real axis

$$A_1(1,\omega) = \mathrm{Im}[G_1(1,\omega - i0) - G_1(1,\omega + i0)] \tag{4.18}$$

is denoted as spectral function.

With the fermionic distribution function $f(\omega) = [\exp(\beta\omega) + 1]^{-1}$ the equilibrium **single–particle occupation number**

$$\langle a_1^\dagger a_{1'}\rangle = \delta_{11'} \int \frac{d\omega}{2\pi} f(\omega)\, A_1(1,\omega) \tag{4.19}$$

is related to the spectral function.

The thermodynamical potential $p(T,\mu) = \int_{-\infty}^{\mu} n(T,\mu')\, d\mu'$ can be derived from the equation of state $n(\mu,T) = \Omega_0^{-1}\sum_1 \langle a_1^\dagger a_1\rangle$, where Ω_0 is the normalization volume. The generalization from a one-component system considered here where the index α can be dropped, to a multicomponent system is straightforward.

The single–particle Green function $G_1(1,z)$ is related to the single–particle self–energy $\Sigma_1(1,z)$ according to the Dyson equation

$$G_1^{-1}(1,z) = z - E(1) - \mu_1 - \Sigma_1(1,z) \ . \tag{4.20}$$

A perturbation expansion for the self–energy can also be given in terms of irreducible diagrams and seems to be more effective because partial summations have already been performed. On the other hand, a simple relation to the spectral function can be derived in the form

$$A_1(1,\omega) = \frac{2\,\mathrm{Im}\Sigma_1(1,\omega - i0)}{[\omega - E(1) - \mu_1 - \mathrm{Re}\Sigma_1(1,\omega)]^2 + [\,\mathrm{Im}\Sigma_1(1,\omega - i0)]^2}\,. \tag{4.21}$$

This way, the determination of thermodynamic properties is traced back to the evaluation of the self–energy.

To include the formation of bound states, a **cluster decomposition of the self–energy** can be performed. Using Feynman diagrams, we have

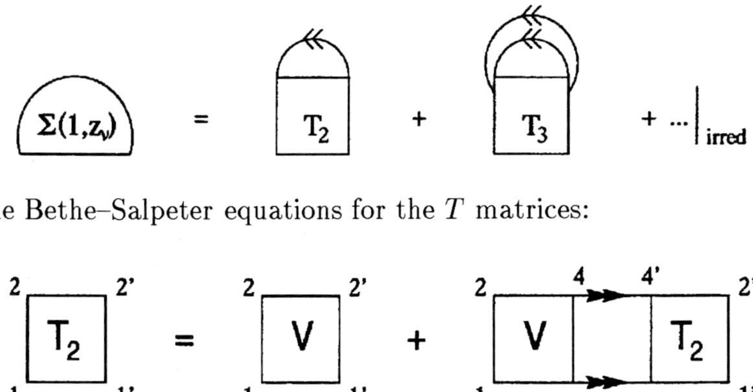

with the Bethe–Salpeter equations for the T matrices:

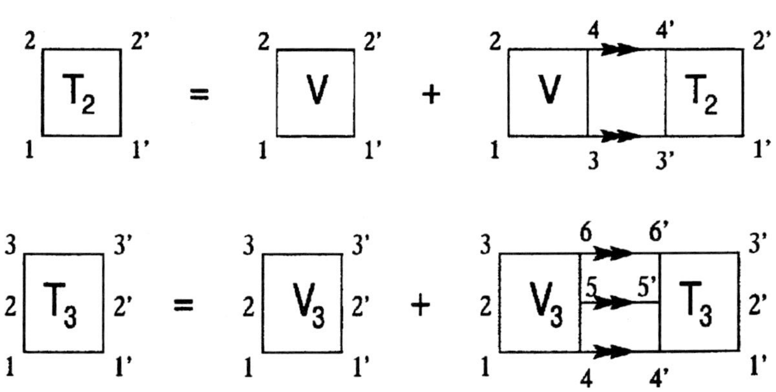

The three–particle interaction V_3 is given by the sum of all pair interactions.

Self–consistent approximations are obtained from partial summations. We discuss two examples:

(i.) In *Hartree–Fock approximation*, the free particle Green functions are replaced by the quasi-particle Green functions with a mean-field energy shift. The consistent solution of the T_2 matrix contains in addition also mean-field effects as Pauli blocking terms to be discussed in the following section.

(ii.) In *Cluster-mean field approximation* [446, 451, 456], mean-field contributions due to bound states are included. Typically, these contributions to the cluster self-energy Σ_A are given by an interaction in first order (Born approximation) with a correlation in the medium, but in addition the full antisymmetrization of the wave function has to be performed. For instance, considering only two-particle correlations, the instantaneous part of the two-particle self-energy is found from

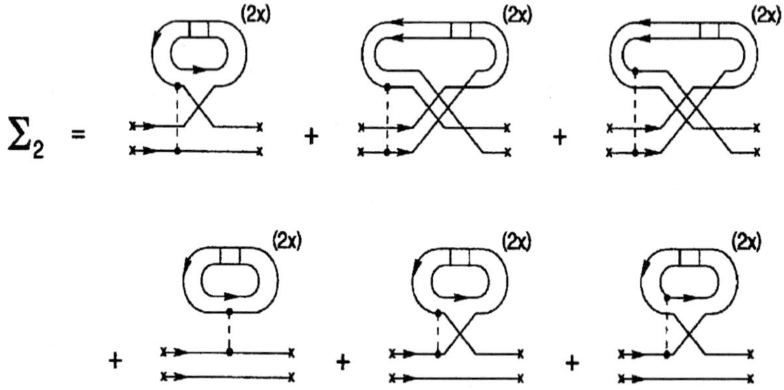

in addition to the ordinary Hartree–Fock contributions. The box in the backward two-particle propagator denotes a T_2 matrix, some low order diagrams have to be subtracted to avoid double counting. The peculiarity of the instantaneous part of Σ_A is that a closed equation for the single-frequency n-particle propagator can be derived. We will discuss this cluster-mean field approximation below. There we discuss the different contributions in the above diagram in a more familiar form.

Both approximations can be improved by the conditions of self-consistency for the single-particle as well as the A-particle propagator. Then, we have also to subtract classes of diagrams to avoid double counting. At present, this highly sophisticated approach has been solved only in some approximations (see Refs. [446, 451]). Instead of an analysis of Feynman diagrams, the cluster-mean field approximation can also be derived from the equations of motions of cluster Green functions and has been denoted as *Self-Consistent RPA* [123, 497].

4.2.4 Clusters in Equilibrium

In the low-density limit, the medium effects on the A-particle Schrödinger equation can be neglected. The cluster decomposition of the self-energy gives an expansion with respect to the fugacity $\exp(\beta\mu)$ (virial expansion) of the density. For simplicity we consider a one-component system, the generalization to a multi-component system with different chemical potentials μ_α is straigtforward. Restricting ourselves to two-particle contributions only, we have the ordinary Beth-Uhlenbeck formula

$$n(T,\mu) = \sum_1 f_1[E(1)] + \sum_{\alpha p} f_2(E_{2\alpha p})$$
$$+\sum_{\alpha p}^{(b)} \int \frac{dE}{\pi} f_2\left(\frac{p^2}{4m} + E\right) \frac{d}{dE}\delta_{2\alpha p}(E)\,. \tag{4.22}$$

Here we introduced the distribution function

$$f_A(E) = \left\{\exp\left[\beta(E - A\mu)\right] - (-1)^A\right\}^{-1}\,, \tag{4.23}$$

which for $A = 1$ describes the Fermi distribution, for $A = 2$ the Bose distribution.

Interesting is that in addition to the contribution of single-particle states (energy $E(1) = p_1^2/2m$) and two-particle bound states (energy $E_{2\alpha p}$), also a contribution due to scattering states arises characterized by the scattering phase shift $\delta_{2\alpha p}(E)$. The contribution of scattering states to the correlated density has the following consequence. If a bound state disappears due to the weakening of the interaction, e.g., a smooth behaviour of the density is obtained. This effect is due to the contribution of resonances, in accordance with the Levinson theorem. From the self-energy, the spectral function and the occupation numbers may be evaluated [488].

Expressions for occupation numbers and the equation of state can be found. As an example, the **generalized Beth-Uhlenbeck formula** was obtained.

In this approach, correlations in the medium were neglected when evaluating the T_2–matrix (quasi–particle mean field). Furthermore, the spectral function (4) was expanded with respect to small $\mathrm{Im}\Sigma_1(1,\omega)$ [488]:

$$n(T,\mu) = \sum_1 f_1[E^{\mathrm{MF}}(1)] + \sum_{\alpha,p>P_{\mathrm{Mott}}}^{(b)} f_2(E_{2\alpha p})$$
$$+ \sum_{\alpha p} \int \frac{dE}{\pi} f_2\left(\frac{p^2}{4m} + E\right) \tag{4.24}$$
$$\frac{d}{dE}\left[\delta_{2\alpha p}(E) - \sin\delta_{2\alpha p}(E)\cos\delta_{2\alpha p}(E)\right] \, .$$

In–medium effects are represented by the shifts of the single particle energies $E^{\mathrm{MF}}(1) = E(1) + \sum_2 V(12,12)_{\mathrm{ex}} f_1(2)$, so that the free particle contribution to the density is replaced by the contribution of quasiparticles. Furthermore, the two–particle energies $E_{2\alpha p}$ and scattering phases $\delta_{2\alpha p}(E)$ are modified by the medium as it will be discussed in the following subsection. The summation over p is restricted to that region where bound states in the medium will exist, i.e., for momenta $p \geq P_{\mathrm{Mott}}$.

The scattering term contains a contribution ($\sin\delta\cos\delta$) which compensates the lowest order of interaction already incorporated in the quasiparticle energies.

A generalization to arbitrary cluster sizes can be done with the following result

$$n(\beta,\mu) = \sum_{A=1}^{\infty} A \sum_{\alpha}^{(b)} g_{A\alpha} \int \frac{d^3p}{(2\pi\hbar)^3} f_A(E_{A\alpha p}) \tag{4.25}$$

Here only bound states are taken into account. $g_{A\alpha}$ is the degeneration factor. This expression reflects the chemical picture, considering the system as a mixture of different components (clusters). However, the neglection of scattering contributions cannot always be justified.

The inclusion of higher clusters into the systematic treatment of a many–particle system needs rules for a selection of the relevant diagrams within the perturbative expansion. Within the chemical picture, each bound state is considered as a new "elementary" species. Elementary particles and bound states should be treated in an equivalent way [267]. The summation over all the excited clusters necessary to evaluate Eq. (4.25) can approximately be performed introducing a density of states.

4.2.5 Mott Effect and Phase Transitions

Here we investigate the influence of the medium on the single particle states and correlated states. In a first approximation, we consider the medium as uncorrelated, i. e. in the quasiparticle approximation. Then we have the advantage that the contribution due to the medium is simply expressed in terms of the single–particle (fermion or boson) distribution function. Formally this approximation is given by a single–particle expression for the entropy operator so that all correlation functions are evaluated using Wick's theorem. The single–particle entropy has to be determined such that the temperature and the effective chemical potential reproduce the thermodynamic parameters such as density in a correct way.

As a trivial case, the ordinary Hartree–Fock approximation for the single–particle properties is reproduced. The single–particle energies are shifted according to

$$E^{\mathrm{HF}}(1) = p_1^2/(2m_1) + \sum_2 V(12,12)_{\mathrm{ex}} f_1(2) \ . \tag{4.26}$$

The shift of the single–particle energies has been investigated for nuclear matter. As a result, semiempirical parametrizations have been performed, such as the so–called Skyrme forces [551]. This procedure allows to evaluate nucleonic density profiles in nuclei according to the local density approximation.

Furthermore, evaluating for symmetric nuclear matter, a phase instability region is obtained in this approximation. It should be mentioned that also the transition to superfluidity can be described within the mean–field approximation if the anomal averages like $\langle a_1 a_{\bar{1}} \rangle$ are nonvanishing [456].

Interesting is the behaviour of two–particle states within the Hartree–Fock approximation. In that approximation the Schrödinger equation reads in momentum representation

$$\begin{aligned} &[E^{\mathrm{HF}}(1) + E^{\mathrm{HF}}(2)]\psi_{2\alpha p}(12) \\ &+ \sum_{1'2'} [1 - f_1(1) - f_1(2)] V(12,1'2') \psi_{2\alpha p}(1'2') \\ &= E_{2\alpha p} \psi_{2\alpha p}(12) \ . \end{aligned} \tag{4.27}$$

The influence of the medium is contained in two terms: The single–particle self–energy shift is given in Hartree–Fock approximation as well as the Pauli blocking terms $[f_1(1) + f_1(2)]V(12, 1'2')$. Latter ones can be interpreted in that way that interactions are effective only from states $\{1', 2'\}$ into states $\{12\}$ if these states are free. Both medium contributions can be joined into an effective contribution to the Hamiltonian

$$\Delta H^{\mathrm{HF}}(12, 1'2') = \sum_3 f_1(3)\,[V(13, 13)_{\mathrm{ex}}\delta_{11'}\delta_{22'} \qquad (4.28)$$
$$+V(23, 23)_{\mathrm{ex}}\delta_{11'}\delta_{22'} + V(12, 1'2')\,(\delta_{13} + \delta_{23})] \;.$$

It should be mentioned that this hamiltonian is not hermitean, but leads to an eigenvalue problem with real eigenvalues.

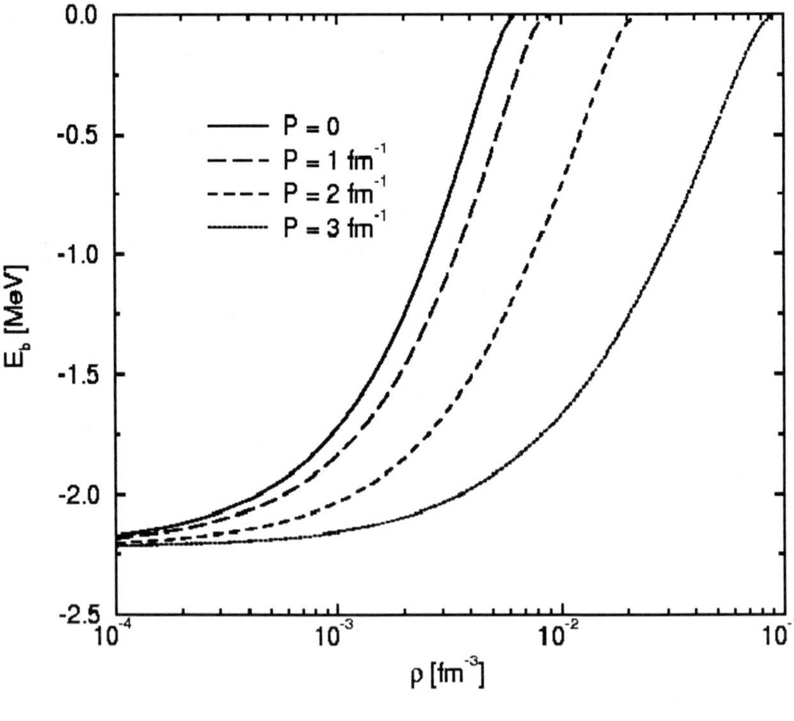

Fig. 4.2 Deuteron bound state energies as function of nuclear matter density ρ at temperature $T = 10$ MeV for different center of mass momenta P.

The solution of the Schrödinger equation now depends on density and temperature. As a simple example, we will give the result for a separable interaction potential [423] describing the nucleon-nucleon interaction

$$V_\alpha^{LL'}(k,k') = \sum_{i,j=1}^{N} w_{\alpha i}^L(k)\, \lambda_{\alpha ij}\, w_{\alpha j}^{L'}(k') \,. \tag{4.29}$$

Here k denotes the relative momentum, L is the angular momentum. The solution of the Bethe–Salpeter equation [488] for the T–matrix can be found as

$$T_\alpha^{LL'}(k,k',p,z) = \sum_{ijl} w_{\alpha i}^L(k)\, [1 - J^\alpha(p,z)]_{ij}^{-1}\, \lambda_{\alpha jl}\, w_{\alpha l}^{L'}(k') \,, \tag{4.30}$$

where the quantity J is given by

$$J_{ij}^\alpha(p,z) = \int \frac{d^3k}{(2\pi)^3} \sum_{nL} w_{\alpha n}^L(k)\, \lambda_{\alpha in}\, w_{\alpha j}^L(k) \tag{4.31}$$
$$\times \frac{1 - f_1\left[E^{\rm HF}(\vec{p}/2+\vec{k})\right] - f_1\left[E^{\rm HF}(\vec{p}/2-\vec{k})\right]}{E^{\rm HF}(\vec{p}/2+\vec{k}) + E^{\rm HF}(\vec{p}/2-\vec{k}) - \mu_1 - \mu_2 - z} \,.$$

Medium effects occur in terms of single–particle mean–field shifts and the Pauli blocking term. The consequences for the two–particle bound state energy and the scattering phase shifts are derived from the T–matrix according to

$$\tan \delta_{2\alpha p}(E) = \frac{{\rm Im} T_\alpha(p,E)}{{\rm Re} T_\alpha(p,E)} \,. \tag{4.32}$$

The results are shown in Figs. 4.2 and 4.3.

Due to the *Pauli blocking*, the deuteron binding energy at fixed temperature is decreasing with increasing density. At the so–called Mott density which depends also on the total two–particle momentum, the bound state disappears. At the same time, the medium dependent scattering phase shift in the triplett channel jumps by π in accordance to the Levinson theorem. The consequences for the in–medium nucleon–nucleon cross sections are shown in [490]. With the help of the T–matrix, the self–energy and the spectral function are evaluated.

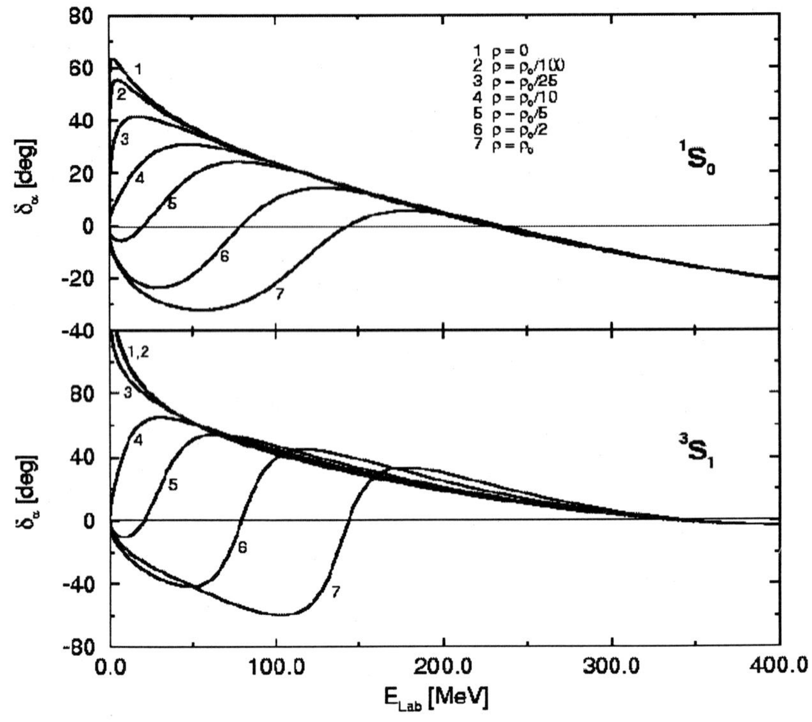

Fig. 4.3 Singlet and triplet nucleon–nucleon scattering phase shifts for different nuclear matter densities ρ and temperature $T = 10$ MeV as function of relative energy.

In addition to a quasi–particle peak, the spectral function shows further structures due to correlation effects. Within a *self–consistent calculation*, the sharp satellite structure of the bound state contribution is smoothed out. Consequently, also the disappearence of bound states with increasing density (Mott effect) is a smooth transition since the bound states are broadened in a self–consistent approximation.

With the energy eigenvalue of the two–particle problem obtained from the Schrödinger equation and the scattering phase shifts, we can evaluate the equation of state (4.24). The first contribution on the right hand side can be interpreted as the contribution of free particles to the density, the second one as the contribution of bound states (b), and the third one as the contribution of scattering processes (sc). The contribution of scattering processes cannot be neglected if bound states or resonances are near the continuum edge of

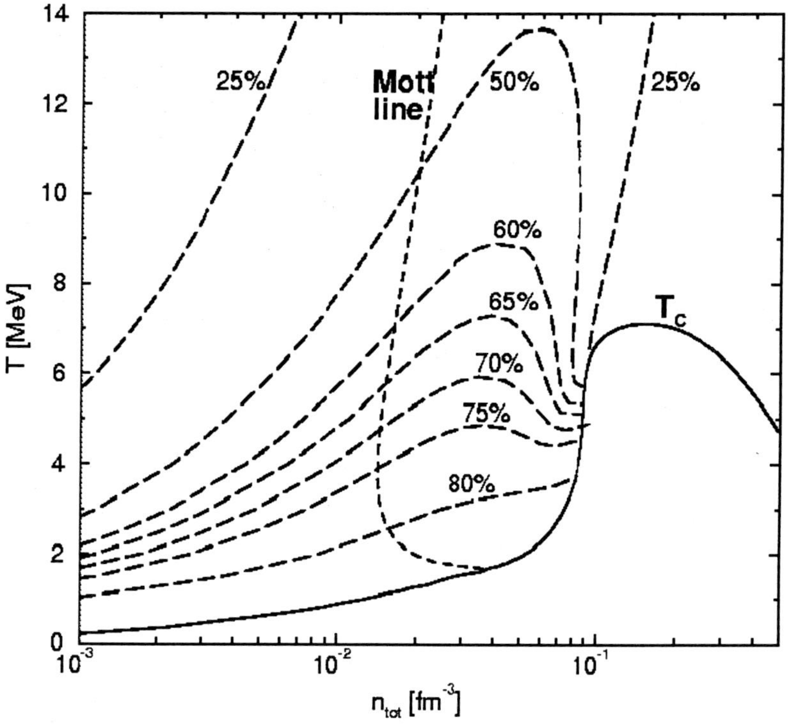

Fig. 4.4 Temperature–density plane of symmetric nuclear matter showing lines of equal concentration of correlated nucleons, the Mott line, and the critical temperature of the onset of superfluidity.

scattering states. Especially near the Mott transition, where bound states are dissolved due to density effects, the contribution of scattering processes is of importance. The applicability of the ideal mixture is extended if the quasiparticle picture is used (shifted single particle and cluster energies).

From the generalized Beth–Uhlenbeck formula, the composition of nuclear matter is obtained as shown in Fig. 4.4.

Two–particle correlations are present also above the Mott density, because bound states with finite total momentum are destructed by the Pauli blocking effect at higher densities. Furthermore, the disappearence of a bound state is a smooth process, resonances will take over the contribution acording to the Levinson theorem. Whereas the Mott density is at about $\rho_0/20$, two–particle correlations continue to contribute up to $\rho_0/3$, with $\rho_0 = 0.17$ fm^{-3} being the nuclear matter density.

Taking the two–particle correlation into account, the phase instability region is partially decreased, in comparison with the single–particle mean–field result [451].

4.2.6 Cluster–Mean–Field Approximation

We note that the quantum statistical approach allows for the systematic inclusion of further contributions, especially many–particle effects, to describe the interaction of the clusters with its surroundings.

The inclusion of correlations in the medium can be performed in mean–field approximation in the following way: In the Hamiltonian of the A–particle cluster, we include the medium by averaging over the other particles. If exchange terms are dropped, the averaged Hamiltonian has the form

$$V(12,1'2')a_1^\dagger a_{1'}\langle a_2^\dagger a_{2'}\rangle \tag{4.33}$$

where the full cluster decomposition of $\langle a_2^\dagger a_{2'}\rangle$ has to be performed.

The inclusion of exchange terms is more involved. For the A–particle problem, the effective wave equation reads [451]

$$\begin{aligned}
&[E(1)+\dots E(A)-E_{A\alpha p}]\psi_{A\alpha p}(1\dots A)\\
&+\sum_{1'\dots A'}\sum_{i<j}^{A} V_{ij}^{A}(1\dots A,1'\dots A')\psi_{A\alpha p}(1'\dots A')\\
&+\sum_{1'\dots A'} V_{\text{nucl.matter}}^{A,\text{MF}}(1\dots A,1'\dots A')\psi_{A\alpha p}(1'\dots A')=0\ ,
\end{aligned} \tag{4.34}$$

with $V_{12}^A(1\dots A,1'\dots A') = V(12,1'2')\delta_{33'}\dots\delta_{AA'}$. The effective potential $V_{\text{nucl.matter}}^{A,\text{MF}}(1\dots A,1'\dots A')$ accounts for the influence of the medium on the cluster bound states and takes the form

$$V^{A,\mathrm{MF}}_{\mathrm{nucl.matter}}(1\ldots A,1'\ldots A') = \sum_i \Delta(i)\delta_{11'}\ldots\delta_{AA'} + \sum_{i,j}{}' \Delta V^A_{ij}(1\ldots A,1'\ldots A') \,, \tag{4.35}$$

with

$$\Delta V^A_{12} = -\Big\{\frac{1}{2}(\tilde{f}_1(1)+\tilde{f}_1(1'))V(12,1'2')+ \sum_{B=2}^{\infty}\sum_{\alpha p}\sum_{\bar{2}\ldots\bar{B}}\sum_{\bar{2}'\ldots\bar{B}'} f_B(E_{B\alpha p})\sum_j^B V^B_{1j}(1\bar{2}'\ldots\bar{B}',1'\bar{2}\ldots\bar{B}) \psi^*_{B\alpha p}(2\bar{2}\ldots\bar{B})\psi_{B\alpha p}(2'\bar{2}'\ldots\bar{B}')\Big\}\,\delta_{33'}\ldots\delta_{AA'} \,, \tag{4.36}$$

$$\tilde{f}_1(1) = f_1(1) + \sum_{B=2}^{\infty}\sum_{\alpha p}\sum_{2\ldots B} f_B(E_{B\alpha p})|\psi_{B\alpha p}(1\ldots B)|^2, \tag{4.37}$$

$$\Delta(1) = \sum_2 (V(12,12)_{\mathrm{ex}}\tilde{f}_1(2) - \sum_{B=2}^{\infty}\sum_{\alpha p}\sum_{2\ldots B}\sum_{1'\ldots B'} f_B(E_{B\alpha p}) \times \sum_{i<j}^{m} V^B_{ij}(1\ldots B,1'\ldots B')\psi_{B\alpha p}(1\ldots B)\psi^*_{B\alpha p}(1'\ldots B') \,, \tag{4.38}$$

where

$$f_A(E) = \left\{\exp\beta(E-A\mu) - (-1)^A\right\}^{-1} . \tag{4.39}$$

Note that within the cluster–mean field approximation the effective potential $V^{A,\mathrm{MF}}_{\mathrm{nucl.matter}}$ remains energy independent, i.e. instantaneous. The quantity $\tilde{f}_1(1)$ describes the effective occupation of the state 1 due to free and bound states, whereas the exchange terms are obtained from the additional terms in ΔV^A_{12} and $\Delta(1)$ thus realizing the Pauli exclusion principle.

Of course, the self–consistent solution of the cluster in a clustered medium is a rather involved problem which has not been solved until now. Starting from an uncorrelated medium, first step iterations have been attempted, including only some bound states, but neglecting all scattering states.

The cluster–mean field may be considered as a generalisation of the ordinary mean field, where in addition to the mean field produced by the single–particle states also the mean field produced by clusters (bound states) is taken into account. We can evaluate the modification of bound state energies as well as wave functions in this approximation. We obtain an optimized set of states which may be of use in evaluating self–energies and spectral functions in a fully self–consistent way, as a prerequisite in evaluating correlation functions.

The thermodynamic properties of a correlated many–particle system are obtained from a thermodynamic potential, which can be derived from the correlation functions. The evaluation of thermodynamic instabilities and phase transitions obtained from the equation of state depend on the approximations made. Compared with a single mean–field approach where all correlations are neglected, the region of phase instability is reduced if correlations in many–particle systems are taken into account.

4.3 Clusters in Non–Equilibrium Processes

Whereas the equilibrium state of a correlated system is decribed in terms of the distribution functions of clusters or, simply, by the composition, in non–equilibrium states also processes are of importance which lead to a time variation of the cluster distributions. We can consider elastic collisions which lead to relaxation in momentum space (thermalization) and inelastic collisions (excitations, rearrangements, formation of clusters and break–up reactions) which relax the composition to chemical equilibrium.

4.3.1 Correlated Medium in Non–Equilibrium States

The state of a many–particle system at time t is described by the statistical operator $\rho(t)$. The statistical operator follows as the solution of the von–Neumann equation

$$\frac{\partial}{\partial t}\rho(t) + \frac{i}{\hbar}[H, \rho(t)] = 0 \qquad (4.40)$$

with the appropriate initial condition. Having the non–equilibrium statistical operator to our disposal, the time–dependent mean value of an operator $\mathcal{O}$ is evaluated according to $\langle \mathcal{O} \rangle^t = \mathrm{Tr}\,[\rho(t)\mathcal{O}]$.

To construct $\rho(t)$ for the non–equilibrium case, we start with the A–particle distributions at time $t_0 \leq t$ which characterize the nonequilibrium state of the system:

$$n_1(11', t_0) = \langle a_1^\dagger a_{1'} \rangle^{t_0} , \tag{4.41}$$

$$n_2(12, 1'2', t_0) = \langle a_1^\dagger a_2^\dagger a_{2'} a_{1'} \rangle^{t_0} , \tag{4.42}$$

etc., 1 denotes a complete set of single–particle quantum numbers.

With a given set of relevant A-particle distributions we can determine the boundary conditions for the statistical operator $\rho(t)$. For this we construct a relevant statistical operator $\rho_{\mathrm{rel}}(t)$ which reproduces the given mean values of the relevant distributions, but is not the solution of the von-Neumann equation.

Before doing so, we note that it is possible to introduce another orthonormal basis ν_1 of single–particle states with $\psi_{\nu_1}(1) = \langle 1|\nu_1\rangle$, of two–particle states $\psi_{\nu_2}(12) = \langle 12|\nu_2\rangle$, etc., so that

$$n_1(\nu_1; \nu_1'; t_0) = \sum_{11'} \psi_{\nu_1}^*(1) n_1(11'; t_0) \psi_{\nu_1'}(1') , \tag{4.43}$$

$$n_2(\nu_2; \nu_2'; t_0) = \sum_{121'2'} \psi_{\nu_2}^*(12) n_2(12, 1'2', t_0) \psi_{\nu_2'}(1'2') , \tag{4.44}$$

etc.

In particular, it is possible to introduce a basis system of quantum states which are optimal stationary, i.e. for which the damping is small. This means that we are looking for a basis in the Fock space that nearly diagonalizes the Hamiltonian. For homogeneous systems (no external fields) one can introduce the total momentum $\vec{p}$ of the A–particle cluster as one of the quantum numbers of such an optimized basis system. Further quantum numbers such as spin or the internal quantum number of a bound state, but also the relative momenta of a scattering state, shall be denoted as α. We propose to use the cluster–mean field basis $\psi_{A\alpha p}(1, \ldots A, t_0)$ introduced in the previous section as such an optimized basis system. In the previous section thermodynamic equilibrium was considered, i.e. the average was performed with the equilibrium statistical operator. Now, the average will be performed with the relevant statistical operator $\rho_{\rm rel}(t_0)$.

This statement means that the influence of the medium will not be characterized by the equilibrium distribution functions $f_A(E) = \{\exp[\beta(E - A\mu)] - (-1)^A\}^{-1}$ but by a non–equilibrium distribution which has to be constructed in a self–consistent way so that the averages (4.43) at time t_0 are correctly reproduced.

For brevity we introduce the A–particle operators $a^\dagger_{(A)} = a^\dagger_{(1\ldots A)} = a^\dagger_1 \ldots a^\dagger_A$ and $a_{(A')} = a_{(1'\ldots A')} = a_{A'} \ldots a_{1'}$. According to Zubarev [592], the relevant statistical operator $\rho_{\rm rel}(t_0)$ follows from the maximum of information entropy

$$S(t_0) = -{\rm Tr}\{\rho_{\rm rel}(t_0) \ln \rho_{\rm rel}(t_0)\} \tag{4.45}$$

at given mean values of the relevant A–particle distributions

$${\rm Tr}\{\rho_{\rm rel}(t_0) a^\dagger_{(A)} a_{(A')}\} = \langle a^\dagger_{(A)} a_{(A')}\rangle^{t_0}. \tag{4.46}$$

The solution is a generalized Gibbs state $\rho_{\rm rel}(t_0) = \exp[-S(t_0)]$ with

$$\begin{aligned} S(t_0) &= \sum_{A=1}^{\infty} s_A(1\ldots A, 1'\ldots A', t_0) a^\dagger_{(A)} a_{(A')} \\ &\quad - \ln {\rm Tr} \exp\left\{\sum s_A(1\ldots A, 1'\ldots A', t_0) a^\dagger_{(A)} a_{(A')}\right\} \end{aligned} \tag{4.47}$$

The Lagrange parameters $s_A(1\ldots A, 1'\ldots A', t_0)$ are introduced to fulfil the self–consistency conditions (4.46). They have the meaning of thermodynamic parameters.

A main problem is the determination of these thermodynamic parameters $s_A(1\ldots A, 1'\ldots A', t_0)$ for given mean values $\langle a^\dagger_{(A)} a_{(A')}\rangle^{t_0}$, i.e., the evaluation of the equation of state for the generalized Gibbs ensemble at time t_0. For brevity we will drop the variable t_0 in the following. For the relevant statistical operator, it can be considered as a parameter.

In principle, the Matsubara technique known from the equilibrium grand canonical ensemble can be applied also to the generalized Gibbs ensemble because it has an exponential form. Thus, the reduced densities can be derived from the relevant statistical operator within perturbation theory. A self–energy can be introduced, and the A–particle problem can be treated in cluster–mean field approximation. Within a certain approximation we can evaluate the correlation functions and obtain relations between the reduced density matrices and the thermodynamic parameters.

To specify the approximation we consider the solution of the A–particle problem in cluster–mean field approximation. In general, we have a discrete spectrum of energy eigenvalues describing bound states and a continuum describing scattering states. The corresponding states form a complete basis for the A–particle Hilbert space. Now, we consider a scattering state $(A\alpha p)$ with asymptotic incoming states $(A_1\alpha_1 p_1)$ and $(A_2\alpha_2 p_2)$ both α_1, α_2 belonging to the bound state spectrum. Then we project out the free motion from the scattering states so that

$$|\psi^{\mathrm{corr}}_{A\alpha p}\rangle = |\psi_{A\alpha p}\rangle - z|\psi_{A_1\alpha_1 p_1}\rangle|\psi_{A_2\alpha_2 p_2}\rangle \tag{4.48}$$

holds with the renormalization factor $z = \langle\psi_{A_1\alpha_1 p_1}|\langle\psi_{A_2\alpha_2 p_2}|\psi_{A\alpha p}\rangle$. This procedure is more rigorously formulated in terms of Green functions, where the T matrix is introduced after separation of the uncorrelated propagators.

With these states we suppose that in the homogeneous case the relevant statistical operator can be given in the form

$$\rho^{\mathrm{corr}}_{\mathrm{rel}}(t_0) = Z^{-1}_{\mathrm{rel}}(t_0)\exp\left[-\sum_{A\alpha p} s_{A\alpha p}(t_0)a^{\dagger}_{A\alpha p}a_{A\alpha p}\right], \tag{4.49}$$

where we used the abbreviation $a_{A\alpha p} = \sum_{1\dots A}\psi^{\mathrm{corr}}_{A\alpha p}(1\dots A)a_A\dots a_1$. Now, the thermodynamic parameters $s_{A\alpha p}(t_0)$ have to be determined from the given mean values

$$\begin{aligned} f_{A\alpha}(p,t_0) &= n^{\mathrm{corr}}_{A\alpha p}(t_0) \\ &= \sum_{1\dots A,1'\dots A'} \psi^{\mathrm{corr}}_{A\alpha p}(1\dots A)^* n^{\mathrm{corr}}_A(1\dots A, 1'\dots A', t_0) \\ &\times\psi^{\mathrm{corr}}_{A\alpha p}(1'\dots A'). \end{aligned} \tag{4.50}$$

Here we introduced the correlated part of the A–particle distribution such as

$$n^{\mathrm{corr}}_2(12, 1'2') = n_2(12, 1'2') - n_1(11')n_1(22')_{\mathrm{ex}}. \tag{4.51}$$

Furthermore we will use the approximation

$$\begin{aligned}\mathrm{Tr}(\rho_{\mathrm{rel}}^{\mathrm{corr}}(t_0)a_{A\alpha p}^{\dagger}a_{A\alpha p}) &= \left(\exp\left[s_{A\alpha p}(t_0)\right] - (-1)^{A}\right)^{-1} \\ &= n_{A\alpha p}^{\mathrm{corr}}(t_0)\end{aligned} \tag{4.52}$$

which immediately gives a relation to determine the thermodynamic parameters. Of course, the evaluation of the average over the relevant statistical operator can be improved systematically. In combination with the effective wave equation to determine the states $\psi_{A\alpha p}^{\mathrm{corr}}$ in the medium, a rather complex system of equations has to be solved in a self–consistent way.

With the relevant operator $\rho_{\mathrm{rel}}^{\mathrm{corr}}(t')$ we can construct the non–equilibrium statistical operator in the usual way [591] and derive equations for the time dependence of the thermodynamic parameters $s_{A\alpha p}(t)$. The method can be developed as in [591, 592]. With the time evolution operator $U(t,t')$, obtained from

$$i\hbar\frac{\partial}{\partial t}U(t,t') = H(t)U(t,t') \tag{4.53}$$

with $U(t,t) = 1$, we find

$$\rho(t) = \epsilon\int_{-\infty}^{t} \mathrm{e}^{\epsilon(t'-t)}U(t,t')\rho_{\mathrm{rel}}^{\mathrm{corr}}(t')U^{\dagger}(t,t')dt' \tag{4.54}$$

as the solution of the equation of motion

$$\frac{\partial}{\partial t}\rho(t) + \frac{i}{\hbar}[H,\rho(t)] = -\epsilon(\rho(t) - \rho_{\mathrm{rel}}^{\mathrm{corr}}(t))\ . \tag{4.55}$$

Here, the initial condition given by the relevant statistical operator is introduced into the von Neumann equation in form of an infinitesimal source term, with $\lim \epsilon \to 0$ after the thermodynamic limit.

4.3.2 Derivation of Kinetic Equations

For the distribution function

$$f_{A\alpha}(\vec{r},\vec{p},t) = \int \frac{d^3k}{(2\pi\hbar)^3} e^{i\vec{k}\cdot\vec{r}/\hbar} \langle a^\dagger_{A\alpha,p-k/2} a_{A\alpha,p+k/2}\rangle^t \tag{4.56}$$

we find the equation of motion [455]

$$\frac{\partial}{\partial t} f_{A\alpha}(\vec{r},\vec{p},t) = -D^{\text{Vlasov}}_{A\alpha}(\vec{r},\vec{p},t) + I^{\text{coll}}_{A\alpha}(\vec{r},\vec{p},t) \ . \tag{4.57}$$

The Vlasov $D^{\text{Vlasov}}_{A\alpha}(\vec{r},\vec{p},t)$ term is immediately obtained from the relevant statistical operator. It describes the reversible time evolution as known from time–dependent Hartree–Fock approximation, but contains also the mean field due to clusters in the medium. Introducing the scattering part of the cluster–cluster interaction as

$$\begin{aligned} V^{\text{cluster,elastic}}_{A\alpha p,A'\alpha' p'}(q) &= \sum_{ij} V(ij, i+q, j-q) \\ &\times \sum_{1\ldots A} \sum_{1'\ldots A'} \psi^*_{A\alpha,p+q}(1\ldots i+q\ldots A)\psi^*_{A'\alpha',p'-q}(1'\ldots j-q\ldots A') \\ &\times \psi_{A'\alpha' p'}(1'\ldots j\ldots A')\psi_{A\alpha p}(1\ldots i\ldots A) \end{aligned} \tag{4.58}$$

the Vlasov term in the cluster–mean field approach reads

$$\begin{aligned} D^{\text{Vlasov}}_{A\alpha}(\vec{r},\vec{p},t) &= \int \frac{dk}{(2\pi\hbar)^3} e^{i\vec{k}\cdot\vec{r}/\hbar} \\ &\times \frac{1}{i\hbar}\text{Tr}\{\rho^{\text{corr}}_{\text{rel}}(t)[H, a^\dagger_{A\alpha,p-k/2} a_{A\alpha,p+k/2}]\} \\ &= \int \frac{dk}{(2\pi\hbar)^3} e^{i\vec{k}\cdot\vec{r}/\hbar} \{\frac{i}{\hbar m}\vec{p}\cdot\vec{k} f_{A\alpha}(\vec{p},\vec{k},t) + \frac{1}{i\hbar}\sum_{A'\alpha'}\int \frac{d^3p' d^3q}{(2\pi\hbar)^6} \\ &\times V^{\text{cluster,elastic}}_{A\alpha p,A'\alpha' p'}(q)[f_{A\alpha}(\vec{p}+\vec{q}/2,\vec{k}-\vec{q},t) f_{A'\alpha'}(\vec{p}'-\vec{q}/2,\vec{q},t) \\ &- f_{A\alpha}(\vec{p}-\vec{q}/2,\vec{k}-\vec{q},t) f_{A'\alpha'}(\vec{p}'+\vec{q}/2,\vec{q},t)]\} \ . \end{aligned} \tag{4.59}$$

In addition to the mean field produced by the single–particle distribution ($A'=1$), also the clusters ($A'>1$) contribute to the mean field. In general, the cluster-mean field will depend on the position $\vec{r}$.

For the collision term we will neglect non-locallity in space so that only the distributions at the position $\vec{r}$ are of relevance, which are given by the diagonal part in the momentum space. Then, the collision term has the form

$$I_{A\alpha}^{\rm coll}(\vec{r},\vec{p},t) = -\int dE \int dE' \int_{-\infty}^{t} dt' \exp\{(i(E-E') \\ +\epsilon)(t'-t)/\hbar\} {\rm Tr}\left\{\frac{i}{\hbar}[H, a^\dagger_{A\alpha p} a_{A\alpha p}]\delta(E-H) \right. \\ \left. \left(\frac{i}{\hbar}[H,\rho_{\rm rel}^{\rm corr}(t')] + \frac{\partial}{\partial t'}\rho_{\rm rel}^{\rm corr}(t')\right)\delta(E'-H)\right\} \tag{4.60}$$

and gives rise to irreversible behaviour. It describes reaction and scattering processes between bound and single–particle states. In the evaluation of the collision term we will neglect non–locality in time, i.e. $\rho_{\rm rel}^{\rm corr}(t') \approx \rho_{\rm rel}^{\rm corr}(t)$. Performing the integrals over t' and E' we get the Kubo–Greenwood type formula

$$I_{A\alpha}^{\rm coll}(\vec{r},\vec{p},t) = \frac{1}{\pi\hbar}\int dE\ {\rm Tr}\left\{[H, a^\dagger_{A\alpha p} a_{A\alpha p}]{\rm Im}\frac{1}{E-H+i\epsilon} \right. \\ \left. \times\ [H,\rho_{\rm rel}^{\rm corr}(t)]{\rm Im}\frac{1}{E-H+i\epsilon}\right\} . \tag{4.61}$$

For the further evaluation, a cluster decomposition of H with respect to different scattering channels is performed. In fact, we decompose H in the Fock space as

$$H = \sum_A H^{(A)} , \tag{4.62}$$

$$H^{(A)} = \sum_{1\dots A, 1'\dots A'} |1\dots A\rangle\langle 1\dots A|H|1'\dots A'\rangle\langle 1'\dots A'| . \tag{4.63}$$

Here, the cluster–mean field basis $\psi_{A\alpha p}(1,\dots,A)$ can be introduced. The scattering states should be further treated separating the contribution of the free components. This way, an effective interaction between the component clusters $\psi^{\rm corr}_{A\alpha p}(1,\ldots,A)$ is introduced,

$$\begin{aligned}\langle 1\ldots A|H|1'\ldots A'\rangle = & \sum_{A_i\alpha_i p_i, A'_j\alpha'_j p'_j} \\ & \psi^{*\rm corr}_{A_1\alpha_1 p_1}(1\ldots A_1)\psi^{*\rm corr}_{A_2\alpha_2 p_2}(A_1+1\ldots A_1+A_2)\ldots \\ & \langle \psi^{\rm corr}_{A_1\alpha_1 p_1}\psi^{\rm corr}_{A_2\alpha_2 p_2}\ldots|H|\psi^{\rm corr}_{A'_1\alpha'_1 p'_1}\psi^{\rm corr}_{A'_2\alpha'_2 p'_2}\ldots\rangle \\ & \psi^{\rm corr}_{A'_1\alpha'_1 p'_1}(1'\ldots A'_1)\psi^{\rm corr}_{A'_2\alpha'_2 p'_2}(A'_1+1\ldots A'_1+A'_2)\ldots \end{aligned} \quad (4.64)$$

what gives the possibility to introduce a cluster–cluster interaction V_{A_i,A'_i}. Now we can try to formulate an optical theorem after introducing the channel–T matrices, and then to find reaction rates.

Considering only the single–particle distribution, the usual Vlasov term and Boltzmann collision term are deduced [455].

4.3.3 Example: Deuteron Formation Rate in Hot Matter

As an example for a kinetic equation in correlated matter, we demonstrate the simplest case for reactive collisions, the formation and break–up reaction of a two–particle bound state [34]. In particular, we consider deuterons in dense nuclear matter. As well–known, to describe these processes we have to solve the three–particle problem. Now, in dense matter, the main obstacle is that the formation of bound states requires the notion of few–body reactions within the medium. Even the simplest case, i.e., the abundances of deuterons that are determined by the deuteron formation via $NNN \to dN$ (N nucleon, d deuteron) and break–up, $dN \to NNN$, reactions, requires a proper treatment of the effective three–body problem.

The essentials of the three–body problem for the isolated system are well known, see, e.g., Ref. [171]. In the following we utilize the AGS–formalism [8] suitably modified to treat the three–body problem within nuclear matter. For this we first recapitulate the non–equilibrium approach including single–particle and two–particle states. Our aim is to derive an expression for the time scale at which the deuteron concentration in dense matter relaxes to its equilibrium value.

We consider a homogeneous system and drop the variable $\vec{r}$. Following a general density matrix approach as given in Ref. [455] the time evolution

(4.57) of the distribution functions for nucleons $f_{1\alpha}(\vec{r},\vec{p},t) \equiv f_N(\vec{p},t)$ and deuterons $f_{2\alpha}(\vec{r},\vec{p},t) \equiv f_d(\vec{p},t)$ read

$$\frac{\partial}{\partial t} f_N(\vec{p},t) = \mathrm{Tr}\left\{\rho(t)\frac{i}{\hbar}[H, n_{Np}]\right\} = -D_N^{\mathrm{Vlasov}}(\vec{p},t) + I_N^{\mathrm{coll}}(\vec{p},t)\ , \quad (4.65)$$

$$\frac{\partial}{\partial t} f_d(\vec{p},t) = \mathrm{Tr}\left\{\rho(t)\frac{i}{\hbar}[H, n_{dp}]\right\} = -D_d^{\mathrm{Vlasov}}(\vec{p},t) + I_d^{\mathrm{coll}}(\vec{p},t)\ , \quad (4.66)$$

where $n_{Np} = a^\dagger_{Np} a_{Np}$ and $n_{dp} = a^\dagger_{dp} a_{dp}$. We have introduced bosonic (two–particle) operators that are given through

$$a_{dp} = \sum_{12} \psi_{dp}(12) a_1 a_2\ , \quad (4.67)$$

and the hermitian conjugate $a^\dagger_{dp}$. For the two–particle system of interest here ψ_{dp} is given by the solution of the respective wave equation in mean–field approximation with the eigenvalues E_{dp} [488]. The Vlasov terms $D^{\mathrm{Vlasov}}_{A\alpha}$ describe the reversible time evolution and are related to time dependent Hartree–Fock calculations as, e.g., explained in Ref. [455].

The collision terms $I^{\mathrm{coll}}_{A\alpha} = I^{\mathrm{E}}_{A\alpha} + I^{\mathrm{R}}_{A\alpha}$ correspond to the irreversible behavior and describe elastic scattering (E) and inelastic (reaction) processes (R), respectively, between the constituents of the system. The elastic processes do not change the internal quantum numbers of the particles. They determine the time scales for thermal relaxation. To be more specific we consider the situation where the collision rate is sufficiently high compared to the reaction rate, so that each of the components are close to their thermal equilibrium distributions. The small deviations of the chemical composition from equilibrium are then treated within the linear response theory. The time scale of the relaxation to chemical equilibrium is set by the reaction processes that will be considered in the following.

The inelastic processes that are related to excitation as well as to bound state formation and disintegration change the abundances of the components characterized by the internal quantum numbers and determine the time scale of chemical equilibration. The reaction condsidered here is the nucleon deuteron break–up $I_{NNN,Nd}(p,t)$ (and the reversed ones), i.e.

$$I_N^{\mathrm{R}}(\vec{p},t) \approx I_{NNN,Nd}(\vec{p},t)\ , \quad (4.68)$$

$$I_d^{\mathrm{R}}(\vec{p},t) \approx I_{dN,NNN}(\vec{p},t)\ . \quad (4.69)$$

The collision terms on the rhs. of Eqs. (4.68) and (4.69) each contain gain and loss terms due to deuteron break–up or formation reactions. Collisions of higher clusters (e.g., *dd*) that require a suitable treatment of the effective four–body problem are left for further evaluations.

The collision integral that involves three–particle processes has been given in Born approximation [455] or evaluated in impulse approximation [108]. In both cases the influence of the surrounding medium on the elementary cross section that enter into the collision integrals has been neglected. To evaluate the integral $I_d^{\rm R}(p,t)$ in a dense correlated medium we use linear response theory, see [592], considering a situation of non–equilibrium which is near to the equilibrium state.

$$I_d^{\rm R}(p,t) = -\sum_{p'p''} \langle \dot{n}_{dp}; \dot{n}_{dp'} \rangle (n_{dp'}; n_{dp''})^{-1} \, \delta f_d(p'',t), \tag{4.70}$$

where $\delta f_d(p,t) = f_d(p,t) - f_d^0(p)$ denotes the fluctuations from the equilibrium distribution, and $n_{dp} = a_{dp}^\dagger a_{dp}$. The Kubo scalar product $(A;B)$ that appears in Eq. (4.70) is given by

$$(A;B) = \frac{1}{\beta} \int_0^\beta d\tau \ {\rm Tr}\{\rho_0 \, A(-i\tau) \, B\} \tag{4.71}$$

and its Laplace transform, i.e. the correlation functions $\langle A(\epsilon \to 0^+); B \rangle$, by

$$\langle A(\epsilon); B \rangle = \int_{-\infty}^0 dt \ e^{\epsilon t} \ (A(t); B) \ . \tag{4.72}$$

Following standard many–body techniques (see, e.g., Refs. [454, 227, 147]) the correlation function is evaluated using Green functions,

$$\langle A(\epsilon); B \rangle = -\frac{1}{\beta} \int \frac{d\omega}{2\pi} \frac{1}{\epsilon + i\omega} \frac{1}{\omega} \left[G_{AB}(\omega + i0^+) - G_{AB}(\omega - i0^+) \right] \tag{4.73}$$

where $G_{AB}(z)$ is the analytic continuation of the Matsubara Green function $G_{AB}(z_\mu)$ that will be discussed in the next subsection.

For homogeneous matter, where $\langle \dot{n}_{dP}; \dot{n}_{dP'} \rangle$ and $(n_{dp'}; n_{dp''})$ are diagonal in momenta p, the response equation (4.65, 4.70) is given by

$$\frac{\partial}{\partial t} f_d(p,t) = \langle \dot{n}_{dp}(\epsilon \to 0^+); \dot{n}_{dp} \rangle (n_{dp}; n_{dp})^{-1} \, \delta f_d(p,t)$$
$$\equiv \frac{1}{\tau_{dp}} \, \delta f_d(p,t). \qquad (4.74)$$

The limit $\epsilon \to 0^+$ implied through Eq. (4.74) has to be taken after the thermodynamic limit.

Here we have introduced the momentum dependent life–time (formation–time) of the deuteron fluctuations τ_{dp} in the surrounding medium, which is of central interest. Note that the disintegration (formation) of deuterons requires the explicit treatment of three–particle equations in nuclear matter, which will be derived in the following subsection.

4.3.4 In–Medium Three–Body Equations and Deuteron Breakup Rate

In order to evaluate the equilibrium correlation functions occurring in Eq. (4.74) we can use the method of finite temperature Green function [34]. The thermodynamic three–particle Green function is defined by

$$G_3(123, 1'2'3', t-t') = -i \, \langle T \; a_{(3)}(t) a^{\dagger}_{(3')}(t') \rangle_0 \; , \qquad (4.75)$$

where T implies Wick time ordering. The operators $a_{(3)} = a_1 a_2 a_3$ are three–particle operators, depending on time according to the Heisenberg picture. The average is performed over the equilibrium statistical operator.

Due to the Kubo–Martin–Schwinger boundary condition [227, 267, 147] one can introduce the Fourier transform $G_3(z_\mu)$ at discrete Matsubara frequencies $z_\nu = \pi\nu / i\hbar\beta$ with odd numbers ν. In mean–field approximation the following Bethe–Salpeter equation for the three–particle Green function at finite temperatures and densities is obtained

$$G_3(z_\nu) = G_3^{(0)}(z_\nu) + R_3^{(0)}(z_\nu) \tilde{V}_3 G_3(z_\nu) \; , \qquad (4.76)$$

which is the central input to derive Faddeev type equations in a medium. The notation will be explained in the following. The proper symmetrization is treated separately. The Green function of the noninteracting system is

$$G_3^{(0)}(z_\nu) = N_3 R_3^{(0)}(z_\nu), \tag{4.77}$$

where with the Hartree–Fock energies (4.26) and $\bar{f}_1 = 1 - f_1$

$$R_3^{(0)}(123, 1'2'3'; z_\nu) = \frac{\delta_{11'}\delta_{22'}\delta_{33'}}{z_\nu + 3\mu - E^{\rm HF}(1) - E^{\rm HF}(2) - E^{\rm HF}(3)}\,, \tag{4.78}$$

$$\begin{aligned} N_3(123, 1'2'3') &= \delta_{11'}\delta_{22'}\delta_{33'}(f_1(1)f_1(2)f_1(3) + \bar{f}_1(1)\bar{f}_1(2)\bar{f}_1(3)) \\ &= \delta_{11'}\delta_{22'}\delta_{33'}(1 - f_1(i) - f_1(j)) \\ &\times (1 - f_1(k) + f_2(E^{\rm HF}(i) + E^{\rm HF}(j)))\,. \end{aligned} \tag{4.79}$$

holds, for all permutations $\{ijk\} = \{123\}$.

Note that $[N_3, R_3^{(0)}] = 0$. The interaction kernel in $\tilde{V}_3$ in Eq. (4.76) is given by

$$\tilde{V}_3(123, 1'2'3') = \sum_{k=1}^{3} \tilde{V}_3^{(k)}(123, 1'2'3')\,, \tag{4.80}$$

$$\tilde{V}_3^{(k)}(123, 1'2'3') = (1 - f_1(i) - f_1(j))V_2(ij, i'j')\delta_{kk'}\,, \tag{4.81}$$

with $\{ijk\} = \{123\}$ cyclic, and $\tilde{V}_3 \neq \tilde{V}_3^\dagger$. If we introduce a potential $V_3 = N_3^{-1}\tilde{V}_3$ we may write instead of Eq. (4.76)

$$G_3(z_\nu) = G_3^{(0)}(z_\nu) + G_3^{(0)}(z_\nu)V_3 G_3(z_\nu)\,, \tag{4.82}$$

which looks formally as the equation for the isolated case [171].

Using Eq. (4.80) we may write the potential $V_3 = \sum_k V_3^{(k)}$ in terms of (e.g., $k = 1$)

$$\begin{aligned} V_3^{(1)}(123, 1'2'3') &= (1 - f_1(1) + f_2[E^{\rm HF}(2) + E^{\rm HF}(3)])^{-1} \\ &\times V_2(23, 2'3')\delta_{11'}\,. \end{aligned} \tag{4.83}$$

In Eq. (4.80) we have already introduced the channel notation that is convenient to treat systems with more than two particles [171].

If the pair and the odd particle are uncorrelated in channel γ, we may define a channel Green function $G_3^{(\gamma)}(z_\nu)$. In this case, only the interaction within the pair of channel γ is taken into account,

$$G_3^{(\gamma)}(z_\nu) = \frac{1}{-i\beta} \sum_\lambda iG_2(\omega_\lambda)\, G_1(z_\nu - \omega_\lambda)\,. \tag{4.84}$$

The summation is done over the bosonic Matsubara frequencies $\omega_\lambda = \pi\lambda/(i\hbar\beta)$, λ even. The equation for the channel Green function is derived in the same way as for the total three particle Green function given in Eqs. (4.76) and (4.82):

$$G_3^{(\gamma)}(z_\nu) = G_3^{(0)}(z_\nu) + G_3^{(0)}(z_\nu) V_3^{(\gamma)} G_3^{(\gamma)}(z_\nu)\,. \tag{4.85}$$

Introducing the notation $\bar{V}_3^{(\gamma)} = V_3 - V_3^{(\gamma)}$ we arrive at the following equation for $G_3(z_\nu)$ expressed through the channel Green functions $G_3^{(\gamma)}(z_\nu)$:

$$G_3(z_\nu) = G_3^{(\gamma)}(z_\nu) + G_3^{(\gamma)}(z_\nu) \bar{V}_3^{(\gamma)} G_3(z_\nu)\,. \tag{4.86}$$

As a generalization of the channel transition operator $U_{\alpha\beta}$ to calculate cross sections [463] to finite densities, we define with Eqs. (4.82), (4.85) and (4.86)

$$G_3(z_\nu) = \delta_{\alpha\beta} G_3^{(\alpha)}(z_\nu) + G_3^{(\alpha)}(z_\nu) U_{\alpha\beta}(z_\nu) G_3^{(\beta)}(z_\nu)\,, \tag{4.87}$$

$$U_{\alpha\beta} = (1 - \delta_{\alpha\beta}) G_3^{(0)-1} + \sum_{\gamma \neq \alpha} V_3^{(\gamma)} G_3^{(\gamma)} U_{\gamma\beta}\,. \tag{4.88}$$

This is the AGS–type (or Faddeev–type) equation valid to treat three–particle correlations at finite temperatures and densities in mean–field approximation. The channel transition operator U can be related to the channel T matrix $T_3^{(\gamma)}$, defined as

$$G_3^{(\gamma)} = G_3^{(0)} + G_3^{(0)} T_3^{(\gamma)} G_3^{(0)}\,, \tag{4.89}$$

or using Eq. (4.85)

$$T_3^{(\gamma)} = V_3^{(\gamma)} + V_3^{(\gamma)} G_3^{(0)} T_3^{(\gamma)}, \qquad V_3^{(\gamma)} G_3^{(\gamma)} = T_3^{(\gamma)} G_3^{(0)} . \tag{4.90}$$

Then it is possible to write a second, more useful version of the AGS–type equations

$$U_{\alpha\beta} = (1-\delta_{\alpha\beta})(N_3 R_3^{(0)})^{-1} + \sum_{\gamma\neq\alpha} T_3^{(\gamma)} N_3 R_3^{(0)} U_{\gamma\beta} . \tag{4.91}$$

In the low density case [33] where $N_3 > 0$ in (4.80), we find for $U^*_{\gamma\beta} = N_3^{1/2} U_{\gamma\beta} N_3^{1/2}$

$$U^*_{\alpha\beta} = (1-\delta_{\alpha\beta}) R_3^{(0)-1} + \sum_{\gamma\neq\alpha} T_3^{*(\gamma)} R_3^{(0)} U^*_{\gamma\beta} . \tag{4.92}$$

The equation for the channel operator $T_3^{*(\gamma)} = N_3^{1/2} T_3^{(\gamma)} N_3^{1/2}$ (and so for V_3) is then

$$T_3^{*(\gamma)} = V_3^{*(\gamma)} + V_3^{*(\gamma)} R_3^{(0)} T_3^{*(\gamma)} . \tag{4.93}$$

Inserting all definitions, the explicit form of the effective potential arising in this equation reads for $f_1^2(1) \ll f_1(1)$

$$\begin{aligned} &V_3^{*(3)}(123, 1'2'3') \\ &\simeq (1-f_1(1)-f_1(2))^{1/2} V_2(12,1'2')(1-f_1(1')-f_1(2'))^{1/2} \delta_{33'} . \end{aligned} \tag{4.94}$$

Utilizing this approximation, Eq. (4.92) has been solved numerically using a separable ansatz for the strong nucleon–nucleon potential [33].

We are now able to evaluate the correlation function $\langle \dot{n}(\epsilon \to 0^+); \dot{n} \rangle$ according to Eq. (4.74). The Green function needed in Eq. (4.74) is given by (see [34])

$$\begin{aligned} G^{(\gamma)}_{\dot{n}_{dp}\dot{n}_{dp}}(\omega_\mu) = \frac{4}{-i\beta} \sum_\lambda \ \mathrm{Tr}\left\{ U_{\gamma 0}\, G_3^{(0)}(\omega_\mu + z_\lambda)\, U_{0\gamma}\, G^{(\gamma)}_{3,dp}(z_\lambda) \right\} \\ +(\omega_\mu \leftrightarrow -\omega_\mu) . \end{aligned} \tag{4.95}$$

To perform the Matsubara summation that is present in Eq. (4.95), we now use the spectral representation of the Green functions that have been given for the quasi–particle approximation in Eqs. (4.77) and (4.84). The resulting expression for $G_{\dot{n}_k \dot{n}_p}(\omega_\mu)$ may be cast into the following form,

$$G^{(\gamma)}_{\dot{n}_{dp}\dot{n}_{dp}}(\omega_\mu) = 4i \sum_{123c} \frac{\langle 123|U_{\gamma 0}|c\psi^{(\gamma)}_{dp}\rangle\langle c\psi^{(\gamma)}_{dp}|U_{0\gamma}|123\rangle}{\omega_\mu + (E^{(\gamma)}_{dp} + E^{\rm HF}(c)) - E_0} \tag{4.96}$$
$$\times[\bar{f}_1(1)\bar{f}_1(2)\bar{f}_1(3)f_1(c)f_2(E^{(\gamma)}_{dp})$$
$$-f_1(1)f_1(2)f_1(3)\bar{f}_1(c)(1+f_2(E^{(\gamma)}_{dp}))]$$
$$+(\omega_\mu \leftrightarrow -\omega_\mu)\ ,$$

with $E_0 = E^{\rm HF}(1) + E^{\rm HF}(2) + E^{\rm HF}(3)$, and $c \in \{1'2'3'\}$ depending on the channel $\gamma \in \{123\}$, resp. The terms in brackets of Eq. (4.97) are usually referred to as Pauli factors of the gain and loss terms. Since presently we are interested in the time scale of fluctuations, we may consider, e.g., the loss term that is given by

$$\langle \delta\dot{n}^{(\gamma)}_{dp}; \delta\dot{n}^{(\gamma)}_{dp}\rangle = 4\sum_{123c} |\langle 123|U_{0\gamma}|\psi^{(\gamma)}_{dp}, c\rangle|^2 \bar{f}_1(1)\bar{f}_1(2)\bar{f}_1(3)f_1(c)f_2(E^{(\gamma)}_{dp})$$
$$\times\ 2\pi\delta(E^{(\gamma)}_{dp} + E^{\rm HF}(c) - E_0)\ . \tag{4.97}$$

For identical particles that are considered here proper symmetrization has to be taken into account. To this end we connect the result given in Eq. (4.97) with the transition matrix T_0, that is evaluated between properly symmetrized and normalized states ϕ_0, ϕ_d and again satisfies the equivalent three–body equation, see Refs.[171] for the isolated case. It is given by

$$\langle \phi_0|T_0|\phi_d\rangle \equiv \sqrt{2}\sum_\gamma \langle 123|U_{0\gamma}|\psi^{(\gamma)}_{dp}, c\rangle\ . \tag{4.98}$$

To be more specific we also separate spin and momentum degrees of freedom and introduce amplitudes

$$\langle m_1 m_2 m_3|M_0(\vec{p}_1\vec{p}_2\vec{p}_3, \vec{p}_d\vec{p}_c; E)|m_d m_c\rangle \equiv \langle \phi_0|T_0|\phi_d\rangle\ . \tag{4.99}$$

Using this amplitudes the life–time may be written in the following way

$$\tau_d^{-1} = \frac{4}{3!}\int d^3p_N \int d^3p_1 d^3p_2 d^3p_3 \ \mathrm{Tr}(M_0\rho_i M_0^\dagger)\ \bar{f}_1(1)\bar{f}_1(2)\bar{f}_1(3)f_1(c)$$
$$\times 2\pi\delta(E-E_0)\ . \tag{4.100}$$

The total energy is given by $E = E_d + p^2/2m_d + E^{\mathrm{HF}}(p_N)$, where E_d is the in-medium deuteron binding energy and $E^{\mathrm{HF}}(p_N)$ is the energy of the odd nucleon. The factor $1/3!$ prevents overcounting due to the six possible ways of arranging the identical particles among the three momenta, the trace is over spin projections only and ρ_i is the initial spin density matrix.

We may now introduce the in–medium break–up cross section σ_0^* in the center of mass system, which coincides with the usual in–vacuum definition in the zero density limit [33]. It is given by

$$\sigma_0^*(E) = \frac{(2\pi)^3}{|\vec{v}_d - \vec{v}_N|}\frac{1}{3!}\int d^3p' d^3q'\ \mathrm{Tr}(M_0^*\rho_i M_0^{*\dagger})$$
$$\times\ 2\pi\delta(E^* - E_0^*)\ . \tag{4.101}$$

The cross section is evaluated in the center of mass system introducing Jacobi coordinates $\vec{p}'$, $\vec{q}'$, and $|\vec{v}_d - \vec{v}_N|$ is the relative velocity of the incoming particles. The center of mass scattering energy is $E^* = 3q^2/4m^* + E_d^*$, where we have used effective mass approximation for the nucleon self energy [33]. Due to the medium effects the deuteron binding energy changes, which is calculated consistently with the two–body input into the Faddeev equation that lead to the amplitudes M_0^*. To evaluate the life–time of the deuteron in medium we introduce the cross section defined in Eq. (4.101) into Eq. (4.100). The remaining integration is over the momentum $\vec{p}_N$ of the odd nucleon. The equation for the life–time of the deuteron is then given by

$$\tau_{dP}^{-1} = \frac{4}{(2\pi)^3}\int d^3p_N\ |\vec{v}_d - \vec{v}_N|\ \sigma_0^*(E^*)\ f(\varepsilon^*)\ , \tag{4.102}$$

and $\varepsilon^* = (3\vec{q}/2 + \vec{p}/2)^2/2m^*$.

The cross section entering into Eq. (4.102) is given in Fig. 4.5, see [34]. The parameters for the nucleon-nucleon interaction are taken from Refs. [71, 422], which lead to a good overall description of the elastic and break–up cross

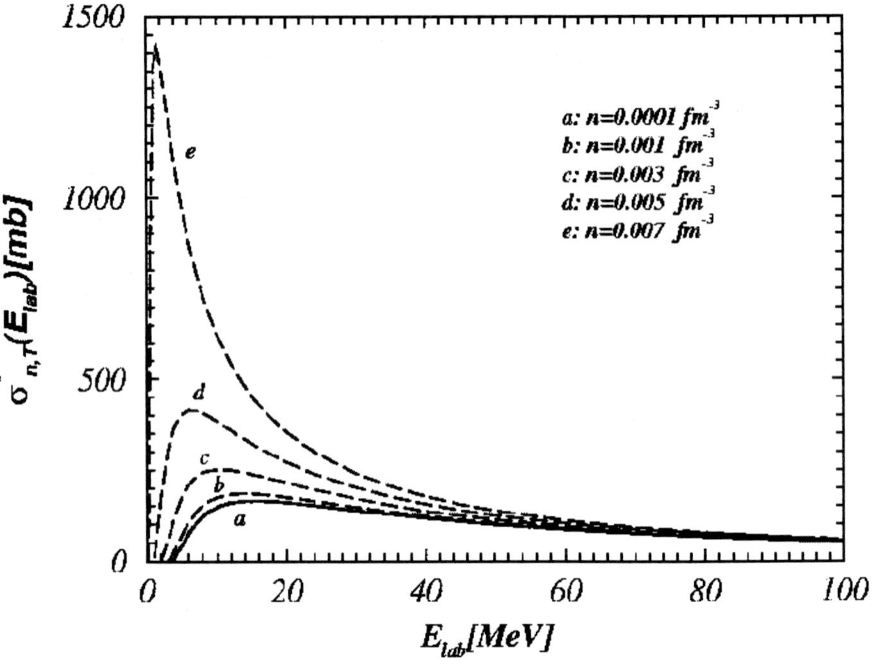

Fig. 4.5 Break–up cross section at temperature $T = 10$ MeV. Free cross section is shown as solid line and reproduces the experimental data, see Ref. [33]. Other lines are due to different nuclear densities, see text.

sections as well as the differential elastic cross section up to $E_{\rm lab} = 50$ MeV [33, 501]. To calculate the break–up cross section we use the optical theorem.

The solid lines represent the isolated break–up cross section. As shown previously it reproduces the experimental data [33, 501]. The dashed lines show the break–up cross section, $\sigma^{*}_{n,T}(E_{\rm lab})$ for densities $n = 0.1,\ 1,\ 3,\ 5,\ 7 \times 10^{-3}$ fm^{-3}, respectively, as a function of the laboratory energy $E_{\rm lab}$. The Mott transition occurs at the density $n \simeq 8 \times 10^{-3}$ fm^{-3} for the temperature $T = 10$ MeV considered.

The medium effects significantly modify the isolated break–up cross section. Two qualitative features are observed. First, the break–up threshold is shifted towards lower scattering energies with increasing density of the nuclear matter. This kinematic effect is due to the decrease of the deuteron binding energy with increasing density (see Fig. 4.6). Second, the cross section in-

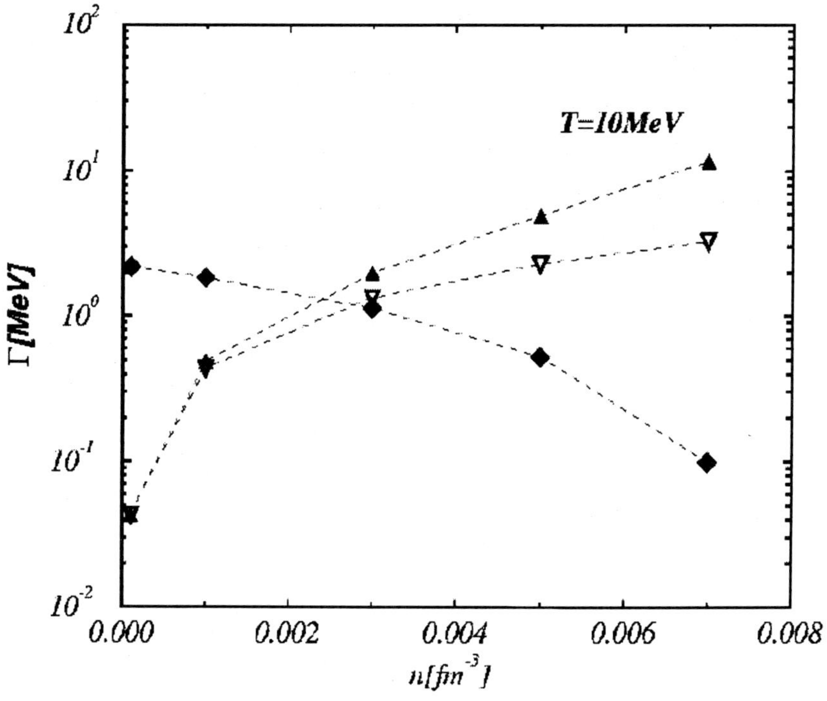

Fig. 4.6 Width of the deuteron in hot and dense nuclear matter at $T = 10$ MeV depending on the nuclear density and $P = 0\ \mathrm{fm}^{-1}$. The full triangles show the full calculation, the empty triangles show the one that uses the vaccuum cross section for the break–up reaction, free masses and deuteron binding energy. The deuteron binding energy is shown by the full diamonds.

creases considerably with increasing density. The maximum is enhanced by one order of magnitude for the largest density value considered. For densities larger than the Mott densities the deuteron disappears as a bound state. Nevertheless, a resonance in the continuum of scattering states is expected to occur the width of that should match with the time of life of the bound state at the Mott density where the bound state disappears.

4.4 Conclusions

We have considered isolated clusters of interacting particles as bound states to be described by the Schrödinger equation. A promising method to solve the bound states of the A–particle system is the local density approximation. Also other methods have been applied successfully, including computer simulations of the quantum mechanical A–particle problem.

The properties of a cluster are modified, if the cluster is imbedded in a medium. In equilibrium, a cluster–mean field approximation has been presented to describe the cluster wave function in an optimal way. In addition to the bound states, also scattering states have to be investigated, to obtain a smooth transition via resonances to bound states. Whereas in the low–density limit the composition is given by an ideal mixture of free and bound states (chemical picture), in dense matter the clusters are dissolved due to the Mott effect.

The non–equilibrium evolution of a clustered quantum system is treated on the basis of a density matrix approach. The non–equilibrium statistical operator is constructed on the basis of a non–equilibrium cluster–mean field approximation. Kinetic equations for the cluster distributions are derived, containing a Vlasov term given by the mean field of all clusters, and a collision term describing elastic as well as inelastic collisions. Reaction rates such as the dimer breakup reaction rate can be derived, which are modified in a dense system. As an example the deuteron break–up reaction cross section in dense nuclear matter is considered.

5 Monte Carlo Modeling of First–Order Phase Transitions

A. Milchev

5.1 Introduction

One of the most important applications of Monte Carlo methods in condensed matter physics is the study of phase transformation processes [41, 49, 53, 52]: one is interested in the determination of the parameter values for which a phase transition takes place; frequently one has to distinguish whether the transition is first order [51] or second order [528]; and one wishes to characterize precisely (i.e. quantitatively) the properties of the transition. For a second order transition this implies estimation of the critical exponents and critical amplitudes [528] whereas for a first order transition one determines the jumps in the first derivatives of the free energy F at the transition, such as order parameter or energy (latent heat), for instance.

First order phase transitions play an important role in many fields of physics [180, 51, 204]. Well known examples are field–driven transitions in magnets, adsorbates, condensation and crystal melting, the electron – hole liquid – gas transition in semiconductors [436], the nematic–isotropic transition in liquid crystals or, at much higher energy scales, the deconfining transition in in hot quark–gluon matter and the various transitions in the evolution of early universe [204].

In order to get deeper insight into this variety of processes, which reflect the rather complex behavior of many–body systems, one usually combines experimental studies with analytical model treatment. Both approaches have their shortcomings – while an idealized simplified model is frequently obscured in laboratory investigations due to various (undesirable) side effects, analytic results are derived, as a rule, at the cost of severe approximations, only. In this sense computer experiments, as the Monte Carlo simulations, which, in principle, allow one to test a model exactly, provide a valuable tool for research and complement favourably the other approaches. In a number of cases computer simulations, serving as critical test for existing theories, have offered significant revision of our understanding of phase transformations as, for example, the acknowledged absence of clear–cut border between nucleation and spinodal decomposition in first order relaxation phenomena, which we shall consider in more detail below.

Although there already exists a vast literature on the subject [180], many properties of first order phase transitions still remain to be investigated in details. Examples are finite–size scaling (FSS) [35], the shape of energy or order parameter distributions [36, 37], partition function zeros [302] etc., which are all closely interrelated. Even the Monte Carlo methods for modeling first order phase changes themselves bear many nontrivialities and pitfalls when meaningful information is to be extracted from statistical observations.

Usually new approaches are tested on a simple and possibly exactly solvable model before such approaches become a standard tool for investigation of more complex models. In the development of Monte Carlo methods for the investigation of first order phase transitions, therefore, as a rule one resorts to familiar models of many body condensed matter systems of the type of the Ising model [359, 217] and its generalization – the Potts model [429, 573], or to the so called ϕ^4–model (or, continuous Ising model) [174, 368]. In all these cases one deals with a system of atoms (spins) which occupy the sites of a d–dimensional lattice whereby the atoms (spins) reside in several states and interact by pair (or multiplet) forces.

In the q–state Potts model [429, 573] each lattice site carries a spin S_i that may take any of the q possible states, corresponding to the degeneracy of the ordered phase, while the disordered state is not degenerated. The q–state Potts Hamiltonian is given by

$$\mathcal{H} = -\sum_{\langle ij \rangle} J\delta_{S_i S_j} \qquad \text{with} \qquad S_i \in \{1, \ldots, q\} \tag{5.1}$$

where $\delta_{S_i S_j}$ is Kronecker's symbol, so that an energy J is gained if two neighboring sites are in the same state, and the summation $\langle ij \rangle$ runs over all nearest–neighbor pairs. In the particular case of $q = 2$ (up– and down–spins only) the Hamiltonian, Eq.(5.1), yields that of the Ising model. Of particular interest is that the order of the phase transition[1] depends on

[1] Recently it has been shown [57] that the order of this phase transition may change at other q–values for sufficiently strong *random* fields.

both q and the dimensionality d of the system: for $d = 3$ the transition is first order when $q \geq 3$. Similarly to the Ising model with respect to second order phase transitions, the two dimensional Potts model is an especially convenient toy for testing new ideas since for $q \geq 5$ it is *exactly* known to exhibit a temperature–driven first order phase transition at a critical temperature $(k_B T_0)^{-1} = \ln(1 + \sqrt{q})$ (here and in what follows k_B denotes the Boltzmann factor). Furthermore, right at the transition also the internal energies of the disordered and ordered phases and the difference of the corresponding specific heats are known exactly [573, 25] and can be compared to simulational results.

The ϕ^4 –model is defined by the Ginzburg–Landau Hamiltonian

$$H(\{\phi_i\}) = \sum_{i=1}^{V} \left[\frac{1}{2}(\nabla\phi_i)^2 - \frac{u^2}{2}\phi_i^2 + g\phi_i^4\right] \tag{5.2}$$

where the spin variables ϕ_i are continuously distributed, $-\infty \leq \phi_i \leq \infty$, and the discretized lattice gradient term $(\nabla\phi_i)^2 = \sum(\phi_{i+1} - \phi_i)^2$ with $i+1$ being the nearest neighbor index in a particular lattice direction. The critical behavior of the ϕ^4–model is described within the universality class of the Ising model [368, 539, 362]. Many of the Monte Carlo studies, quoted below, are based on simulations of these models.

In the following of the present chapter we shall focus briefly on some important methodological problems in Monte Carlo studies of *static* properties, associated with first order phase transition, and on the model description of the *dynamics* of such transformations. In particular, we shall consider in more detail the reason for gradual transition between the two major kinetic mechanisms of discontinuous phase changes – nucleation and spinodal decomposition, whose understanding as been due mainly to numeric experiments.

5.2 Static Properties

5.2.1 Location and Identification of First–Order Phase Transitions – Finite Size Scaling

A typical feature of first order transitions is the existence of *hysteresis* effects. Due to effect of critical slowing down [41], typical for continuous phase changes, however, the observation times during the simulation may prove far too short, and a spurious hysteresis, observed near second order transition, can be distingushed from a 'true' one by a careful finite size scaling analysis, only. Moreover, even if the transition may be clearly identified as first order, it is a problem to locate at which parameter value it really occurs since it can be easily surpassed into the metastable region.

Since a Monte Carlo simulation, as a rule, does not yield thermodynamic potentials as direct observables, a simple prescription is to compare the free energies of the various phases, derived by thermodynamic integration[41, 366] of suitable derivatives

$$\frac{F}{k_B T} = \lim_{T \to \infty} \frac{F}{k_B T} + \int_0^{1/k_B T} E(T') d\frac{1}{k_B T'} \tag{5.3}$$

although the method requires the knowledge of internal energy, $E(T)$, over a broad range of temperatures and of F at some particular temperature, usually $T = 0$, or $T = \infty$. Recently this method has been successfully utilized to determine the phase diagram of gas–liquid condensation in thin–film geometry confined between two parallel plates ("capillary condensation") [55].

If for a reasonable time a hysteresis is not found, a transition may still be of first order, albeit weak. The standard method for localization and characterization of first order phase transitions is to use a finite size scaling analysis [151, 82, 433] in order to derive reliable and efficient criteria. Indeed, while a singular behavior of the free energy of a many body system of linear size L should be possible in the thermodynamic limit ($V = L^d \to \infty$) only, in a finite system no such singularities are possible, rather, for V finite the transition becomes somewhat diffuse and rounded, and is also shifted relative to the critical parameter value where it occurs for $V \to \infty$. For instance, for a first order change, driven by temperature, the specific heat δ-function singularity at T_c appears as a rounded peak of finite height, $C_{\max} \propto L^{\alpha_m}$, and the transition itself is shifted, $T_c(L) - T_c(\infty) \propto L^{-\lambda}$. Now the finite–size behavior of the specific heat can be simply understood from the observation[86] that the energy distribution $P_L(E)$ is a superposition of two Gaussians ($\Delta T = T - T_c$):

$$P_L(E) = g_+ \exp\left[-\frac{(E - E_+ - C_+ \Delta T)^2 L^d}{k_B T^2 C_+}\right] \tag{5.4}$$

$$+ g_- \exp\left[-\frac{(E - E_- - C_-\Delta T)^2 L^d}{k_B T^2 C_-}\right]$$

where, assuming that the low temperature phase has a degeneracy q_- and the high temperature phase – degeneracy q_+, the weights g_+, g_- are:

$$g_+ = q_+ \exp\left[\frac{\Delta T(E_+ - E_-)}{2k_B T T_c} L^d\right], \qquad g_- = q_- \exp\left[-\frac{\Delta T(E_+ - E_-)}{2k_B T T_c} L^d\right] \tag{5.5}$$

Using Eqs.(5.4,5.5) one may take readily the average, $\langle E\rangle = \int E P_L(E) dE$, and from $C_L = d\langle E\rangle/dT$ the specific heat peak in a finite system is obtained in a straightforward manner

$$C_L = \frac{g_+ C_+^{3/2} + g_- C_-^{3/2}}{g_+\sqrt{C_+} + g_-\sqrt{C_-}} + \frac{q_+ q_- \sqrt{C_+ C_-}}{k_B T^2} L^d \left[\frac{(E_+ - E_-) + (C_+ - C_-)\Delta T}{g_+\sqrt{C_+} + g_-\sqrt{C_-}}\right]^2 \tag{5.6}$$

so that a maximum of height [86]

$$C_{\max} \approx \frac{(E_+ - E_-)^2}{4 k_B T_c^2} L^d + \frac{C_+ + C_-}{2} \tag{5.7}$$

occurs at temperature $T_c(L)$. The value of this maximum is expected for a first order phase transition to scale with the system size $\propto L^d$ – Fig.5.1. The phase change occurs where the *weights* of the two Gaussians are equal, $g_+\sqrt{C_+} = g_-\sqrt{C_-}$, which yields [86]

$$\frac{T_c(L) - T_c}{T_c} = \frac{k_B T_c}{E_+ - E_-} \ln\left(\frac{q_-}{q_+}\sqrt{\frac{C_-}{C_+}}\right) L^{-d} . \tag{5.8}$$

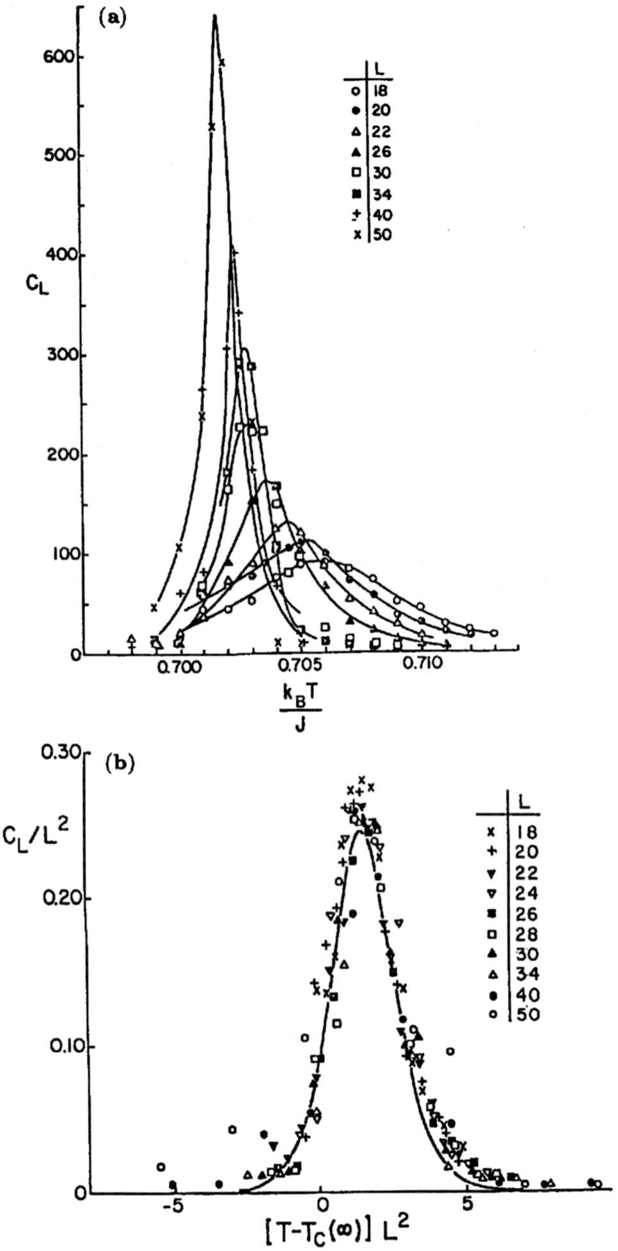

Fig. 5.1 (**a**) Temperature variation of the specific heat for various lattice sizes, for the 10–state Potts model on a square lattice. (**b**) Scaling of the specific heat data, plotting C_L/L^2 versus $[T - T_c(\infty)]L^2$. In this model $q_- = 10, q_+ = 1$. The solid curve results from Eq.(5.6). From Ref. [52].

Since the width of the specific heat rounding can be defined by requiring that the argument of the exponential in Eq.(5.5) is ± 1, one gets

$$\frac{\Delta T}{T_c} = \frac{2k_B T}{E_+ - E_-} L^{-d} . \tag{5.9}$$

It is thus clear that the exponents $\alpha_m = \lambda = d$, i.e., at first order transitions various second derivatives of the thermodynamic potentials, such as specific heat, susceptibility or compressibility, should scale with system size as the *linear dimension of the system to the power of its volume.* The scaling law differs thus significantly from the case of continuous phase transformations where it is determined by the respective critical exponents. Similar probability distributions can be derived for any order parameter, ψ, Fig.5.2, one should only keep in mind that they are valid as long as the system size exceeds considerably the correlation length, $L \gg \xi$, but this limit is always

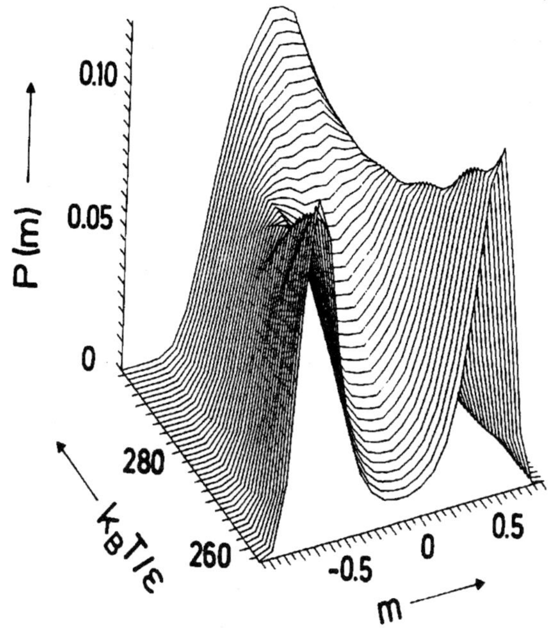

Fig. 5.2 Order parameter distribution $P(m)$ of a symmetrical binary mixture of two types of polymers chains of length $N = 128$ shown as a function of m and T in the critical region. Here m is defined in terms of the number $n_A(n_B)$ of A–chains (B–chains) $m = (n_A - n_B)/(n_A + n_B)$. The shape of the peaks away from criticality is very nearly Gaussian. From Ref. [55].

reached for first order phase transitions with a discrete number of ordered states since the correlation length remains finite at T_c. With the ansatz for the probability distribution of the order parameter one can calculate all its moments of interest. Especially useful for location and identification of the nature of a phase change is the fourth order cumulant, $g_L(T)$, which is given by the ratio of the fourth– and second moments of the order parameter (or, the energy E)

$$g_L(T) = \frac{n}{2}\left(1 + \frac{2}{n}\frac{\langle\psi^4\rangle}{\langle\psi^2\rangle^2}\right) \tag{5.10}$$

where n denotes the order parameter dimensionality (number of components). A phenomenological theory [555] which has been nicely illustrated as an example of the probability distribution of the 3–state Potts in three dimensions [55] has shown that for first order phase changes $g_L(T)$ exhibits a characteristic minimum at $(T_{\min} - T_c) \propto L^{-d}$ and the value of the minimum behaves like $g_L(T_{\min}) \propto L^{-d}$. Moreover, all cumulants $g_L(T)$ for different $L \gg \xi$ approximately intersect at a common crossing point, $(T_{cross} - T_c) \propto L^{-2d}$, with a universal value $g_L(T_{cross}) = 1 - n/(2q)$ (q is again the degeneracy of the ordered phase).

5.2.2 Interface Tension

A quantity of central importance for the kinetics of first order phase transitions is the surface tension σ associated with the interface between coexisting phases [180, 51] – Fig.(5.3). Although it is known that for systems of dimensionality $d \leq 3$ there exists a typical long wavelength interface instability due to capillary waves[73], and in the absence of fields, stabilizing the position of the interface, its width in a system with linear size L parallel to the interface diverges for $L \to \infty$ as $L^{(3-d)/2}$ (or, $\propto \sqrt{\ln L}$ in $d = 3$), there still exists well defined interface free energy. For example, this instability also occurs for the two–dimensional lattice–gas (Ising) model [590] where the interface energy, $\sigma = 2J - k_B T \ln[\coth(J/(k_B T))]$, has been calculated exactly [399].

In numerical caculations of the interface free energy one could simulate a system, containing an interface (its thickness is typically $\approx \xi$, the correlation length), and another one without interface, at otherwise identical conditions – Fig.5.4. Studying the excess energy E_s of the system of Fig.(5.4b) in

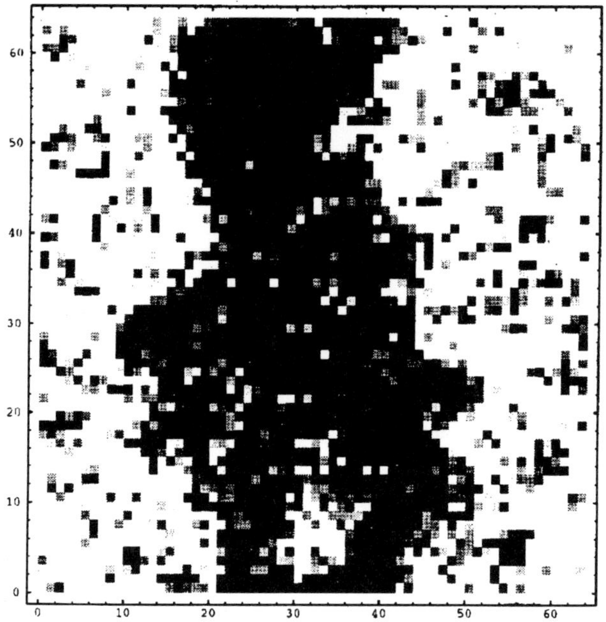

Fig. 5.3 Interfaces in a typical mixed phase configuration in the ϕ^4–model with for $r^2 = 1.40$ and $L = 64$. Values of $\phi > 0.5 (< -0.5)$ are depicted black (white), values of $| \phi_i | < 0.5$ are depicted by varying gray shades. For this configuration the magnetization was $m \approx -0.2$. From Ref. [223].

comparison with that of Fig.(5.4c), one could find the interface free energy σ by numerical integration – Eq.(5.3). If the linear dimensions $L, M \gg \xi$ one would include possible capillary wave fluctuation contributions to the interface free energy by this method.

This method is useful for temperatures far below critical. Near the critical point, where ξ becomes very large and E_s – very small, it is difficult to apply since the fluctuations of the bulk energy (of order $\sqrt{ML^{d-1}C}$) strongly increase as the bulk specific heat C tends to diverge. A method has been suggested [48], however, which yields directly the surface tension near T_c by sampling the (nonzero) minimum of the order parameter distribution at $\rho_{\min} = \frac{1}{2}(\langle \rho_+ \rangle + \langle \rho_- \rangle)$ – Fig.(5.4d). This minimum is due to configurations where interfaces form spontaneously due to thermal fluctuations in the finite

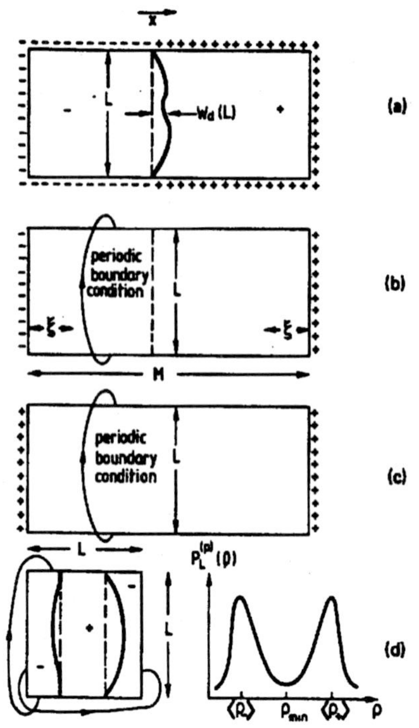

Fig. 5.4 (a) Boundary conditions for an Ising model in $d = 2$ which lead to the formation of an interface below the critical point separating phases with negative and positive magnetization. (b) Standard boundary conditions for computer simulation of a system with interface. The linear dimension $M \gg \xi$ where ξ is the bulk correlation length. (c) Boundary for a reference system without interface. (d) Finite system with fully periodic boundary conditions and its order parameter distribution function $P_L^{(p)}(\rho)$. The minimum corresponds to a situation with two interfaces, while the maxima correspond to pure phases with order parameters $\langle\rho_-\rangle$ and $\langle\rho_+\rangle$.

system. For a lattice gas, for instance, the probability distributions in the pure phases, similar to Eq.(5.4), will be a Gaussian

$$P_L^{(p)}(\rho) = \frac{L^{\frac{d}{2}}}{\sqrt{2\pi k_B T \chi}} \exp\left[-\frac{(\rho - \langle\rho\rangle)^2}{2k_B T \chi} L^d\right] \tag{5.11}$$

where χ denotes the order parameter "susceptibility" (for the liquid–gas system – the compressibility $\chi = (\partial\rho/\partial\mu)_T$). Neglecting configurations which contain domains describing states with mixed phases, a state at the coexistence curve where both coexisting phases are equally likely to occur, would then have a probability distribution

$$P_L^{(p)}(\rho) \simeq \frac{L^{\frac{d}{2}}}{2\sqrt{2\pi k_B T \chi_-}} \exp\left[-\frac{(\rho - \langle\rho_-\rangle)^2}{2k_B T \chi_-} L^d\right] + \frac{L^{\frac{d}{2}}}{2\sqrt{2\pi k_B T \chi_+}} \exp\left[-\frac{(\rho - \langle\rho_+\rangle)^2}{2k_B T \chi_+} L^d\right] \tag{5.12}$$

where χ_+ and χ_- are the values of susceptibility at the two branches of the coexistence curve. According to Eq.(5.12), a state with order parameter $\rho_{\min}$ would have a probability of order $L^{d/2}(2\pi k_B T\chi)^{-1/2}\exp[-(\langle\rho_-\rangle - \langle\rho_+\rangle)^2 L^d/(2\pi k_B T\chi)]$, i.e., the probability of a *homogeneous* state with order parameter $\rho_{\min}$ decreases *exponentially fast* with the volume L^d of the system. Evidently the probability of the minimum will be reduced by the Boltzmann factor involving the free energy cost, $\Delta F = 2L^{d-1}\sigma$, needed to create two interfaces: $P_L^{(p)}(\langle\rho_{\min}\rangle) \simeq P_L^{(p)}(\langle\rho_+\rangle)\exp(-2L^{d-1}\sigma/k_B T)$. The size dependence of the preexponential factor is much weaker than the exponential variation: $P_L^{(p)}(\langle\rho_+\rangle) \simeq \frac{1}{2}L^{d/2}(2\pi k_B T\chi_+)^{-1/2}$. This result suggests that $\sigma/k_B T$ can be found by extrapolating $\ln P_L^{(p)}(\langle\rho_{\min}\rangle)/2L^{d-1}$ or $\ln P_L^{(p)}(\langle\rho_{\min}\rangle)/P_L^{(p)}(\langle\rho_+\rangle)$ as linear function of $\ln L/L^{d-1}$ for $L \to \infty$. This method has been checked for the Ising model in two dimensions [48] where the simulational data converge nicely to the exactly known result.

If this method is employed at temperatures far below the critical point (or, for strong first order phase transitions), however, in order to derive accurate results the system has to travel many times between the two peaks through mixed phase configurations, which has been demonstrated above to happen with vanishing probability, implying an exponential divergence of the autocorrelation time $\tau \propto \exp(2L^{d-1}\sigma/k_B T)$ for sufficiently large system size L. This property is sometimes called *supercritical* slowing down. To overcome the slowing down problem, a so called *multicanonical* method has been proposed which is basically a reweighting approach and can, in principle, be combined with any legitimate updating algorithm. In multicanonical simulations [31, 222] of field driven transitions one tackles the problem of simulating a canonical distribution, Eq.(5.4), or Eq.(5.11), by simulating an auxiliary distribution, $P^{mc}(\rho) = P_L^{(p)}(\rho)\exp[-f(\rho)]$, where the reweighting factor $\exp[-f(\rho)] \equiv w(\rho)^{-1}$ is adjusted iteratively in such a way that $P^{mc}(\rho) = const.$ between the two peaks. This gives the mixed phase configurations the same statistical weight as the pure phases. Canonical expectation

parameter ψ , one can show that $[l(t)]^d \propto (t^x \xi^{1-xz})^d$ for $L \to \infty$ where again z is the dynamic critical exponent and the growth exponent x within the Ornstein–Zernike theory is $x = 1/z = 1/2$ so that one recovers the well known Cahn–Allen [6] theory of domain growth. Similar analysis shows then that the statistical error in the determination of the average size of the newly formed domains after a temperature quench depends on the number of observations n at time t (as in Eq.(5.16)) but not on lattice size L.

5.3 Dynamic Properties

The suitability of the Monte Carlo method for the study of dynamic properties of various stochastic models stems from the fact that Monte Carlo sampling is a *kinetic process in itself* by construction [41]. One may attempt to choose the details of this kinetic process in such a way that a realistic description of the desired system is achieved. It may be then applied to various systems as binary and ternary alloys, adsorbed surface layers, highly anisotropic magnets, molecular crystals or polymers. An important feature is that the simulation is *not* restricted to small deviations from thermal equilibrium, as most well established analytical methods are. Thus nonlinear relaxation phenomena, like nucleation or spinodal decomposition, are easily accessible to study. This property makes Monte Carlo sampling a powerful method in materials science which complements and tests existing laboratory investigations and theoretical interpretations. In the following we shall briefly consider the basic concepts of phase separation in alloys on the ground of the recent insight into the nature of these phenomena gained largely from intensive Monte Carlo studies.

We consider a solid binary alloy whose phase diagram in temperature–density coordinates has a miscibility gap between the two branches $c^{(1)}$ and $c^{(2)}$ of the coexistence curve (binodal) which merge in the critical point, T_c, c_c, and separate the one phase region from the two phase coexistence region – Fig.5.5. In thermal equilibrium the system is macroscopically homogeneous in the one phase region, but inhomogeneous in the two phase region: at concentration c, according to the so called *lever rule*, a fraction $\frac{c-c^{(1)}(T)}{c^{(2)}(T)-c^{(1)}(T)}$ of the volume is in domains of concentration $c^{(2)}$ and the remaining fraction $\frac{c^{(2)}(T)-c}{c^{(2)}(T)-c^{(1)}(T)}$ has the composition $c^{(1)}$.

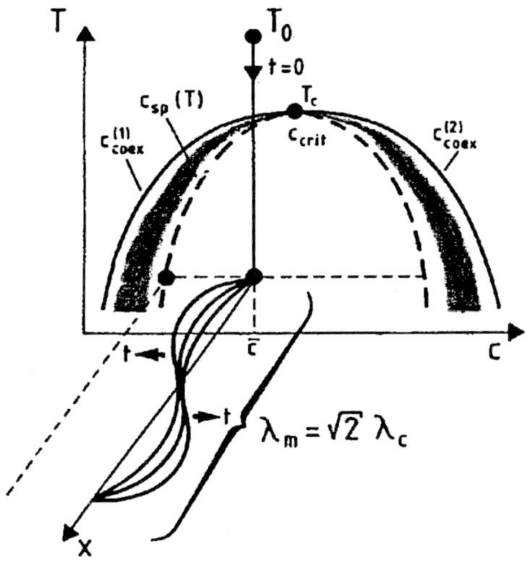

Fig. 5.5 Schematic phase diagram with a miscibility gap below a critical temperature T_c, characterized by a binodal (solid line) and spinodal (dashed line), which mark the regions of metastability and instability for given concentration c and temperature $T \leq T_c$, respectively. In between both lines a region of gradual transition from nucleation to spinodal decomposition is marked as a shaded area. The growth of density fluctuations at length scales $\approx \lambda_{max}$ is also shown.

After an initial quench from a temperature T_0 in the one phase region to a final state at temperature T in the two phase region a growth of initially microscopic inhomogeneities is observed and the system evolves towards a phase separated equilibrium state. In studying this evolution it is common [79, 180] to distinguish in the two phase region between a *metastable* region, enclosed between the binodal and the so–called "spinodal curve" $c_{sp}(T)$, defined as an inflection point of the (mean field) free energy in the T, c plane, $(\partial^2 F/\partial c^2)_T = 0$, and an *instable* region within the spinodal itself (where $(\partial^2 F/\partial c^2)_T \leq 0$) with the branches of the binodal and the spinodal merging at the critical point, as shown in Fig.5.5. With respect to the early stage dynamics of phase separation it is then usually assumed that the decay of metastable states proceeds by the formation of *localized* large–amplitude *heterophase* fluctuations (droplet formation, nucleation) whereas in the unstable region this decay proceeds by amplification of initially weak *delocalized* long wave–length fluctuation (spinodal decomposition). At a later stage the inhomogeneous concentration distribution, thus generated, coarsens in order to minimize the interface – Figs.5.6 and 5.7.

Simulational results meanwhile suggest that such a sharp distinction be-

Fig. 5.6 Spinodal decomposition in the two–dimensional Ising model with conserved magnetisation $m = 0$. Typical configurations for quenches at $T = 0.5, 1.0, 2.0$, and 3.0 (from left to right) after: $0, 2^4, 2^6, 2^8, 2^{10}, 2^{12}$ and 2^{14} MC steps. From Ref. [371].

Fig. 5.7 The same as in Fig.5.6 for $2^{15}, 2^{17}, 2^{18}, 2^{19}, 2^{20}, 2^{21}$ and 2^{22} MC steps.

tween different mechanisms of phase formation is based on mean field type arguments only and cannot be really justified – rather the transition from nucleation to spinodal decomposition is gradual, the spinodal curve is smeared over a concentration interval which is itself shifted away from the mean field position of the spinodal towards the coexistence curve, as shown in Fig.(5.5). It should also be emphasized at this point that weak delocalized long–wavelength fluctuations cannot be actually identified in terms of the atomic concentration variable c_i which changes discretely ($c_i = 0$ to $c_i = 1$) from one lattice site to the next, but require the introduction of coarse–grained concentration field $c(\mathbf{r},t)$, defining $c(\mathbf{r},t) \equiv L^{-d} \sum_{i \in L^d} c_i(t)$, $\mathbf{r}$ being the center of mass of the cell L^d.

It becomes thus evident that the free energy F of "homogeneous" states in the two phase region will depend on the length scale L over which short wave length fluctuations have been integrated, moreover, this coarse–grained free energy describes homogeneous states only if $L \ll \xi$, the interfacial width, otherwise the states which contribute to F are phase separated on local scale, and F should tend smoothly to a double–tangent construction between the pure phase minima as $L \to \infty$. Hence there is no unique method for calculating the "true" location of the spinodal curve.

5.3.1 Spinodal Decomposition

Using the coarse–grained concentration variables $c(\mathbf{x},t)$ one can derive the governing equation describing spinodal decomposition from the continuity equation, $\partial c(\mathbf{x},t)/\partial t + \nabla \cdot \mathbf{j}(\mathbf{x},t) = 0$, since in the total volume V the average concentration $\bar{c} = V^{-1} \int c(\mathbf{x},t) d\mathbf{x}$ remains constant. The standard assumption relates the concentration current $\mathbf{j}$ to the mobility $M(T)$ and the local chemical potential difference $\mu(\mathbf{x},t)$ as $\mathbf{j}(\mathbf{x},t) = -M(T)\nabla\mu(\mathbf{x},t)$. Since in equilibrium $\mu = (\partial F/\partial c)_T$, one generalizes this relation to an inhomogeneous situation far from equilibrium, defining a free energy functional $\mathcal{F}$

$$\frac{\mathcal{F}}{k_B T} = \int d\mathbf{x} \left\{ \frac{f[c(\mathbf{x},t)]}{k_B T} + \frac{r^2}{2} [\nabla c(\mathbf{x},t)]^2 \right\} \tag{5.17}$$

where f is the *coarse–grained* free energy density, and r is related to the range of the interactions between atoms in lattice sites i and j of the original (non coarse–grained) lattice gas Hamiltonian, e.g., if $\mathcal{H} = \sum_{i \neq j} [\varphi_{AA}(\mathbf{x}_i -$

$\mathbf{x}_j)c_ic_j + \varphi_{BB}(\mathbf{x}_i - \mathbf{x}_j)(1-c_i)(1-c_j) + \varphi_{AB}(\mathbf{x}_i - \mathbf{x}_j)(c_i(1-c_j) + c_j(1-c_i))]$ one defines r as

$$r^2 = \frac{\sum_j (\mathbf{x}_i - \mathbf{x}_j)^2 [\varphi_{AA}(\mathbf{x}_i - \mathbf{x}_j) + \varphi_{BB}(\mathbf{x}_i - \mathbf{x}_j) + \varphi_{AB}(\mathbf{x}_i - \mathbf{x}_j)]}{2d \sum_j [\varphi_{AA}(\mathbf{x}_i - \mathbf{x}_j) + \varphi_{BB}(\mathbf{x}_i - \mathbf{x}_j) + \varphi_{AB}(\mathbf{x}_i - \mathbf{x}_j)]} \tag{5.18}$$

The term $(r^2/2)(\nabla c)^2$ in Eq.(5.17) accounts for the free energy cost of existing inhomogeneities (interfaces) in the system. As mentioned above, in reality f will depend on L and is hard to obtain explicitly in practice, therefore, f is frequently taken as the mean field free energy of a mixture

$$\frac{f(c)}{k_BT} = c\ln c + (1-c)\ln(1-c) + 2\frac{T_c^{MF}}{T}c(1-c) \tag{5.19}$$

$$\text{with} \quad 4k_BT_c^{MF} = \sum_j (\varphi_{AA} + \varphi_{BB} - 2\varphi_{AB})$$

or its Landau expansion (f_0, a, b being suitable coefficients) around critical concentration c_c,

$$f(c) = f_0 + a(c-c_c)^2 + b(c-c_c)^4 + \ldots \tag{5.20}$$

where $a \propto (T/T_c^{MF} - 1) < 0$, $b > 0$. From Eq.(5.17) $\mu(\mathbf{x})$ follows as a functional derivative $\mu(\mathbf{x}) \equiv \delta F/\delta c(\mathbf{x}) = (\partial f/\partial c)_T - r^2 k_BT\nabla^2 c(\mathbf{x})$ which, inserted into the continuity equation, yields the Cahn–Hilliard equation[79]

$$\frac{\partial c(\mathbf{x},t)}{\partial t} = M(T)\nabla^2 \left[\left(\frac{\partial f}{\partial c}\right)_T - r^2 k_BT\nabla^2 c(\mathbf{x},t)\right]. \tag{5.21}$$

The central assumption of the Cahn–Hilliard theory is that in the initial stages of unmixing the fluctuation $\delta c(\mathbf{x},t) \equiv c(\mathbf{x},t) - \bar{c}$ is everywhere small in the system (however, since L cannot be made arbitrary large, this assumption is typically *not true*). Linearizing the expression for the derivative in Eq.(5.21), one obtains then

$$\frac{\partial \delta c(\mathbf{x},t)}{\partial t} = M(T)\nabla^2 \left[\left(\frac{\partial^2 f}{\partial c^2}\right)_{T,\bar{c}} - r^2 k_BT\nabla^2\right] \delta c(\mathbf{x},t) \tag{5.22}$$

which, by introducing the Fourier transform $\delta c_{\mathbf{k}}(t) \equiv \int d\mathbf{x} \exp(i\mathbf{kx})\delta c(\mathbf{x},t)$, is easily solved to

$$\delta c_{\mathbf{k}}(t) = \delta c_{\mathbf{k}}(0) \exp[R(\mathbf{k})t] \,, \qquad R(\mathbf{k}) \equiv -M(T)k^2[f''_{T,\bar{c}} + r^2 k_B T k^2] \tag{5.23}$$

Usually one studies the equal–time structure factor $S(\mathbf{k},t)$ at time t after the quench which, according to Eq.(5.23), shows exponential growth for $0 < k < k_c$, $k_c \equiv (2\pi)/\lambda_c = \left[-f''/(r^2 k_B T)\right]^{1/2}$, with a maximum at a fixed wave vector $k_{\max} = k_c/\sqrt{2}$,

$$S(\mathbf{k},t) \equiv \langle \delta c_{-\mathbf{k}}(t)\delta c_{\mathbf{k}}(t)\rangle_T = S_{T_0} \exp[2R(\mathbf{k})t] \tag{5.24}$$

with $S_{T_0} \equiv \langle \delta c_{-\mathbf{k}}(0)\delta c_{\mathbf{k}}(0)\rangle_{T_0}$ being the structure factor in equilibrium at the initial temperature T_0, (we consider infinitely rapid quench from T_0 to T at $t=0$).

It is clear from Eq.(5.23) that the amplification rate $R(\mathbf{k})$, plotted as $2R(\mathbf{k})k^{-2}$ vs k^2, should be linear, $2R(k)k^{-2} = -D_0\left[1 - (k/k_c)^2\right]$, with a *negative* diffusion constant $D_0 \equiv 2Mf''$ ("uphill diffusion"). In addition, at $k = k_c$ the structure factor should be time independent $S(\mathbf{k}_c,t) = S(\mathbf{k},0)$ so that curves $S(\mathbf{k}_c,t)$, recorded at different times t, should intersect in a common point.

In reality, however, one typically observes a convex curve instead of the linear $2R(\mathbf{k})k^{-2}$ vs k^2 relationship, and there is no common intersection point. Some of the reasons for the failure of the simple linearized Cahn theory of spinodal decomposition are:

- Random statistical fluctuations, existing in the final state, should not be completely disregarded, as this is the case in Eq.(5.21).
- The concentration field $c(\mathbf{x},t)$ may be coupled to another slowly relaxing variable [45].
- Appreciable relaxation of the structure factor occurs during the quench from T_0 to T if the cooling rate $-dT/dt$ is finite[213, 84].

• Nonlinear effects are important already in the early stages of the quench [79, 300, 369].

• Since thermal conductivity in the unmixing system is not infinitely large, during the quench the system behaves to some extent as adiabatically isolated until the temperature T is established everywhere [371].

The last point is covered in Chapter 17 so we shall consider the other points briefly in the following.

Fluctuations in the final state are usually represented by a random force $\eta(\mathbf{x}, t)$ which is assumed to be delta–correlated random noise and whose mean–square amplitude $\langle\eta^2\rangle_T$ is linked to the mobility $M(T)$ by a fluctuation – dissipation relation [99]

$$\langle\eta(\mathbf{x}, t)\eta(\mathbf{x}', t')\rangle_T = \langle\eta^2\rangle_T \nabla^2 \delta(\mathbf{x} - \mathbf{x}')\delta(t - t') \; ,$$
$$\langle\eta^2\rangle_T = 2k_B T M(t) \; . \tag{5.25}$$

In the framework of Eq.(5.22) one then obtains

$$\frac{dS(\mathbf{k}, t)}{dt} = -2M(T)k^2 \left[(f''_{T,\bar{c}} + r^2 k_B T k^2)S(\mathbf{k}, t) - k_B T\right] \; . \tag{5.26}$$

It is seen that the effective diffusion constant

$$D_{eff}(\mathbf{k}, t) \equiv \frac{1}{k^2}\frac{d \ln S(\mathbf{k}, t)}{dt} \tag{5.27}$$

now yields

$$D_{eff}(\mathbf{k}, t) = -2M(T) f''_{T,\bar{c}} \left(1 - \frac{k^2}{k_c^2}\right) + \frac{2M(T)k_B T}{S(\mathbf{k}, t)} \; . \tag{5.28}$$

The solution of Eq.(5.26) can be represented as

$$S(\mathbf{k}, t) = S_{T_0}(\mathbf{k}) \exp[2R(\mathbf{k})t] + S_T(\mathbf{k}) \left(1 - \exp[2R(\mathbf{k})t]\right) \tag{5.29}$$

where

$$S_T(\mathbf{k}) = \frac{k_B T}{f''_{T,\bar{c}} + r^2 k_B T k^2} \, . \tag{5.30}$$

For $t \to \infty$, Eq.(5.28) would reduce to the simple Cahn result quoted above since in the linear theory $S(\mathbf{k}, t \to \infty) \to \infty$, but in this limit the theory is never applicable due to nonlinear effects. On the other hand, at $t = 0$ Eq.(5.28) leads to linear relation between $D_{eff}(\mathbf{k}, 0)$ and k^2 again (see Eq.(5.30) for T_0), the point where $D_{eff}(\mathbf{k}, 0)$ changes sign is not k_c, however, but shifted to a larger value. However, even with thermal fluctuations included, the linearized theory still produces an exponential growth of the structure factor, and a common intersection point is still present.

Suppose now the concentration field couples to a *slowly relaxing variable* $A(t)$, whose fluctuations in the absence of any coupling would decay exponentially, $\propto \exp(-\gamma t)$. Such coupling may typically occur in glasses [100] or in fluid polymer mixtures near their glass transition [363]. The decay of concentration fluctuations will be affected if (i) their maximum amplification rate $R_{\mathrm{max}} = R(k_{\mathrm{max}})$, is of the order of γ, and (ii) if the coupling between the variables $A(t)$ and $c(\mathbf{x}, t)$ is sufficiently strong. For a strength of this coupling expressed in terms of a parameter $(1 - D_\infty/D_0)$ [45], Fig.5.8, one can see that the unmixing mode $R(k) = \Gamma^-(k)$ leads to a pronounced convex curvature. In this case the slow variable will also relax with a broad spectrum of rates rather than with a single rate γ. In addition, fluctuations in the final state need also to be accounted for.

In the case of *finite quench rates*, when the mobility $M(T)$ follows Arrhenius law with sufficiently large activation energy, most of the relaxation may occur in the one phase region even before the spinodal is crossed. It can be demonstrated by stepwise cooling [84] that even in the Cahn–Hilliard approximation there is no longer a unique intersection point in the $S(\mathbf{k}, t)$ curves and the effective diffusion coefficient depends very weakly on time but is again rather curved.

Probably one of the most essential features of spinodal decomposition even immediately after the quench are, however, the *nonlinear effects*, as already the first computer simulations [354] in the nearest–neighbor Ising model have shown. However, one could demonstrate this directly on the model, based on the free energy functional, Eq.(5.17), used by Cahn and Hilliard as a starting point of their theory. The results of such direct simulation [369]

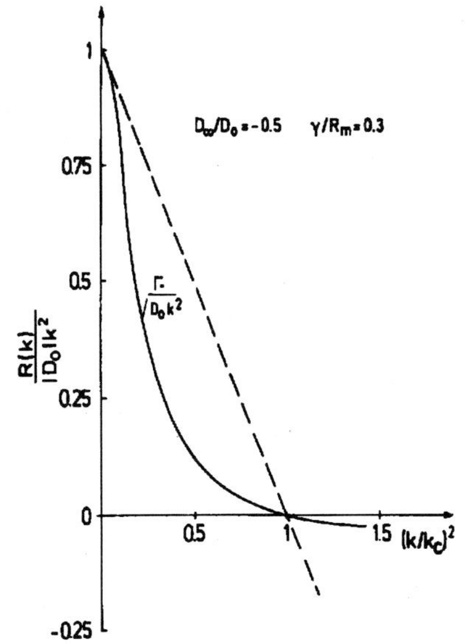

Fig. 5.8 Cahn plot, $R(k)/k^2$ plotted vs k^2/k_c^2: Broken straight line indicates absence of coupling to a slow variable, solid line represents a model with coupling. From Ref. [45].

of a lattice version of the Cahn – Hilliard model are shown in Fig.(5.9c). Evidently, the Hamiltonian of a *continuum* Ising model on a square lattice, where the coarse–grained concentration variable at each lattice site i is in the interval $-\infty < c_i < \infty$, can be transformed from

$$\frac{\mathcal{H}}{k_BT} = \sum_i \left[A(c_i - c_c)^2 + B(c_i - c_c)^4\right] + \sum_{i \neq j} \frac{C}{2}(c_i - c_c)^2 \qquad (5.31)$$

into Eq.(5.17), if one expands $c_j = c_i + (\mathbf{x}_j - \mathbf{x}_i) \cdot \nabla c_i$, and identifies $\frac{C}{2}\sum_{i,j}\left[(\mathbf{x}_j - \mathbf{x}_i) \cdot \nabla c_i\right]^2 \equiv \frac{r^2}{2}(\nabla c)^2$, replacing also $\sum_i$ by an integral $\int d\mathbf{x}$. Thus the lattice in Eq.(5.31) has nothing to do with the original lattice of the solid (if one considers binary alloy crystals) and is rather a lattice of coarse–grained cells with $L \sim r$, provided r is assumed to be of the order of interatomic spacing. The variables in this model may be rescaled [366] choosing

$$\widetilde{c}_i = \frac{c_i - c_c}{\sqrt{-\frac{A+dC}{2B}}}, \quad \widetilde{\alpha} = \frac{A+dC}{2B}, \quad \widetilde{\beta} = -\frac{C(A+2dC)}{B}. \qquad (5.32)$$

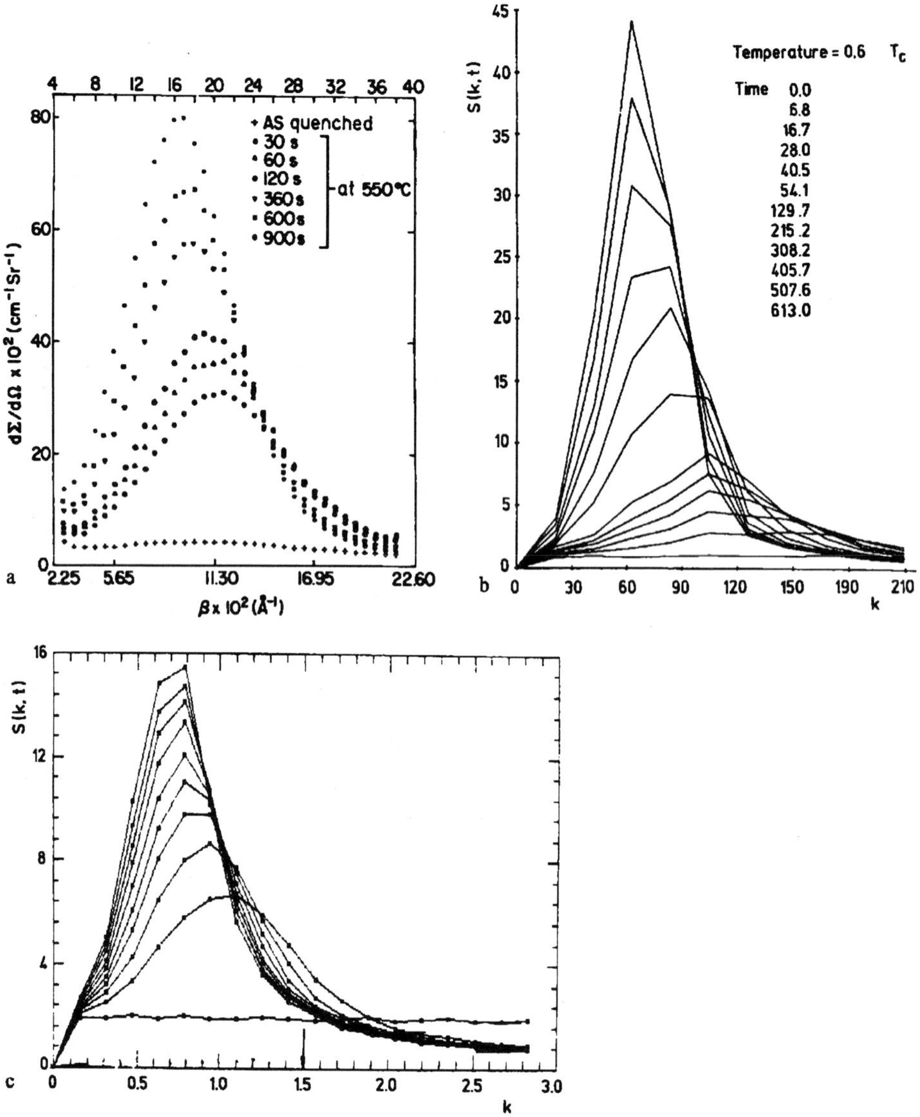

Fig. 5.9 (a) Neutron small scattering intensity [506] vs scattering vector **k** (β: scattering angle) for an Au–60at.%Pt alloy quenched to 550^{0}C. (b) Time evolution of the structure factor $S(k,t)$ according to a Monte Carlo simulation of a three-dimensional nearest neighbor Ising model at critical concentration and $T = T_c$. Due to the periodic boundary conditions for the $30 \times 30 \times 30$ lattice, k is only defined for discrete multiples of $2\pi/30$ which are connected by straigth lines on the plot [354]. (c)Time evolution of $S(k,t)$ for the Hamiltonian, Eq.(5.31), on a 40×40 lattice with $A/B = -2.292, \sqrt{B}/C = 0.972$ at times $t = 0, 10, 20, ..., 90$ Monte Carlo steps per site and spherical averaging for the same value of $\mid \mathbf{k} \mid$. The arrow indicates $k_{max} = k_c/\sqrt{2} = \sqrt{-A/C}$. From Ref. [369].

so that the scaled variables $\widetilde{\alpha}$ and $\widetilde{\beta}$ are now related to r and $(1 - \frac{T}{T_c})$. The parameters used in Fig.5.9 still correspond to fairly short range r so that nonlinear effects appear very important during the earliest stages of the quench already.

As a matter of fact, though, one should point out that the linearized Cahn – Hilliard – Cook theory still may become valid for systems in which interactions are sufficiently long–range – the mean field theory then becomes correct since statistical fluctuations are suppressed (at the same time nucleation in metastable states becomes impossible because *heterophase* fluctuations become also suppressed). This is seen by applying the Ginzburg criterion for the validity of mean field theory for critical phenomena to spinodal decomposition [50]: nonlinear terms in $\delta c(\mathbf{x},t)$ can be neglected if the mean square amplitude of the growing fluctuations is small in comparison to the concentration difference over which $f(c)$ in Eqs.(5.17,5.19) has appreciable nonlinearities:

$$\langle[\delta c(\mathbf{x},t)]^2\rangle_{T,L} \ll [\bar{c} - c_{sp}(T)]^2 \ . \tag{5.33}$$

One can estimate the quantity $\langle[\delta c(\mathbf{x},t)]^2\rangle_{T,L}$ as being equal to the expression $\langle[\delta c(\mathbf{x},t)]^2\rangle_{T,L} \approx \langle[\delta c(\mathbf{x},0)]^2\rangle_{T,L} \exp[2R_{\max}t]$ where $R_{\max}$ is the maximal growth rate, defined above, and the initial mean–square amplitude is related to the correlation function of the concentration fluctuations in the initial state at T_0. Using again the coarse grained variable $c(\mathbf{x}) = L^{-d}\sum_{i\in cell} c_i$

$$\begin{aligned}\langle[\delta c(\mathbf{x},0)]^2\rangle_{T,L} = \langle[\delta c(\mathbf{x})]^2\rangle_{T_0,L} &= L^{-2d}\sum_{i,j}\left[\langle c_i c_j\rangle - \bar{c}^2\right] \\ &= L^{-d}\int d\mathbf{x}\left[\langle c(0)c(\mathbf{x})\rangle_{T_0} - \bar{c}^2\right]\end{aligned} \tag{5.34}$$

where one of the sums is cancelled making use of the translational invariance of the correlation function. The mean field result near criticality for the latter gives at distances $|\,\mathbf{x}\,| \ll \xi$ a power law decay, $\langle c(0)c(\mathbf{x})\rangle_{T_0} \approx r^{-2}x^{2-d}$ so that one gets $\langle[\delta c(\mathbf{x},0)]^2\rangle_{T_0,L} \approx r^{-2}L^{2-d}$, that is, the mean square concentration fluctuation is the smaller the larger the range of the interactions and the coarse–graining length. Since the largest permissible choice of L is $L \approx \lambda_c$, which from Eq.(5.20) near the spinodal yields

$$\lambda_c \propto \frac{r}{\sqrt{1 - \frac{T}{T_c^{MF}}}} \sqrt{\frac{c_c - c_{sp}(T)}{\overline{c} - c_{sp}(T)}} \,, \tag{5.35}$$

one derives the condition for the validity of Cahn's theory as[50]

$$\exp[2R_{\max}t] \ll r^d \left(1 - \frac{T}{T_c}\right)^{\frac{4-d}{2}} \left[\frac{\overline{c}}{c_{sp}(T)} - 1\right]^{\frac{6-d}{2}} \,. \tag{5.36}$$

Evidently, far off the spinodal and sufficiently below T_c for large enough r the inequality (5.36) is fulfilled and the linear theory holds initially, as has been nicely verified by Monte Carlo simulations[199]. Since the long wave length problems of polymer mixtures[44] can be mapped onto long range Ising models [50] if one replaces r^d by $l^d N^{d/2-1}$, where l is the size of a polymer segment and N – the number of segments in the chain, one can thus understand why the validity of the linearized theory has been established mainly in polymer mixtures [44].

5.3.2 Nucleation Theory

Nucleation, that is, the decay of metastable states by forming and growth of droplets, means that the systems moves from a local minimum in phase space (the metastable state without a droplet) toward another stable minimum over a saddle point (the metastable state with one critical droplet). As in the calculation of the free energy from Eq.(5.17)

$$F = -k_B T \ln \int d\{c(\mathbf{x})\} \exp\left[-\frac{\mathcal{F}\{c(\mathbf{x})\}}{k_B T}\right] \tag{5.37}$$

in the functional $\mathcal{F}$, one restricts the phase space to states near the metastable minimum of the coarse–grained free energy density $f(c)$ and to nonuniform solutions $c(\mathbf{x})$ tending to $\overline{c}$ at large distances from the origin. As in the mean field theory of stable or metastable states, where one replaces the actual distribution $\exp\left[-\frac{\mathcal{F}\{c(\mathbf{x})\}}{k_B T}\right]$ by a delta–function $\delta(c - \overline{c})$ with $\overline{c}$ minimizing $\mathcal{F}$, the mean field theory of nucleation [79] is derived by replacing the functional integral by a saddle point solution of $\delta\mathcal{F}\{c(l)\}/\delta c(l) = 0$ where $c(l)$ is a spherically symmetric concentration profile of the critical droplet. In the mean field critical region, i.e., for $r^d(1 - T/T_c)^{(4-d)/2} \gg 1$ this yields a free energy barrier $\mathcal{F}^*$ to form one critical droplet [50, 546]

$$\frac{\mathcal{F}^*}{k_B T_c} \propto r^d \left(1 - \frac{T}{T_c}\right)^{\frac{4-d}{2}} \left[\frac{\bar{c} - c^{(1)}}{c^{(2)} - c^{(1)}}\right]^{-(d-1)}, \quad \text{for } \bar{c} \to c^{(1)} \tag{5.38}$$

and,

$$\frac{\mathcal{F}^*}{k_B T_c} \propto r^d \left(1 - \frac{T}{T_c}\right)^{\frac{4-d}{2}} \left[\frac{c_{sp}(T) - \bar{c}}{c^{(2)} - c^{(1)}}\right]^{\frac{6-d}{2}}, \quad \text{for } \bar{c} \to c_{sp}(T)\,. \tag{5.39}$$

Thus for $r^d(1-T/T_c)^{(4-d)/2} \gg 1$ the barrier $\mathcal{F}^*/k_b T_c \gg 1$ even if one comes close to the spinodal. The ultimate limit of metastability is reached when $\mathcal{F}^*/k_B T_c \approx 1$, i.e., the width of the region where the spinodal is *smeared out* is determined by

$$\frac{\delta c_{sp}}{c^{(2)} - c^{(1)}} \propto \left[r^d \left(1 - \frac{T}{T_c}\right)^{\frac{4-d}{2}}\right]^{\frac{-2}{6-d}} \tag{5.40}$$

$$\propto (\text{for } d = 3\ )\ r^{-2} \left(1 - \frac{T}{T_c}\right)^{-\frac{1}{3}}\,.$$

Eq.(5.40) is also obtained from the inequality (5.36) on the *unstable* side if one inserts $\bar{c} - c_{sp}(T) = \delta c_{sp}$ and $t = 0$. Also for nucleation the Ginzburg criterion tells us under which conditions the mean field treatment above is valid (by using the saddle–point approximation one neglects fluctuations): the fluctuations along the radial droplet concentration profile must be smaller than the order parameter difference between $l = 0$ and $l = \infty$ as described by the profile itself,

$$\langle[\delta c(\mathbf{x})]^2\rangle_{T,L} \ll [c(l \to \infty) - c(l = 0)]^2\,. \tag{5.41}$$

For $\bar{c} \to c^{(1)}$ one may put $L \approx \xi$, $c(l = 0) \approx c^{(2)}$, $c(l \to \infty) = \bar{c} \approx c^{(1)}$ and then (5.41) yields the standard Ginzburg criterion. On the other hand, from

Eq.(5.35) and $L \approx \xi$ taking $c(l=0) - c(l \to \infty) \propto c_{sp}(T) - \bar{c}$ one gets from (5.40)

$$1 \ll r^d \left(1 - \frac{T}{T_c}\right)^{\frac{4-d}{2}} \left[\frac{c_{sp}(T) - \bar{c}}{c^{(2)} - c^{(1)}}\right]^{\frac{6-d}{2}} . \tag{5.42}$$

Comparison of Eq.(5.39) and Eq.(5.42) shows that as long as $\mathcal{F}^*/k_BT \gg 1$ the mean field treatment of nucleation is in fact self–consistent. Comparison of Eq.(5.42) and Eq.(5.36) shows also that the validity criteria are essentially the same on both sides of the spinodal curve. Monte Carlo simulations of metastable states in Ising models with long–range interactions [197] clearly confirm this.

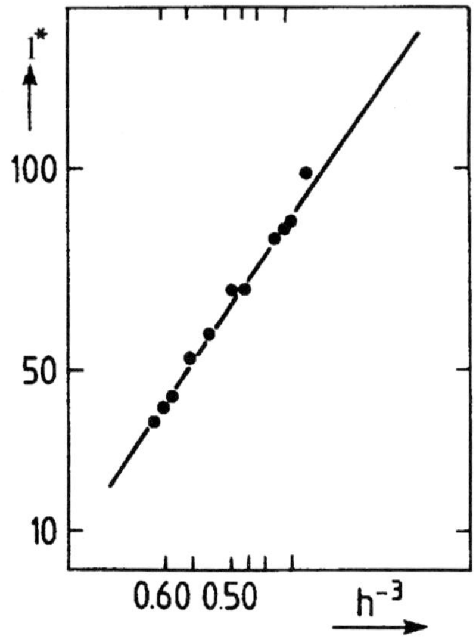

Fig. 5.10 Critical droplet size l^* obtained from the simulation of a three dimensional Ising model versus $1/h^3$ for the temperature $T/T_c = 0.59$. From Ref. [198].

Thus one could argue that states are metastable if $\mathcal{F}^*/k_BT \gg 1$ while they are unstable if $\mathcal{F}^*/k_BT \leq 1$. In metastable states the formation of a cluster of size l of the new phase is basically determined by two factors (see Chapter 4): the growth rate, $R(l)$, which gives the rate of change (by growing or shrinking) of l–clusters, and the free energy of formation, $\mathcal{F}_l$, (the excess free energy in comparison to the ambient phase). At the critical cluster size, l^*, the competition between the negative bulk and the positive surface energy term, and, therefore, $\mathcal{F}_{l^*}$, itself, reach a maximum. If nonlinear terms, allowing for the coalescence of clusters (coagulation), are neglected, one can describe nucleation by means of the continuity equation for the cluster concentrations $\overline{n}_l$ [39]

$$\frac{\partial \overline{n}_l(t)}{\partial t} + \frac{\partial J_l}{\partial l} = 0; \quad J_l = -R_l \frac{\partial \overline{n}_l(t)}{\partial l} - R_l \overline{n}_l(t) \frac{\partial}{\partial l}\left(\frac{\mathcal{F}_l}{k_B T}\right) . \quad (5.43)$$

The cluster current J_l consists of two terms: the first is diffusive (in the space of cluster sizes l), and the second ($\propto \partial \mathcal{F}_l / \partial l$) is a drift term which changes sign for $l = l^*$. The drift, therefore, acts against diffusion for subcritical clusters ($l < l^*$) whereas it "assists" size–diffusion for supercritical clusters ($l > l^*$). If $\mathcal{F}_l$ in Eq.(5.43) is expressed in terms of the equilibrium cluster size distribution n_l by $n_l = \exp(-\mathcal{F}_l / k_B T)$ (which is actually a fictitious quantity for a metastable state) it follows from (5.43) $J_l = -R_l n_l \frac{\partial}{\partial l} \frac{\overline{n}_l(t)}{n_l}$ and hence a steady state solution $\partial \overline{n}_l(t)/\partial t = 0$ is obtained for $J_l = J$ independent of l. From $-(\partial/\partial l)\{\overline{n}_l(t)/n_l\} = J(R_l n_l)^{-1}$ one then obtains, using the boundary conditions $\lim_{l \to 0}\{\overline{n}_l(t)/n_l\} = 1, \quad \lim_{l \to \infty}\{\overline{n}_l(t)/n_l\} = 0$,

$$1 = J \int_0^\infty \frac{dl}{R_l n_l}, \quad \frac{\overline{n}_l(t)}{n_l} = J \int_l^\infty \frac{dl'}{R_{l'} n_{l'}}. \quad (5.44)$$

Thus the knowledge of R_l and n_l appears sufficient to calculate both the so called nucleation rate J and the steady state cluster size distribution $\overline{n}_l(t)/n_l$. Since this Becker–Döring theory is based on many simplifying assumptions (only one "coordinate" l is used to describe a cluster [39], coagulation effects are neglected, etc.), a test of the theory would be important in view of its application to various branches of physics and chemistry.

Since it is difficult to derive R_l and n_l from experiments, however, which hampers experimental tests, computer simulations are very helpful. Thus it turns out, that far from the critical temperature, T_c, and close to the coexistence line, the classical nucleation theory is clearly confirmed [198]. From the classical expression for the cluster formation free energy $F_l = (36\pi)^{1/3}\sigma l^{2/3} - hl$ where σ is the surface tension (of a flat surface) and h is the applied magnetic field (the "supersaturation") in the three dimensional Ising model, one has for the critical droplet size $l^* = [2(36\pi)^{1/3}\sigma/(3h)]^3$, for example, which is shown in Fig. 5.10 to be nicely confirmed. For details, pertaining to the nontrivial definition of a "droplet" in such simulations, the interested reader should refer to the corresponding literature [198, 98].

Also at criticality, which is an especially interesting and hardly accessible region, both R_l and n_l have been determined [39] for the Ising model in two

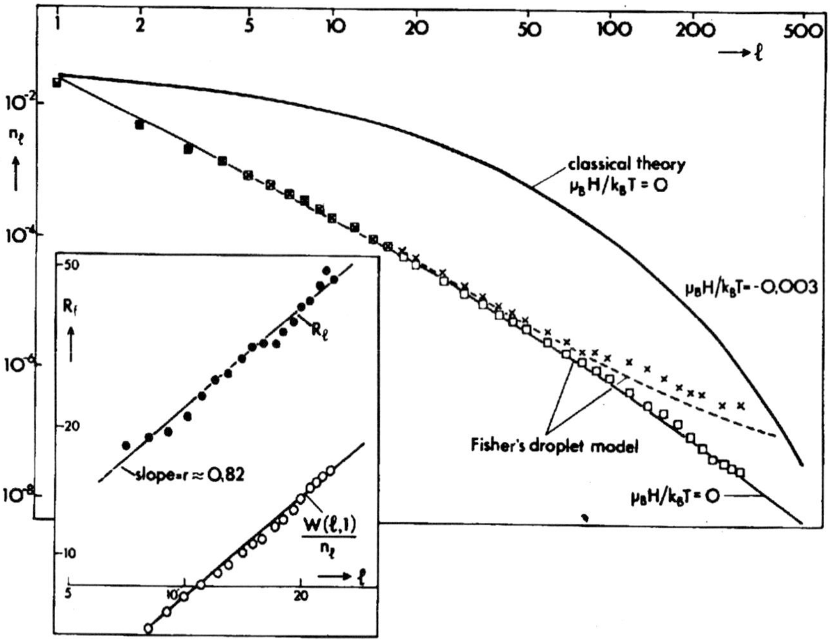

Fig. 5.11 Concentration n_l of clusters containing l reversed spins plotted vs l for the square Ising lattice at $T/T_c \approx 0.96$ and two values of the field. Predictions of both the classical nucleation theory and Fisher's droplet model (neither of which contain any adjustable parameter) are included. The inset shows the reaction rate (for zero magnetic field) and includes also the contribution $W(l,1)/n_l$ of single flip–processes. From Ref. [39].

dimensions – Fig.5.11. There it appears that the observed growth rate R_l for $l \to \infty$ is quite well approximated by a power law, $R(l) = \overline{R} l^{2-\nu z/\beta\delta}$, where $z \approx 2$ is the dynamic exponent, describing critical slowing down, and ν, β, δ – the exponents of the Ising universality class, describing the singular behavior at criticality of the correlation length, $\xi \propto \mid 1 - T/T_c \mid^{-\nu}$, of the order parameter, $(c - c_c)_{\Delta\mu=0} \propto \mid 1 - T/T_c \mid^{\beta}$ and $(c - c_c)_{T=T_c} \propto (\Delta\mu)^{1/\delta}$. The supersaturation $\Delta\mu$ is the reduced chemical potential difference $(\mu - \mu_c)/k_B T$, μ_c being the chemical potential at the coexistence curve. Similarly, the cluster distribution is well fitted by an expression, dominated by a power law, namely the Fisher droplet model [150]: $n(l) \propto l^{-2-1/\delta} \exp[-b(1-T/T_c)l^{1/\beta\delta} - \Delta\mu l]$. From Fig.5.11 is also evident that the "classical" Becker–Döring "capillary" approximation $n_l = n_0 \exp[-2l(h/k_BT) - 2(\sigma/k_BT)\sqrt{\pi}l^{1/2}]$ is a bad approximation, at least for l not too large. Note that in this case this classical formula contains no adjustable parameters – the surface tension σ is known exactly[399] and the coexistence between spin–up and spin–down phase occurs precisely at magnet field $h = \Delta\mu = 0$.

6 Pattern Formation in Cellular Automata Models

R. Mahnke

6.1 Introduction

The first task that faces the scientist who wants to interpret the time evolution of a complex system is the construction of a model. In equilibrium and more ever in non–equilibrium systems many features are likely to be important. Not all of them, however, should be included in the model. Only the few relevant features which are thought to play an essential role in the interpretation of the observed phenomena should be retained. Such a simplified description could be criticized, but it is often very helpful in developing the intuition which is neccessary for the understanding of the behaviour of complex real systems. In many body physics, as pointed out in [172], models such as the van der Waals model of a fluid, the Heisenberg model of ferromagnetism, the spring model of lattice vibrations, the Landau model of phase transitions, the Ising model of spin phenomena and others have played a major role. A simple model, if it captures the key elements of a complex system, may generate highly relevant questions.

Most models in aggregation kinetics are formulated in terms of differential equations. One set of well known *deterministic nonlinear dynamical equations* are the Becker–Döring equations which can be used to model a variety of phenomena in the kinetics of phase transitions including metastability, nucleation and coarsening. The system of equations, first introduced by Becker and Döring in 1935 [26], is connected with the evolution equations by Lifshitz–Slyozov–Wagner (LSW) in the theory of coarsening (Ostwald ripening) [308]. Penrose and others [418, 419] study the behaviour of solutions of the Becker–Döring equations at very late times, long after any possible metastable state has broken down, and show by means of a systematic approximation procedure that in lowest approximation the distribution of the large clusters at these late times obeys the equations of the LSW theory.

The modelling of supersaturated (metastable or unstable) systems can be done by different means. One of the fundamental concepts is based on the classical mechanics of nonequilibrium processes [452, 592]. Nonlinear relaxation processes can by described in terms of the balance equation for the probability distribution of stochastic variables known as Master equation. The *Master equation approach* of nucleation, growth and condensation has been worked out in Chapter 4. The Master equation for the chemical reactions of the attachment and the detachment of elementary particles (monomers) on or out of clusters has been solved with phenomenologically motivated transition probabilities [72, 330, 289].

In the following we are engaged in more computational directed methods. We divide between

- *Cellular automata models*
Chemically reacting systems with phase separations like the well–known Schlögl model can be simulated by probabililtic lattice–gas automaton models where space, time and particle velocities are discrete. Each molecule (occupied site) can undergo elastic and reactive collisions and depending on the updating rule interesting spatio–temporal configurations of the system emerge.

- *Monte Carlo simulation technique*
The formation process of clusters on a two–dimensional lattice can be modelled by a so–called nucleation automaton with a well–explained energy function. The relaxation to a one–big–cluster–state as the equilibrium situation is a searching process on the energy landscape to find a minimum in a very high dimensional situation.

- *Molecular dynamics*
Based on the classical many–particle Hamiltonian with the well–known Lennard–Jones interaction the integration of the equations of motion shows the clustering process quite straightforwardly.

6.2 Description of Cellular Automata Models

Cellular automata are mathematical models for complex systems in which many simple components act together to produce complicated patterns of behaviour. These automata may be considered as discrete dynamical systems and serve as key models to study the origins of chaos (randomness) in physical systems. There are examples known in which a cellular automaton evolving from a single initial state produces a structure so complicated that some features of it seem random. Farmer, Toffoli, Wolfram, Packard and co–workers have been investigated one–dimensional and two–dimensional automata extensively in several ways, see e. g. [140, 181, 571].

Cellular automata have been a wide variety of applications since their invention in the late 1940's. The inventor of cellular automata, John von Neumann, observed that although physics has had amazing success in recent years providing fundamental explanations for many phenomena, there remains a wide class of phenomena that have been unapproachable with the well–established conventional tools. The most intriguing examples are found in biology, where there seems to be a sense in which nature makes something from nothing under the right conditions, as in the genesis of life, biological evolution or learning process in organism. Many people have suggested that there should be physical laws governing such phenomena. J. von Neumann emphasized that the real biological world is too complicated as to be modelled directly, he developed cellular automata to serve as simpler systems that display some of the essential dynamical features that might form the basis for such laws [391].

The approach of J. von Neumann, worked out together with S. Ulam, is based on a universal Turing maschine with the property of self–reproduction. The rules governing the evolution of states are often motivated by biological mechanisms. The abstract idea of cellular automata can be applied in games using two–dimensional lattice representation. In the 60's the most popular game which shows the growth, the decay and the evolution of populations was that of J. Conway named *Life* (Conway's Game of Life) [131]. Cellular automata help to understand the laws that govern complex phenomena by studying the temporal evolution of typical initial conditions under the action of relatively simple local rules. This task is related to synergetics where the cooperation of microscopic simple components produces macroscopic spatial, temporal or functional structures [191].

In natural sciences, nonlinear phenomena are often described by a system of reaction–diffusion equations. It is well known that nonlinear spatio–temporal dynamics of interacting particles, chemical or biological species can generate numerous local and spatially distributed effects far from equilibrium such as steady–state multiplicity, oscillations like limit cycles, propagating fronts, target patterns, spiral waves, pulses as well as stationary spatial pattern like a dense spot surrounded by a dilute phase [125, 129, 342]. In one–dimensional space (coordinate r) the processes can be described generally by

$$\frac{\partial}{\partial t} x(r,t) = f(x(r,t)) + \frac{\partial^2 x(r,t)}{\partial r^2} \tag{6.1}$$

to get the evolution of the quantity $x(r,t)$ at time t and position r. Such reaction–diffusion equations are suitable for continuous systems with not too many degrees of freedom. Using a bistable reaction function f inhomogeneous structures with spherically symmetric regions (droplet–like configurations) can be studied [125, 129, 342].

In contradistinction to the differential equations mentioned above cellular automata may be considered as an alternative and complementary approach to nature based on discretization of space and time. In the simplest case, a cellular automaton consists of a line of sites (cells, compartements) with each site having a value 0 or 1. The sequence of site values is the configuration evolving in discrete time steps. At each time step, the value of each cell is updated according to a definite rule.

Wolfram has given the following five fundamental defining characteristics for cellular automata (CA) [140]:

1. CA consist of a discrete lattice of sites.
2. CA evolve in discrete time steps.
3. Each site takes on a finite set of possible values.
4. The value of each site evolves according to the same updating rules.
5. The rules for the evolution of a site depend on a local neighbourhood of sites around it.

Cellular automata as general discrete models with local interactions may be considered as idealizations of partial differential equations (6.1), in which time and space are assumed discrete, and dependent variables may take on a finite set of possible values only. Self–organization in cellular automata occurs by the preferential generation of special sets of states known as attractors for the evolution [181, 571]. Usually cellular automata are assumed to be entirely deterministic. Of course, one may introduce random noise directly into the cellular automaton rules in analogy to lattice spin systems at nonzero temperature. Such probabilistic cellular automata are found to exhibit phase transitions as a function of noise level. Another question is the influence of different types of updating in respect to the transient and the final configuration states [435]. In the parallel updating one applies the update rules simultaneously to all sites. This type of updating procedure produces the strongest correlations between the sites and is often used in the context of CA traffic flow models (see [385] and Sect. 6.6). The other way of updating, the sequential update, is much more common for practical purposes like computer simulations.

A very simple binary cellular automaton consists of sites on a one–dimensional line having only two different values. Taking $a_i(t)$ to be the value of a at site i and time step t:

$$a_i(t) = \begin{cases} 0 & \text{e. g. spin up} \\ 1 & \text{e. g. spin down} \end{cases} \qquad \text{Boolean variable} \tag{6.2}$$

The local updating rule may written as

$$a_i(t+1) = (a_{i-1}(t) + a_{i+1}(t)) \mod 2 \tag{6.3}$$

showing that the value of a particular site is given by the sum modulo two of the values of its two nearest neighbours on the previous time step. Starting from general initial configurations the evolution of the cellular automaton can be investigated.

Explicit cellular automaton models of natural systems have been constructed and analysed. Aggregation phenomena, such as snowflake growth, follow at the one hand simple local rules but yield on the other hand very complex patterns. Kohyama [261] proposed a class of simple two–dimensional cellular automata with particle conservation for easy simulations of interacting particle systems. The automata are defined by the exchange of states of neighbouring cells, depending on the configuration around the cells. Since cellular automata are discrete in time, space and state, it is difficult to construct models which correspond directly to real physical systems. But as the number of possible cellular automata is huge, one can expect to find rules with physically realistic behaviour as cluster growth by liquid droplet formation.

As already mentioned, Farmer, Wolfram, Toffoli, Packard and co–workers have been investigated one– and two–dimensional cellular automata with two different states (binary situation) very extensively [140, 410, 571]. Despite the simple short–range local rules the variety of different types of automata is enormous. One can speak about a CA zoology. For instance, two–dimensional binary cellular automata with nearest neighbour interaction on a quadratic lattice (neighbourhood of 9 cell are of interest) has about 10^{154} different evolution rules. But nobody with the fastest computer of the world can test all of them, although special purpose computers for CA make millions of updates per second.

Spatially extended binary cellular automata models are known to be a powerful tool to simulate a variety of self–organizing, learning and selection processes in different physical and biological systems. Exploring general features of their behaviour may therefore be helpful to understand complex phenomena found in nature like the emergence of spontaneous nucleation in undercooled systems.

6.3 Deterministic One–dimensional Binary Cellular Automata

In the following we investigate a complete set of patterns generated by deterministic one–dimensional cellular automata. In the notation of cellular automata (CA) the quantity $a_i(t)$ is taken to denote the value of site (cell) i at time t. The site values may evolve by iteration of the mapping (r = range of local generation rule)

$$a_i(t+1) = F(a_{i-r}(t), \dots, a_i(t), \dots, a_{i+r}(t)) \, . \tag{6.4}$$

Let us take into account nearest–neighbour interactions (range $r = 1$) only and that the sites a_i are specified by the values 0 (false, empty, white) und 1 (true, filled, black), then the iteration rule (6.4) reads for the elementary case

$$a_i(t+1) = F(a_{i-1}(t), a_i(t), a_{i+1}(t)) \quad \text{with} \quad a_i = \{0, 1\} \, . \tag{6.5}$$

Since we have $2^3 = 8$ possible values of the three variables a_{i-1}, a_i, a_{i+1}, we may construct in general $2^8 = 256$ distinct cellular automaton rules according to Eq. (6.5). But we restrict ourselves to rules for growing patterns, that means we fix

$$F(0,0,0) = 0\ ; \quad F(0,0,1) = 1\ ; \quad F(1,0,0) = 1\ . \tag{6.6}$$

These restrictions (6.6) (quiescence and boundary conditions for both sides) leave us $256/2^3 = 2^5 = 32$ possible local rules for the generation of growing triangle–like patterns [327]. Other forbiddances of rules were studied, in particular, the so–called reflection symmetry condition which guarantees isotropy in cellular automaton evolution.

In Figs. 6.1 – 6.3 the evolution of all 32 different elementary one–dimensional cellular automata is shown which we calculate by the rule (6.5) together with (6.6). Starting in the zeroth generation with one cell in state 1 (black dot in the uppermost line) we present in Figs. 6.1 – 6.3 the time evolution up to 100 generations, which we have iterated step by step deterministically. If the cell has value 1 we use a black dot (pixel) otherwise (value 0) the place is empty.

At a first glance the generated sequences based on the different local mappings (6.5) show many various structures in the presented evolution patterns, see Figs. 6.1 – 6.3. But a detailed analysis is necessary to answer the question which of the patterns is the most complex one. This can be done by complexity calculations [327, 333].

It is shown in Figs. 6.1 – 6.3 that different production rules (at least eight) can reproduce the same structure. There are mirror symmetrical patterns visible by comparison. Altogether we are able to generate 17 different evolution patterns, 4 of which are mirror–symmetric, from the given 32 iteration rules. To mark each configuration directly we use as identification number the eight digits of the binary representation of the evolution equation (6.5) and from that its decimal number (given in brackets under each triangle

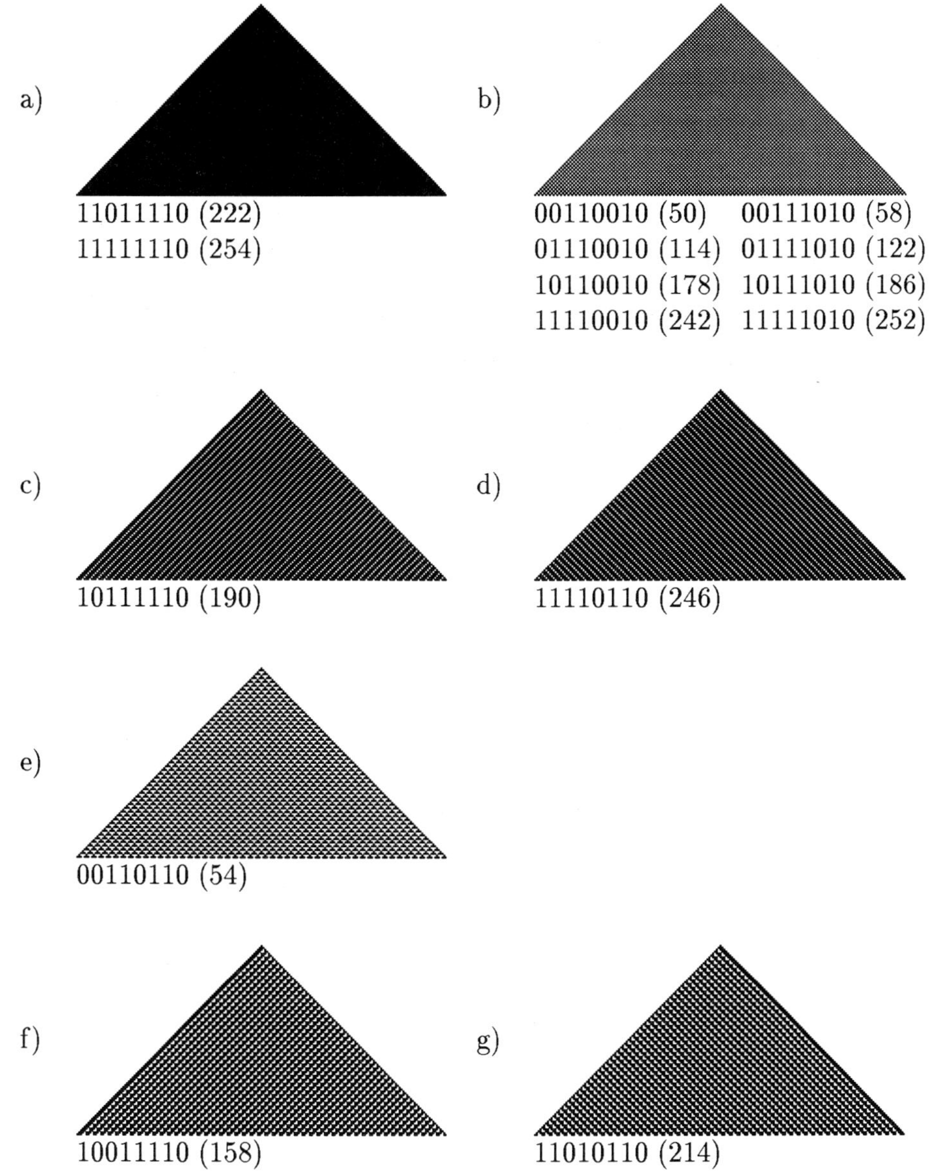
a)
11011110 (222)
11111110 (254)
b)
00110010 (50) 00111010 (58)
01110010 (114) 01111010 (122)
10110010 (178) 10111010 (186)
11110010 (242) 11111010 (252)
c)
10111110 (190)
d)
11110110 (246)
e)
00110110 (54)
f)
10011110 (158)
g)
11010110 (214)

Fig. 6.1 See next page.

h)

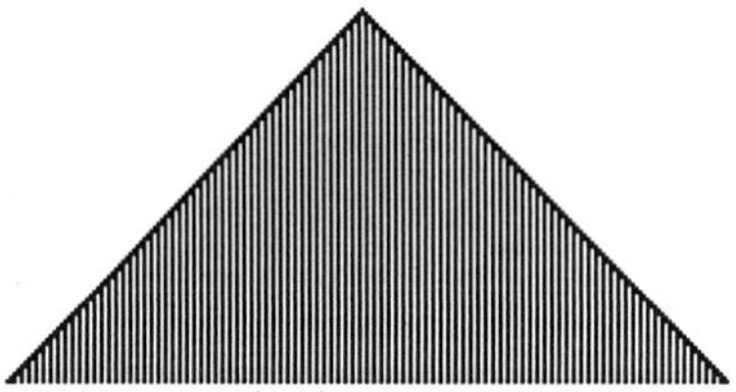

01011110 (94)

i)

10110110 (182)

j)

00010010 (18) 00011010 (26)
01010010 (82) 01011010 (90)
10010010 (146) 10011010 (154)
11010010 (210) 11011010 (218)

k)

00010110 (22)

l)

01111110 (126)

m)

10010110 (150)

Fig. 6.2 See next page.

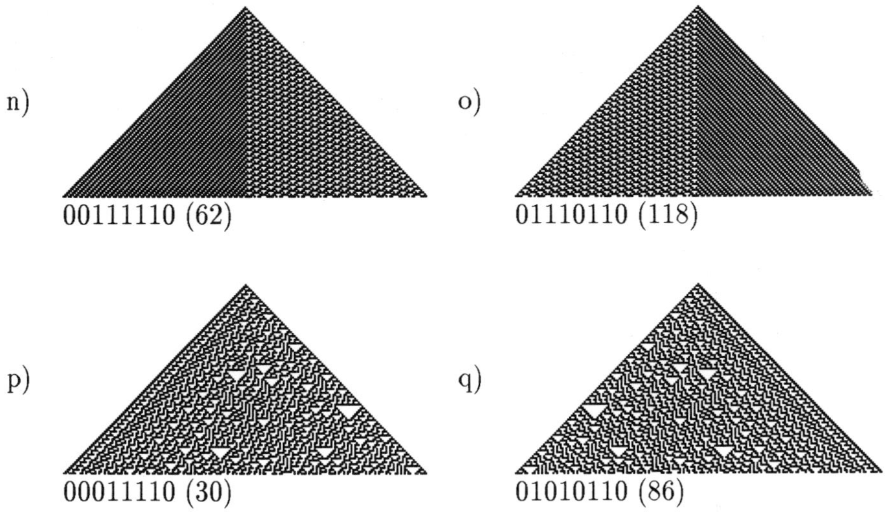

Fig. 6.3 Evolution patterns (a–q) generated by one–dimensional binary deterministic cellular automata. On the basis of all 32 given production rules (6.5, 6.6) the evolution of 0–1–sequences is shown for 100 generations. The local mapping has a binary identification number and its decimal value which is given under each pattern.

in Figs. 6.1 – 6.3). For instance the structure given in the top–left pattern (Fig. 6.1) originates from

$$\left.\begin{array}{l} F(0,0,0)=0 \\ F(0,0,1)=1 \\ F(0,1,0)=1 \\ F(0,1,1)=1 \\ F(1,0,0)=1 \\ F(1,0,1)=0 \\ F(1,1,0)=1 \\ F(1,1,1)=1 \end{array}\right\} \begin{array}{l}\text{Identification code of cellular automaton} \\ \text{(binary number)}\end{array} \tag{6.7}$$

The right hand side of Eq. (6.7) gives the binary identification code. Using the usual transformation formula to the decimal code

$$0\cdot 2^0+1\cdot 2^1+1\cdot 2^2+1\cdot 2^3+1\cdot 2^4+0\cdot 2^5+1\cdot 2^6+1\cdot 2^7=222 \tag{6.8}$$

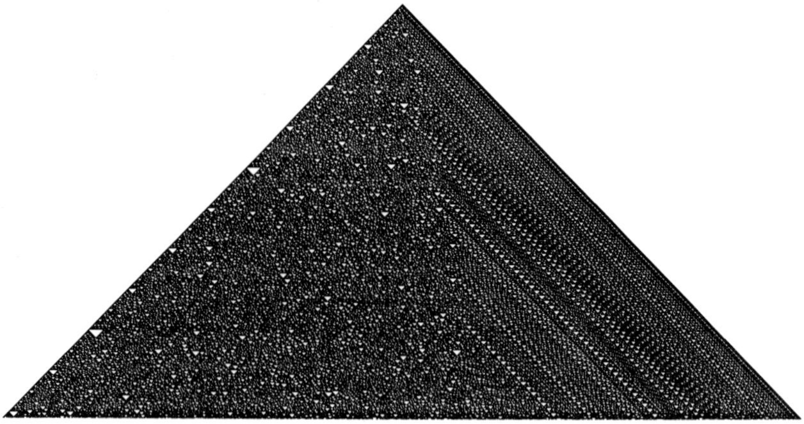

Fig. 6.4 The most complex pattern, cellular automaton 01010110 (86), now generated for 650 time steps, shows in a large scale regular as well as chaotic structures (compare left and right part).

we get the value 222, which was already written under the top–left very simple homogeneous pattern (Fig. 6.1), see also Eq. (6.9).

$$\begin{array}{lc} 0. & 1 \\ 1. & 111 \\ 2. & 11111 \\ \vdots & \vdots \\ n. & 11\cdots\cdots 11 \end{array} \qquad \text{simple regular pattern} \tag{6.9}$$

By human feeling the picture with code 86 (Fig. 6.1) or the mirror symmetric one (code 30) seems to be the most complex pattern. As already pointed out by several authors this pattern, given in Fig. 6.4 up to 650 generations, is a standard example of a chaotic structure created by a one–dimensional deterministic cellular automaton. The analysis of the generated objects concerning their structure is called complexity problem. The approach to complexity connected with the names of Kolmogorov, Chaitin and others is that the complexity of an object must measure the expense or length of all instructions, which are necessary to generate, construct or describe the object. Fig. 6.4 shows such interesting object with high complexity.

6.4 Two–dimensional Cellular Automata

The already discussed one–dimensional cellular automata, see Eq. (6.4), can be extended to two–dimensional CA with square lattices and well–defined neighbourhood structures. In Fig. 6.5 a five–neighbour square is shown. Other examples are nine–neighbour squares on quadratic lattices and triangular or hexagonal lattices as well. The evolution rule in analogy to Eq. (6.4) reads in the case of Fig. 6.5

$$a_{i,j}(t+1) = F\left(a_{i,j-1}(t), a_{i,j}(t), a_{i,j+1}, a_{i-1,j}(t), a_{i+1,j}(t)\right) \ . \qquad (6.10)$$

Sometimes cellular automata are called totalistic if the updated value of the center site depends only on the sum of the values of the sites in the neighbourhood (see Eq. 6.12).

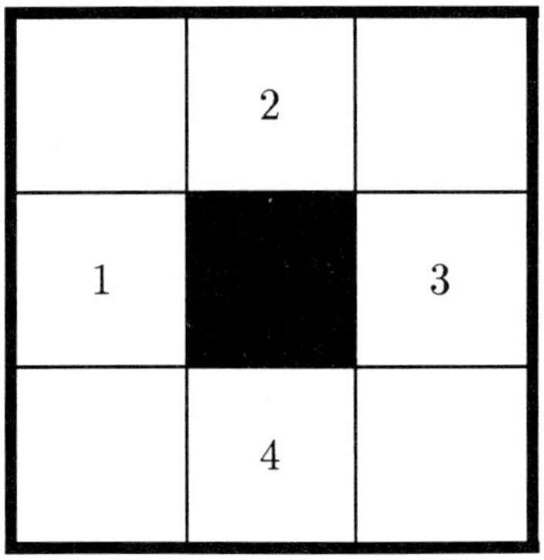

Fig. 6.5 Example of neighbourhood structure for two–dimensional automata known as five–neighbour square.

The number of possible cellular automaton rules increases rapidly with the lattice dimension and the number of states per site. Packard and Wolfram [410] stated that this number is too large to investigate each CA explicitly. For the larger part, one must resort to random sampling with the expectation that the selected rules are typical. Hopefully generic properties are significant independent of precise details of cellular automata construction.

Let us explain a simple two–dimensional flip dynamics with spins. The state of the spin (up or down) at time $t+1$ is determined by four neighboured spins at time t without any memory effect.

$$a_i = \begin{cases} \uparrow & \text{spin up} \\ \downarrow & \text{spin down} \end{cases} \tag{6.11}$$

Based on 16 possible configurations of the neighbourhood the value of the spin in the center may be up or down. Therefore we deal with $2^{16} = 65\,536$ different automata which has to be investigated.

Choosing typical initial states (random configuration with 50% spin up $\uparrow$ and 50% spin down $\downarrow$), an empirical analysis after 30 000 iterations on a quadratic lattice of size 64×64 gives the following result which is in agreement with the universality classes by Wolfram [140] (cf. Sect. 6.5):

- About 2/3 of all automata evolve to homogeneous or periodic final states.
- About 1/3 cellular automata exhibit chaotic behaviour.

In a deterministic cellular automaton, cell by cell has to be updated. In the contrary to the concept of updating of each cell certainly the stochastic (probabilistic) cellular automaton deals with randomly chosen cells and their updating according to the given production rule. The time step is called Monte–Carlo step (MCS) and depends on the size of the lattice. If each cell is chosen once in the average then the system makes one Monte–Carlo step.

Here we consider at first a totalistic two–dimensional automaton which models the linear cluster decay. The totalistic automaton has the simple property that in the transformation rule the new result of a cell depends only on the sum of the states of neighbouring sites. Our decay rule of a site with coordinates i and j are

$$a_{i,j}(t+1) = \begin{cases} 1, \text{ if } & a_{i,j}(t) = 1 \text{ and } \sum a_{n,m} \geq Code \\ 0, \text{ if } & a_{i,j}(t) = 0 \quad \text{or } \sum a_{n,m} < Code \end{cases} \tag{6.12}$$

The sum goes over all of the eight neighbours in a quadratic lattice.

Fig. 6.6 shows a two–dimensional stochastic binary cellular automaton after four Monte–Carlo steps. Starting from a circle filled with atoms ($a_{i,j} = 1$) of radius $r = 80$ cells (overall pixel number = 20 081, total boundary length = 644) the updating according to the rule (6.12) with the decay parameter

Fig. 6.6 Simulation of linear cluster decay by a stochastic totalistic cellular automaton. Starting with a circle of filled sites (a cluster of atoms) the situation of a evaporating droplet with fractal boundary is shown after 4 Monte–Carlo steps.

Code = 7 after a short time the presented situation with values of pixel number = 17 530 and boundary length = 1 432 is found. The fractal dimension (see definition below) decreases from 1.94976 ≈ 2 (Euclidian dimension for an area) to 1.66170 .

The series of pictures in Fig. 6.7 together with the data in the Table shows the influence of the decay parameter *Code*. According to rule (6.12) an atom at position i, j (filled state $a_{ij} = 1$) can only survive if it is surrounded by quite enough other atoms. The number how much are necessary is the parameter *Code*. The decay takes place at the surface (boundary) of the cluster. At that position an atom must evaporate (going over into a state $a_{ij} = 0$) because it has not so many neighbouring molecules. The cluster becomes smaller and smaller with a very rugged boundary. The fractal dimension is a measure of the roughness calculated from the filled area P (total pixel number) and the overall boundary length R by

$$\text{frac. dim.} = \frac{\ln P}{\ln R/4} \equiv D . \qquad (6.13)$$

The area P equals the number of occupied sites (filled cells by pixels). The boundary length R has to be scaled to $R/4$ because each (quadratic) cell has four boundaries. With this scaling the fractal dimension D is in agreement with the power law $P \sim R^D$ which is known as number–size relation.

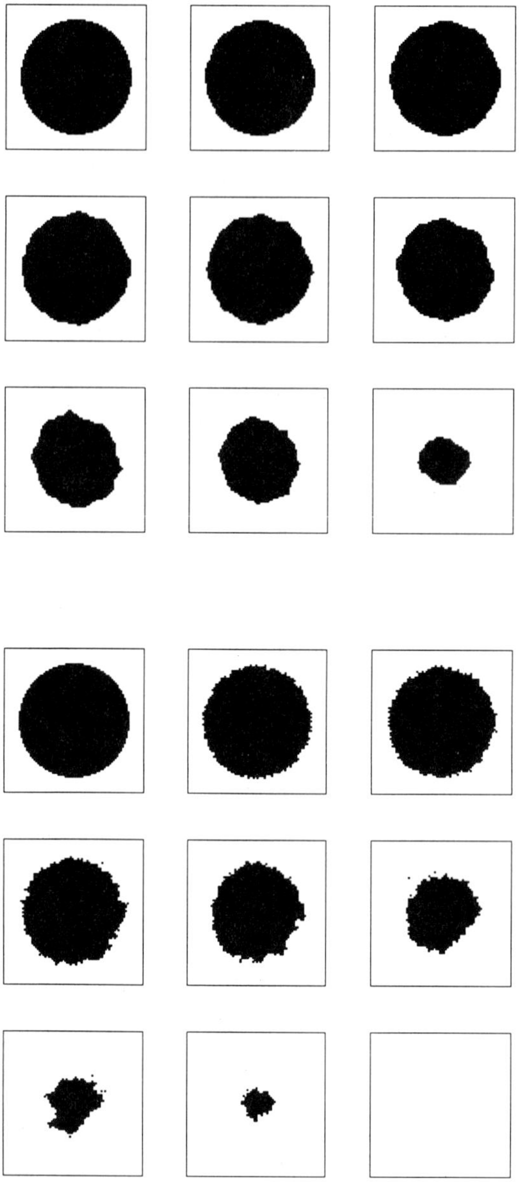

Fig. 6.7 Evolution of a circular cluster by the decay rule (6.12) in dependence on the parameter *Code* for a stochastic two–dimensional cellular automaton. The decay process goes faster for greater values of the parameter (*Code* = 5 (top); *Code* = 6 (bottom)). The data of this CA–simulation are collected in the Table.

Collection of data (area P and boundary length R in pixel, fractal dimension D) for the decay of a spherical cluster as function of time. The configurations generated with the parameter $Code = 5$ and $Code = 6$ (Fig. 6.7) were analysed after 2^n ($n \leq 6$) Monte–Carlo steps including $t = 0$ (initial situation) and $t = 24$ MCS.

MC steps	Code 5			Code 6		
	P	R	D	P	R	D
0 (Start)	2828	240	1.94105	2828	240	1.94105
1	2789	240	1.93766	2706	298	1.83336
2	2735	240	1.93288	2583	298	1.82256
4	2646	238	1.92874	2349	304	1.79225
8	2471	234	1.91996	1906	268	1.79627
16	2100	214	1.92219	1145	210	1.77821
24	1773	202	1.90731	581	170	1.69749
32	1459	182	1.90834	186	90	1.67841
64	506	108	1.88921	0	0	–

6.5 Some Remarks on Universality Classes

Based on the notation (see Eq. (6.4)), the mapping

$$a_i(t+1) = F[a_{i-r}(t), a_{i-r+1}(t), \cdots, a_i(t), \cdots, a_{i+r}(t)] \tag{6.14}$$

with the parameter r (range) and values $a_i(t) \in [0, 1, \ldots, k-1]$ of size k may be written in an alternative form

$$a_i(t+1) = f\left[\sum_{j=-r}^{+r} \alpha_j a_{i+j}(t)\right], \tag{6.15}$$

where the α_j are integer constants, and the function f takes a single integer argument.

In general the value of a given site a_i depends on the last values of a neighbourhood of at most $(2r+1)$ cells. Therefore the region affected by a given site grows by at most r sites in each direction at every time step. After time t, a region of at most $(1+2rt)$ sites may be affected by a given initial site value [140].

In the formalism of cellular automata, we distinguish between additive CA (f is a linear function) and totalistic CA (all $\alpha_j = 1$ in Eq. (6.15)). The state $a_i = 0 \ \forall \ i$ may be considered as zero–configuration (ground state). The statement that this configuration remains invariant under time evolution implies

$$F[0, 0, \cdots, 0] = 0 \, . \tag{6.16}$$

Also it is convenient to consider symmetric rules for which

$$F[a_{i-r}, \cdots, a_{i+r}] = F[a_{i+r}, \cdots, a_{i-r}] \tag{6.17}$$

holds and the automaton evolves to a symmetric state, in which $a_{n+i} = a_{n-i}$ for some n und all i.

The iteration of one–dimensional cellular automata with a binary alphabet $k = 2$ and interaction range $r = 2$, starting from a disordered state (random initial configuration), shows that each pattern which appear falls into one of the four following classes, denoted by Wolfram [140] the four universality classes:

1. Evolution leads to a homogeneous state (Code 60, Fig. 6.8 left top).
2. Evolution leads to a set of simple spatial separated or periodic structures (Code 56, Fig. 6.8 right top).
3. Evolution leads to chaotic patterns (Code 10, Fig. 6.8 left bottom).
4. Evolution leads to complex localized structures, sometimes long–lived (Code 20, Fig. 6.8 right bottom).

Universality implies that many details of the construction of the cellular automata are not relevant in determining its qualitative behaviour. In comparison to non–linear dynamical systems the four classes mentioned above characterize the attractors in cellular automaton evolution. Class 1 (unique homogeneous state) can be considered as a fixed point in phase space. Class 2 is related to limit cycles and class 3 to strange attractors with aperiodic patterns in the framework of non–linear dynamical systems [144, 233, 334].

A very important extension of cellular automata discussed above became known as *lattice gas automata* (LGA). They provide a microscopic approach

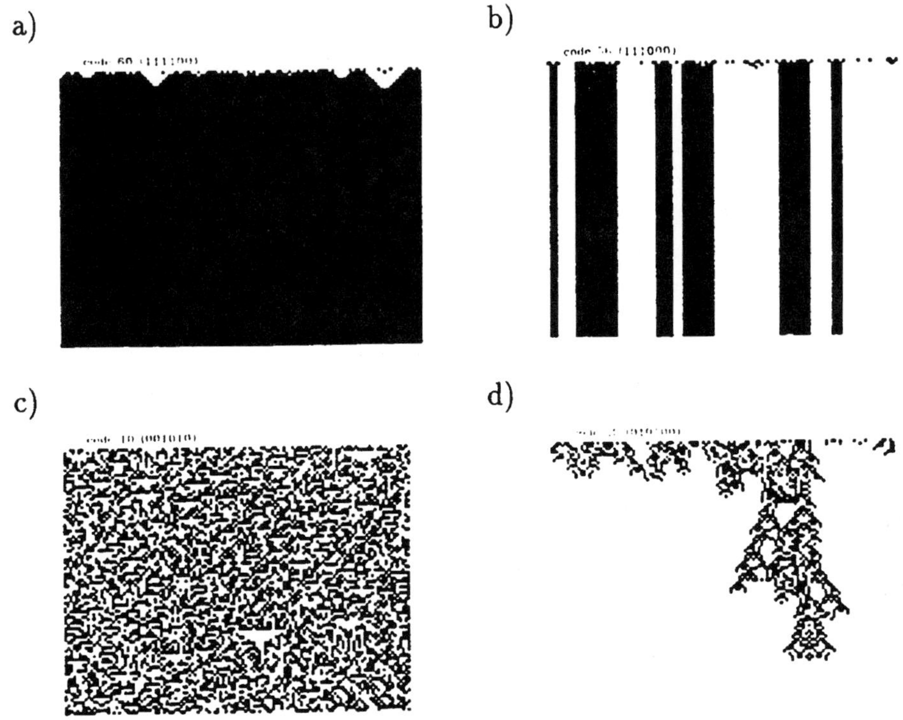

Fig. 6.8 Examples of one–dimensional binary automata, which are typical for the universality classes investigated by Wolfram.

to the dynamics of spatially distributed reacting systems [66, 457]. In a liquid–gas model there is only one kind of particles, but two (thermodynamic) phases. The liquid phase has a high density of particles, while the gas phase is relatively rarefied. The two phases result from a rule that exchange momentum between sites separated by one or more lattice units, which modifies the relationship between pressure and density to allow the coexistence of dense and dilute phases. Starting with a homogeneous random mixture, a phase separation (clustering) takes place [457]. This effect will be shown in the following traffic flow system.

6.6 Traffic Flow by Cellular Automata Models

Recently simulations of traffic flow based on cellular automata (CA) have gained considerable importance. Numerous articles have been published in the past few years investigating discrete models of highway traffic flow, see e. g. current Conference Proceedings [496, 570]. Here we want to show that a very simple probabilistic cellular automaton model can reproduce features of real traffic including the jamming transition from low–density laminar flow to high–density congested flow, where start–stop waves are dominant. The behaviour of that model is very complex and general questions are still under consideration [435, 464]. We mention also other theoretical models like statistical description, hydrodynamic fluids, car–following theory [201, 432] and the stochastic approach formulated by a newly derived Master equation [338, 339] (see Chapter 15).

Modern versions of cellular automata for traffic flow are called *particle hopping models.* Nagel [383] and others [89] pointed out that an initial proposition of a CA model for traffic is from Gerlough in 1956 [169] and has been further extended by Cremer (cf. [102]). Then in 1992, CA traffic models began to receive wide attention after the simple formulation by Nagel and Schreckenberg [385] was published. The one–dimensional road is represented as a string of cells (sites), which are either empty or occupied by exactly one particle. The movement takes place by hopping between cells. If all particles are updated simultaneously (parallel update), then the particle hopping model treated below in detail is formally analogue to one–dimensional cellular automata described before.

Nagel and Schreckenberg [385] introduced in 1992 the following cellular automaton for one–dimensional traffic with periodic boundaries (circular lane) called stochastic traffic cellular automaton (STCA), which has been extended in a large number of papers by different authers. Some of them are given in the References [383, 384, 386, 441, 495]. In this model, the space coordinate i of the road, the time t and the velocity values $v_i(t)$ are discrete variables. STCA is defined on a one–dimensional array of a finite number of sites (length L of the street). Each particle (number N of cars) can have an integer velocity v between 0 and v_{max} (maximal velocity). For an arbitrary configuration, one update of the system consists of the following four consecutive steps, which are performed in parallel for all vehicles [385]:

1. *Acceleration:* If the velocity v of a vehicle is lower than v_{max} and if the

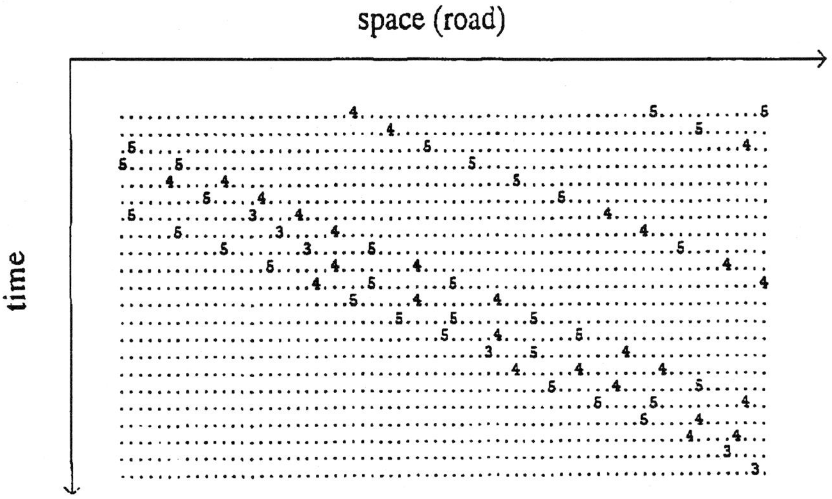

Fig. 6.9 Simulation result for uncongested freeway traffic at a low density of 0.03 cars per site. One of the original diagrams by Nagel and Schreckenberg [385].

distance to the next car (headway = number of empty sites ahead) is larger than $v+1$, the speed is advanced by one ($v \to v+1$).

2. *Slowing down due to other cars:* If a vehicle at site i sees the next car at site $i+j$ (with $j \leq v$), it reduces its speed to $j-1$ ($v \to j-1$).

3. *Randomization by stochastic braking:* With probability p the velocity of each vehicle (if greater than zero) is reduced by one ($v \to v-1$).

4. *Propagation:* Each car moves v sites ahead.

One of the original diagrams by Nagel and Schreckenberg [385] is shown in Fig. 6.9. The overall density ϱ is defined by

$$\varrho = \frac{N}{L} = \frac{\text{number of cars}}{\text{number of sites}} \,. \tag{6.18}$$

For low densities the simulation result consists of (more or less) straight lines of an undisturbed motion with (more or less) constant velocity. Whereas laminar flow at low densities is dominant, we find car clusters (small jams) at higher densities, which are formed randomly due to fluctuations of velocity (see upper line of Fig. 6.10). Here we see the well-known *Stau aus dem Nichts*

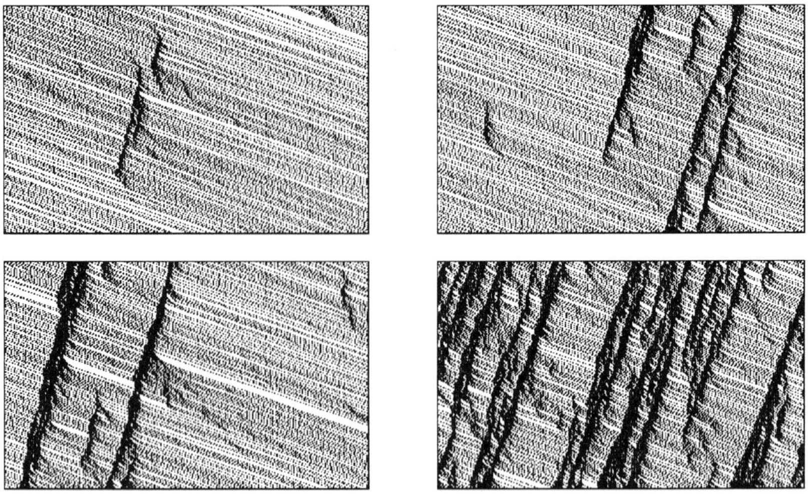

Fig. 6.10 Space–time–simulation results for vehicular motion on a circular road generated by the Nagel–Schreckenberg stochastic traffic cellular automaton (STCA). In dependence on the overall density $\varrho = N/L$ of cars on the highway (top left: 0.08; top right: 0.12; bottom left: 0.15; bottom right 0.3) the traffic behaviour is changing from free flow to heavy congested traffic. The diagrams have been carried out with an implementation due to M. Mahnke [324].

(congestion from nowhere). More cars on the road (density increases) are responsible for very heavy traffic with large car clusters (bottom line of Fig. 6.10).

6.7 Monte Carlo Model of Interacting Monomers and Clusters on a Lattice

Using a binary stochastic model we want to describe drift (directed movement), diffusion (stochastic movement) and chemical reactions in a supersaturated system [340]. Our model consists of a quadratic lattice of sites on a torus, that means a quadratic lattice of sites with periodic boundary conditions. We can use other boundary conditions as well. Having in mind modelling the stream in a pipe (cylinder) or on a road we take two reflecting boundaries together with two periodic boundaries. If one cell is filled by a particle, then the site has the value 1 (black), otherwise it has the value 0 (white). Each molecule can move by drift/diffusion to one of the four neighboured cells by a one–step process, if the cell is empty. Isolated particles (molecules) are called monomers with size $n = 1$. An aggregation of n ($n \geq 2$) neighboured monomers bound by chemical reactions is called a cluster of size n, which is decribed by the boundary length and the number of bindings (chemical bonds) inside the cluster. These quantities describing the shape of a size–n–cluster correspond to the surface and the volume of the cluster in the three–dimensional situation. The data for the size (n_i), the bond number (b_i) and the boundary length (r_i) of each aggregate i can be calculated from the spatial particle configuration on the lattice.

The basic quantity we want to investigate is the discrete cluster ssize distribution in dependence on time, i.e.,

$$\mathbf{N(t)} = (N_1, N_2, ..., N_n, ..., N_N) \tag{6.19}$$

which gives the number N_n of clusters of size n. The free particles (molecules) are called monomers of size $n = 1$. The binding states are clusters of size $n \geq 2$. Investigating a finite system the overall number of particles is fixed. The boundary condition

$$M_0 = \sum_{n=1}^{N} n N_n = \text{const} \tag{6.20}$$

takes into account that the particles are either free or bounded in clusters.

The temporal and spatial development of the cells is given by flip dynamics including a physically motivated energy function. This procedure is called Boltzmann machine, it works as follows. After choosing randomly one filled by a molecule cell and choosing one of the four possible directions of movement in agreement with the given drift probabilities the transition takes place in dependence on the energy change. Based on general Monte Carlo methods in statistical mechanics [52], the algorithm generates a stochastic process of moving particles and interacting monomers and clusters.

Known from atomic as well as cluster physics, the binding energies of spherical charged droplets of different sizes n can be approximated by the well–established Bethe–Weizsäcker expression

$$f_n^{BW} = -c_1\, n + c_2\, n^{2/3} + c_3\, \frac{Q^2}{n^{1/3}}\ . \tag{6.21}$$

In connection to experimental and theoretical investigations of neutral ($Q = 0$) aggregates, e.g. Lennard–Jones clusters, we choose on the basis of Eq. (6.21)

$$f_i \equiv F_i/k_B T = \beta\,(-\alpha\, b_i + (1-\alpha) r_i) \tag{6.22}$$

as the energy of a cluster i with b_i bonds and a boundary length r_i. The parameter β is related to the inverse temperature ($\beta \sim 1/k_B T$), the other parameter α is related to the ratio of volume and surface contributions. The control parameters $\beta > 0$ and $0 \leq \alpha \leq 1$ are constant during each simulation. The binding energy ansatz (6.22) has already been used in an earlier Monte Carlo experiments [332] concerning nucleation, growth and condensation to a single equilibrium cluster regime in a system without drift (see Fig. 6.11).

Since we have a large number of clusters of different sizes (6.19), we define three overall quantities as the sum over all aggregates including monomers. The total number of bindings in the system is $B = \sum b_i$, the total boundary length is $R = \sum r_i$, the total energy summed over each value aacoridng to Eq.(6.22) is

$$E = \sum f_i = \beta\,(-\alpha\, B + (1-\alpha) R)\ . \tag{6.23}$$

During the stochastic Monte Carlo simulation the evolution of the cluster distribution as well as the total number of bonds B, the boundary length R and the energy E can be measured at each time step.

In our simulations we use a lattice of size 100 x 100 sites. We fix the control parameters to $\alpha = 0.5$ and $k_B T \sim 1/\beta = 0.95$.

The series of snapshots of the first Monte Carlo experiment without drift (Fig. 6.11) demonstrates the evolution of the finite system. Starting with the initial distribution of mainly small clusters $\mathbf{N(t=0)} = \{N_1 = 638, N_2 = 103, N_3 = 28, N_4 = 9, N_6 = 1\}$ (Fig. 6.11, top left) the reaction–diffusion system reaches after a very long time a configuration of one big aggregate together with a number of monomers and some small clusters. A typical final situation is shown in the last picture of the series which has the cluster distribution $\mathbf{N(t=15.2)} = \{N_1 = 106, N_2 = 11, N_3 = 1, N_4 = 4, N_{853} = 1\}$ (Fig. 6.11, bottom right).

To extend the first model discussed above (see also [332]) in the second experiment we include the influence of fluids or external fields in which

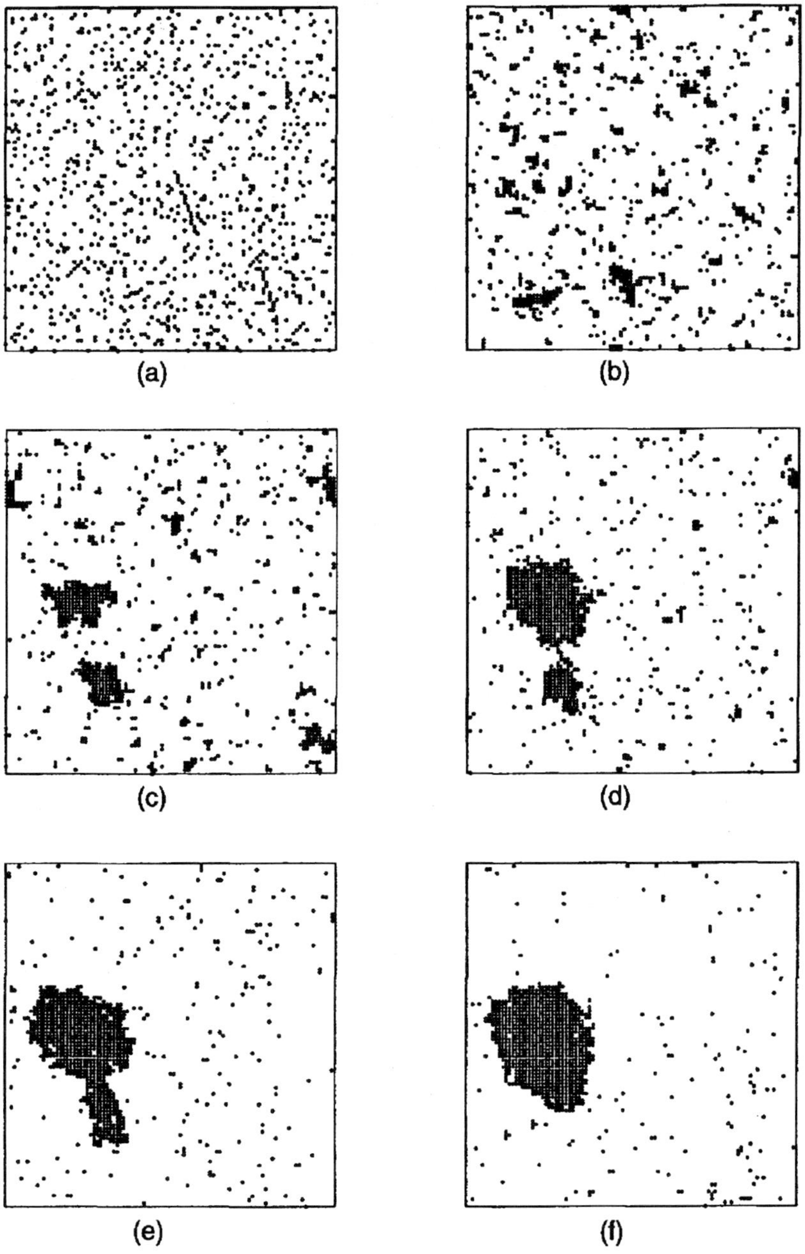

Fig. 6.11 Series of snapshots showing formation of clusters in a box without drift.

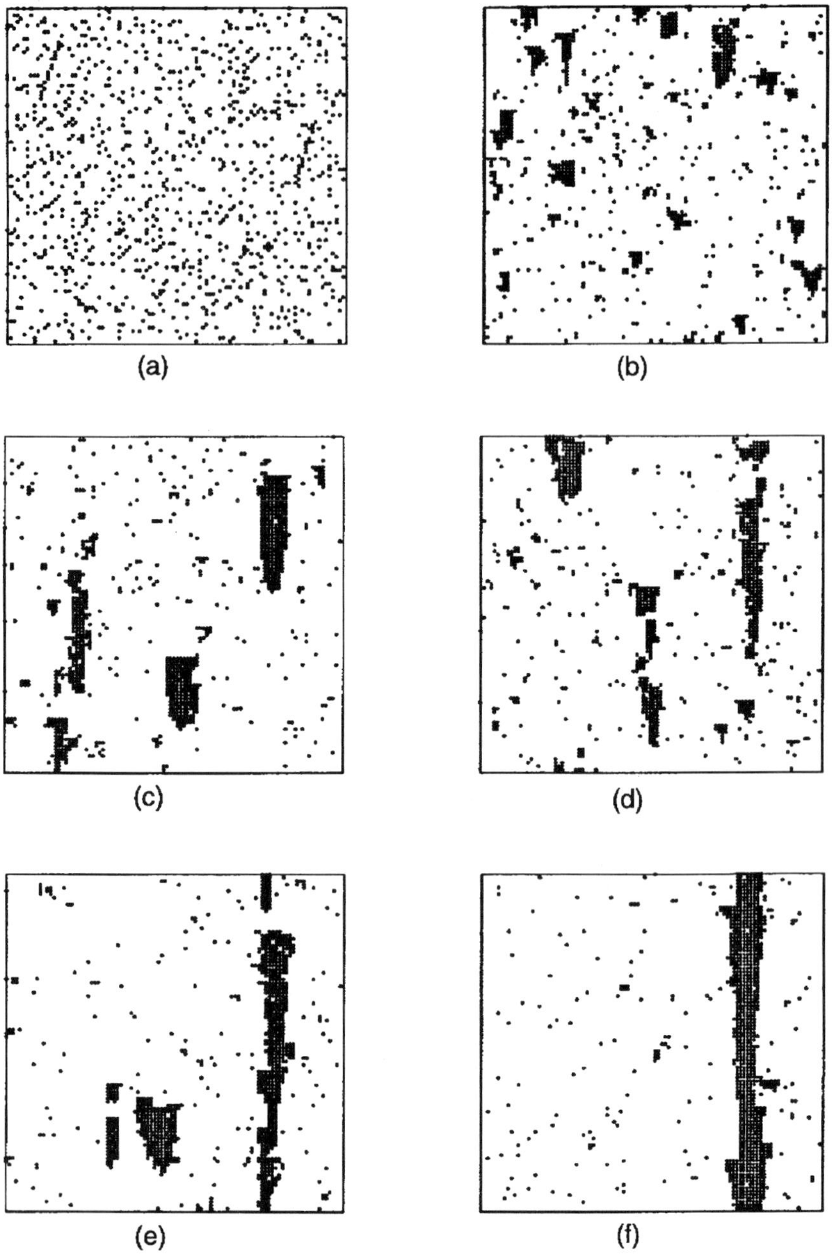

Fig. 6.12 Series of snapshots showing formation of clusters in a flow channel.

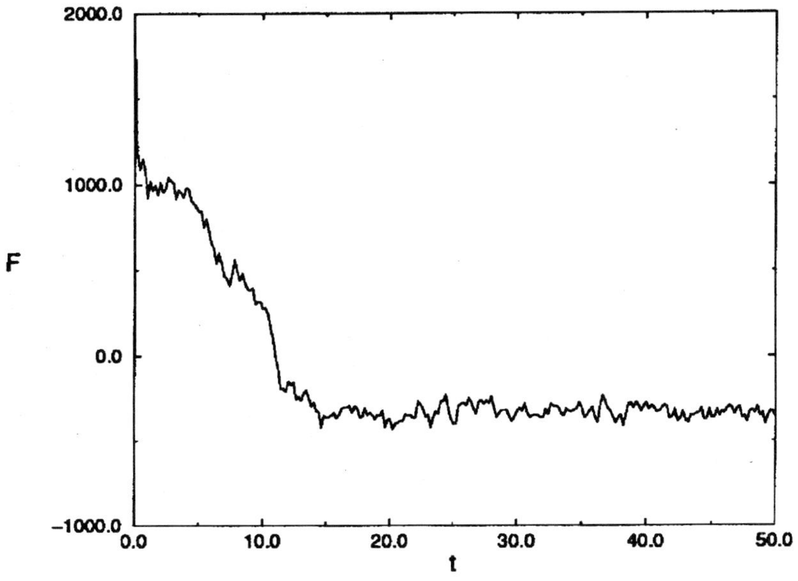

Fig. 6.13 Evolution of the total energy E as function of time. The energy decreases to reach its extremum (minimum value).

the particles are moving. In this so-called drift model we are starting the stochastic process with a random distribution as before. In the initial situation we distribute 1000 monomers randomly over the 10 000 sites, that means that 10 % of the cells are occupied by particles. In Fig. 6.12 (top left) such a random spatial configuration is shown, which has a delta-like cluster distribution $\mathbf{N(t=0)} = \{N_1 = 674, N_2 = 110, N_3 = 17, N_4 = 8, N_5 = 2, N_6 = 1, N_7 = 1\}$. The drift rate of particles is mainly upwards, in probabilities 0.65 to the top, 0.05 to the botton and 0.15 to both sides. The evolution of the cluster configuration (see Fig. 6.12) shows the appearence of small aggregates as binding states, the growth of the stream filaments and the coalescence process up to the final situation with one long filament. The cluster distribution becomes bimodal $\mathbf{N(t=15.2)} = \{N_1 = 110, N_2 = 11, N_3 = 4, N_6 = 1, N_7 = 1, N_{843} = 1\}$.

As it is shown in Fig. 6.13 the system reaches the extremum of the corresponding energy function (Eq.(6.23)) which has a complicated high-dimensional landscape over the space of clusters of different sizes and shapes. In general up-hill and down-hill transitions are possible and the system has

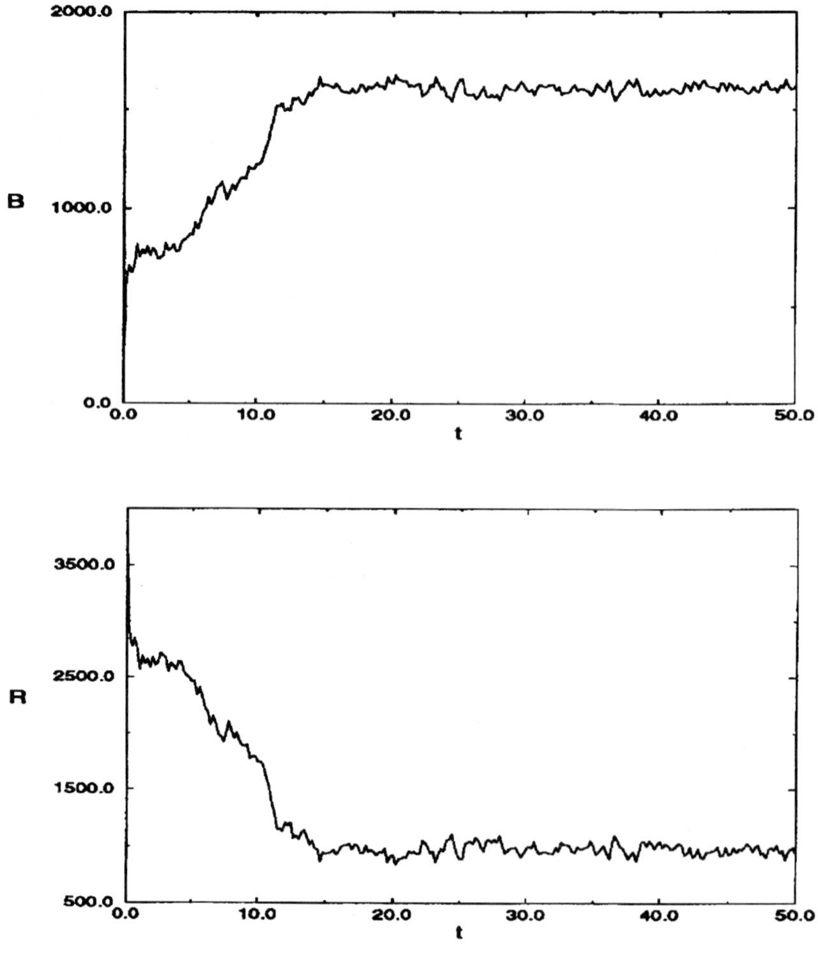

Fig. 6.14 Time evolution of the total number of bindings B (top) and boundary length R (bottom). There are two opposite tendencies, the aggregating system increases the number of bonds and decreases the length of the boundaries.

to find the minimum. The evolution of the cluster configuration in a flow channel (Fig. 6.12) shows the typical behaviour of an aggregation process like in traffic flow. In the beginning the particles condense to a lot of small nuclei. The number of bonds increases rapidly (Fig. 6.14, top). After this nucleation stage the clusters (filaments) grow more or less independently. Since the number of free monomers is decreasing, a competition process, the so–called Ostwald ripening, starts. The number of filaments as well as the energy (Fig. 6.13) and the total boundary length (Fig. 6.14, bottom) are decreasing, only one filament will survive and will grow up to its final size.

In conclusion we emphasize that our model studying the filament structure in a stream of moving particles is related to the cluster formation in open systems [503, 340]. The experimental situation in expanding molecular beams consisting of a condensable vapor and a carrier gas is similar to the treatment which is presented here.

7 Sandpile Model and Self–Organized Criticality

V. B. Priezzhev

7.1 Introduction

In this chapter, we consider a class of phenomena united by a common term, which has been coined recently as "self–organized criticality".

To give an image of this subject, one refers usually to earthquakes [14, 83, 87, 143, 205, 356, 387, 396, 461], forest fires [12, 121, 122, 177, 178, 377], luminocity of stars [360], traffic fluctuations [104, 385, 386, 554], river flows [124], intermittence of biological evolution [15, 59, 153]. The diversity of examples provokes an idea that a total list of them is hardly finite. Indeed, we deal here not so much with distinct physical phenomena but rather with a very general property of dynamical systems. Presumably, the ancient Egyptians watching mysterious causeless changes in the flow of Nile were first who observed self–organized criticality in conditions close to the laboratory ones, as we would say now bearing in mind evidence of the effect. Let us follow them and start with an extremely simplified model of a long quiet river.

Consider a "river" as a one–dimensional dynamical system of L blocks interconnected by springs. Each block is connected to two nearest neighbors. Additionally, each block is connected frictionally to a fixed plate. The flow is provided by an external force F. When the force on a block is larger than the maximal static friction, the block slips. The slip of one block changes the forces on its nearest neighbors. This results in further slips and the chain moves.

Let us put a glass of water on a block in the middle of the chain to imitate an indicator of velocity fluctuations. The question is whether it is possible to draw the chain so accurately that the water in the glass would be never spilled. The answer is surprisingly categorical: not. If the length of the chain

L becomes very large, the probability of spilling tends to unity. We will see below that this simple fact reflects the essence of self–organized criticality.

In a more formal way, the motion of the chain in the steady state is characterized by an average velocity V_{av}. However, the quantity of interest for us is the distribution of the kinetic energy of a block

$$P(\varepsilon) = \text{Prob}(| E - E_{av} | > \varepsilon)$$

measuring the probability that the deviation of the kinetic energy E from its average value E_{av} is greater than ε. In principle, two scenarios of the motion seem possible.

First, the fluctuations of the energy of a block are restricted by a fixed value independent of the number of blocks L and therefore $P(\varepsilon)$ has a typical form

$$P(\varepsilon) \sim \exp\left(-\frac{\varepsilon}{\varepsilon_0}\right) \tag{7.1}$$

where ε_0 is a cutoff independent of L. The exponential decay of fluctuations corresponds to a smooth motion if ε_0 is sufficiently small.

The second scenario is more complicated. The sequences of blocks at rest form clusters of growing size. When the force at the edge of a cluster exceeds the maximal static friction, the motion of the whole cluster is triggered. A block involved into the motion of the large cluster performs a big jump giving rise to a considerable fluctuation of the kinetic energy. If the size distribution of clusters has no characteristic length except L, there are no cutoff parameters in the system in the limit of large L. Therefore, the energy distribution for large L can be written in the form

$$P(\varepsilon) \sim \varepsilon^{-\tau} \tag{7.2}$$

where τ is a constant. The power law (7.2) implies that unlimitedly large fluctuations are possible in the course of motion.

The concept of self–organized criticality claims necessity of the second kind of behavior. More definitely, it claims that [13]

any extended dissipative dynamical system of many degrees of freedom with a driving force tends automatically to a specific state whose large scale behavior is characterized by power–law decay of spatial and temporal correlations.

Now, we are ready to explain both parts of the term "self–organized criticality". The word "criticality" comes from the theory of second–order phase transitions. The critical state in thermodynamics is associated with self–similar fluctuations of observable values. Self–similarity means the absence of a characteristic scale or infinity of the correlation length. This means, in turn, that correlation functions obey power laws at the point of phase transition. To reach the critical point, one should tune one of parameters, say temperature, with high accuracy. In contrast to thermodynamics, dissipative dynamical systems drive themselves to the critical state, or "self–organize" themselves, automatically, without any fine–tuning of parameters.

Thus, the ubiquity of the power laws in Nature is obliged to self–organized criticality which is a universal mechanism of dissipation of energy in large dynamical systems.

The paradigm of self–organized critical systems is a pile of sand [137, 202, 220]. Let us drop grains of sand on the horizontal face of a cube on randomly chosen places, one grain at a time. At some moment, the average slope of the pile reaches a steady state corresponding to the angle that cannot be exceeded no matter how long we carry on adding sand. The stationary state of the sandpile is not completely uniform since variations of the local slopes are possible. If we add a grain of sand, which causes the local slope to exceed the critical angle, an avalanche is triggered. We get a picture quite similar to our spring–block model. The regular input of sand represents the external force of the previous example. The adding of particles leads to avalanches similar to clusters of moving blocks. As a result, we obtain an extremely irregular output of sand or energy having no characteristic intervals between peaks and therefore obeying the power–law distribution.

To examine these observations mathematically, Bak, Tang, and Wiesenfeld [13] proposed a cellular automaton model that borrows the name of its physical prototype. The sandpile model on the two–dimensional square lattice can be defined as follows. Consider a rectangular lattice of side L. Each site i of the lattice is characterized by an integer z_i, the number of particles or the local height at this site. One drops a grain of sand on a site i chosen at random, thereby increasing its height by one:

$$z_i \to z_i + 1 \,. \tag{7.3}$$

If this new height exceeds a maximal stable value, say 4, then the column of sand at the site i gets unstable and topples. The height z_i decreases by 4 and each of four nearest neighbors j of the site i gets one particle:

$$z_i \to z_i - 4, \qquad z_j \to z_j + 1 \, . \tag{7.4}$$

If the toppling occurs at the edge of the lattice, the toppled site gives one particle to each of three neighbors while one grain drops out of the system.

To simplify the toppling rule, Dhar [114] introduced the useful toppling matrix Δ_{ij} with integer elements representing the change in height z_i at site i resulting from the toppling at site j:

$$z_i \to z_i - \Delta_{ij} \, . \tag{7.5}$$

For the two–dimensional square lattice, the toppling matrix has the elements

$$\begin{array}{ll} \Delta_{ij} = 4 & \text{i=j} \\ \Delta_{ij} = -1 & \text{i, j nearest neighbors} \\ \Delta_{ij} = 0 & \text{otherwise} \end{array}$$

It is easy to see that the matrix Δ is nothing else but the discrete Laplacian for the lattice with open boundary conditions.

To watch the evolution of the sandpile in time, we assume that one adds a particle to a stable configuration at each discrete moment of time. If the height z_i reaches 5 somewhere, there is a toppling wherein 4 particles are transferred from the unstable site to its neighbors. The transferred particles may cause instabilities among new sites. The toppling of the latters perturbs next neighbors and a chain reaction propagates up to the moment when all sites get stable again. One assumes the updating to be done concurently, with all sites updated simultaneously. The relaxation processes are assumed to be quick enough to be completed to the next discrete moment of time.

A collection of s distinct sites relaxed during an interval between two successive discrete moments of time forms an avalanche of size s. If a toppling at a given site causes instabilities at all nearest neighbors, the initial site receives 4 particles back and gets unstable after the next updating and therefore topples again. Typically, almost all sites inside a large avalanche undergo multiple topplings. The total amount of topplings in the given time interval is the mass m of the avalanche. The duration t of an avalanche is the number of updatings for the relaxation process to complete.

The formulated rules describe the cellular automaton which is useful for computer simulations. The very first investigations of sandpiles [90, 91, 216, 587] displayed clear power–law dependences for distributions of all basic characteristics of the model:

$$D(s) \sim s^{-\tau} , \tag{7.6}$$

$$D(m) \sim m^{-\kappa} , \tag{7.7}$$

$$D(t) \sim t^{-\alpha} \tag{7.8}$$

and provided rough estimates for the critical exponents τ, κ, α in the size, mass and duration distributions.

Below, we consider the sandpile dynamics in more detail to clarify intrinsic reasons for the self–organized criticality.

7.2 Characterization of Critical States

7.2.1 Recurrent Configurations

The states where all z_i do not exceed 4 are called stable configurations, a state that has any z_i larger than 4 is called unstable. Deepak Dhar [114] introduced operators acting on the stable states in the space of stable configurations $\{C\}$ and discovered remarkable algebraic properties of them. He defined operators a_i by requiring that a_iC be the stable configuration obtained by adding a particle at site i to the configuration C and allowing the system to evolve by topplings to the next stable configuration C':

$$a_iC = C' . \tag{7.9}$$

It is not obvious that the operator a_i exists because the updating procedure may enter a nontrivial cycle. Creutz [103] proved, however, that this is impossible by noting that a toppling in the interior of the lattice does not change the total amount of sand. A toppling on the boundary decreases this sum due to sand falling off the edge, so no cycle can have topplings at the boundary. Next, the sand on the boundary will monotonically increase if there is any toppling one site away. This cannot happen in cycle, thus there can be no topplings one site away from the edges. By induction, there can be no toppling at arbitrary distances from the boundary, thus, there can be no cycle, and therefore the avalanche operator a_i exists.

Consider an unstable configuration in which two sites i and j are both greater than 4. The first toppling at the site i leaves j unstable, and after toppling both i and j one gets a stable configuration in which z_k decreases by $\Delta_{ik} + \Delta_{jk}$. This expression is symmetrical under exchange of i and j. By a repeated use of this argument one obtains that in an avalanche the same final stable configuration is produced irrespectively of whether i or j is toppled first. This fact can be expressed as

$$[a_i, a_j] = 0 \tag{7.10}$$

for all i, j.

The set of all stable configurations $\{C\}$ can be divided into two classes, recurrent and transient. The first subset, denoted by $\{R\}$, includes those stable states that can be obtained from an arbitrary stable configuration by sequential addition of particles. The subset $\{R\}$ is closed under multiple action by operators a_i. Once the sandpile gets into $\{R\}$, it never get out under subsequent evolution. All nonrecurrent stable configurations are called transients and form the subset $\{T\}$ which is the complement of the set $\{R\}$.

Remembering the physical prototype, we see that the recurrent configurations correspond to the steady state of the real sandpile when the critical slope is achieved. So, the subset of recurrent configurations $\{R\}$ represents the self–organized critical ensemble. One can show that all configurations in $\{R\}$ have equal probabilities to occur at a given moment of the infinite evolution.

Figure 7.1 schematically illustrates the configurational space $\{C\}$. The hatched region represents the subset of recurrent configurations $\{R\}$. Each segment of the broken line corresponds to a single addition of a particle followed by relaxation. Assuming that we start with the initial state of the sandpile which is the lattice filled by single particles at each site, we find readily the total number of representatives in $\{C\}$

$$N_{stable} = 4^{L^2} \tag{7.11}$$

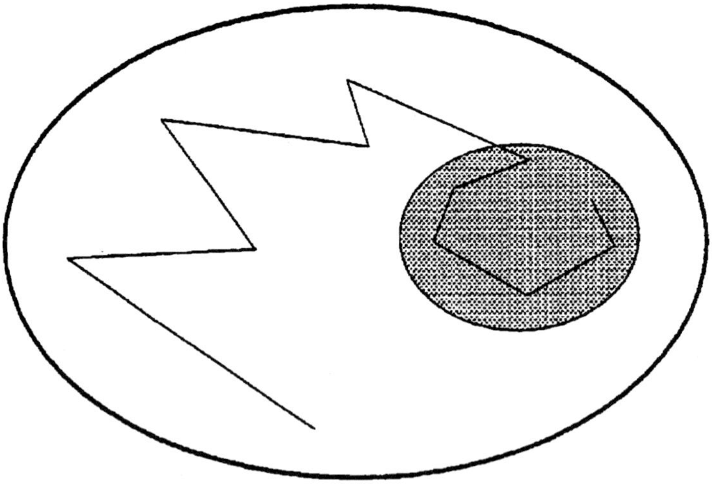

Fig. 7.1 The configurational space of the sandpile model. The external oval confines the set of stable configurations. The hatched region represents the subset of recurrent configurations. A segment of the broken line corresponds to a single addition of a particle followed by relaxation.

as every C is simply the collection of uncorrelated numbers 1, 2, 3, 4.

The determination of the number of recurrent configurations needs some analysis of the algebra of the operators a_i [114].

First, define the inverse operator a_i^{-1}:

$$a_i^{-1}C' = C \tag{7.12}$$

and the identity operator E:

$$E = \prod_j a_j^{\Delta_{ij}}, \tag{7.13}$$

which means that the operator a_i^4 produces the same effect as operators a_j acting at all neighboring sites of i. Therefore, the operator $a_i^4 a_{j_1}^{-1} a_{j_2}^{-1} a_{j_3}^{-1} a_{j_4}^{-1}$, where j_1, j_2, j_3, j_4 are the nearest neighbors of i, does not change the initial configuration.

To proceed, let us note that any recurrent configuration is reachable from a particular one, say, from the maximal stable configuration C_0 which is the lattice filled by 4 at each site. Then, any configuration from the subset $\{R\}$ can be represented by means of the construction:

$$C = (\prod_i a_i^{n_i}) C_0 \,, \tag{7.14}$$

where n_i are integers. Labelling states of $\{R\}$ by a vector $\mathbf{n}$ of L^2 components, we see from Eq.(7.13) that two states are equivalent if the difference of vectors is $\sum_j m_j \Delta_{ij}$ with integers m_j. Thus, the set of vectors $\mathbf{n}$ forms a periodic lattice in the L^2–dimensional space. An elementary cell of this lattice is the L^2–dimensional parallelopiped whose base edges are the vectors Δ_{ij}, $i = 1, 2, 3, ..., L^2$. The volume of the elementary cell is the determinant of Δ. Therefore, the number of nonequivalent recurrent configurations can be expressed in the lapidary form:

$$N_R = \det\Delta \,. \tag{7.15}$$

7.2.2 Burning Algorithm

When the size of the lattice tends to infinity, the discrete Laplacian can be easily diagonalized by a Fourier transform and the determinant in Eq.(7.15) can be found explicitly

$$\det\Delta \sim \exp\left(\frac{L^2}{4\pi^2}\int\int_0^{2\pi} d\alpha\, d\beta \ln\left[4 - 2\cos(\alpha) - 2\cos(\beta)\right]\right) \tag{7.16}$$

$$= (3.21...)^{L^2}$$

Comparing Eq.(7.11) and Eq.(7.16) we can see that the hatched area in Fig.1 is exponentially small with respect to the total configuration space. It means that many stable states admissible in $\{C\}$ get forbidden due to evolution in the recurrent set $\{R\}$.

For instance, consider two neighboring sites i and j occupied by one particle each. This subconfiguration is clearly admissible in the set of stable configurations $\{C\}$. If we want to find it, however, among the recurrent states, we must admit that it appears after a sufficiently long evolution which includes the topplings in both sites i and j. Let the last toppling in i be happened at the moment t_i and the last toppling in j at t_j, assuming $t_i < t_j$ for definiteness. Just before t_j, the site i can contain 1, 2 or 3 particles but not 4 as as it does not topple after t_j. After the toppling in j, the site i receives one particle and increases its height to 2, 3, 4. Therefore, the subconfiguration $z_i = 1, z_j = 1$ is forbidden in the critical state.

Generally, let us denote by $\deg_F(i)$ the number of bonds connecting the site i to the other sites of a subset F of the lattice. The subconfiguration F is said to be forbidden if

$$z_j \leq \deg_F(j) \tag{7.17}$$

for all $j \in F$. If a configuration R is recurrent, it contains no forbidden subconfigurations and is allowed. A typical recurrent configuration on the lattice 10×10 is shown in Fig. 7.2.

Fig. 7.2 A recurrent configuration on the lattice of linear size $L = 10$. The heights are reflected by: white (-1), grey (-2), dark grey (-3), black (-4). The configuration does not contain any forbidden subconfiguration.

Majumdar and Dhar [351] suggested a convenient algorithm to determine whether a given configuration is allowed in the recurrent state. To describe this algorithm, we add an auxiliary site i_0 to our $L \times L$ square lattice and connect it with all boundary sites and with each corner site by two bonds.

Correspondingly, we introduce the new $(L+1) \times (L+1)$ toppling matrix Δ with the additional elements

$$\Delta_{i_0 i_0} = 4L + 1; \Delta_{i_0 j} = \Delta_{j i_0} = -1; \Delta_{i_0 k} = \Delta_{k i_0} = -2 , \qquad (7.18)$$

where j are the boundary sites and k are the corner sites. The sandpile model for the new lattice is quite identical with the original one because the additional site i_0 plays simply the role of a sink collecting particles before they leave the system finally. In this way, we obtain the sandpile model with the single boundary site i_0.

The height z_{i_0} is given by $z_{i_0} = \Delta_{i_0 i_0}$ in all recurrent configurations. Indeed, if $z_{i_0} < \Delta_{i_0 i_0}$, the whole lattice obeys the condition (7.17) forming the maximal forbidden configuration. Now, consider a stable configuration C and drop one particle onto the site i_0. According to Majumdar and Dhar [351], this will generate an avalanche under which each site of the lattice topples exactly once if and only if the configuration C is recurrent. If C is not recurrent, some untoppled sites will remain. All untoppled domains coincide with forbidden subconfigurations.

To show this, consider a boundary site b of a forbidden subconfiguration F. Due to the condition (7.17), the height z_b in the site b does not exceed the number of its nearest neighbors belonging to F. Then, the difference $4 - z_b$ is greater than or equal to the number of its nearest neighbors outside F. Therefore, this site cannot get unstable as a result of the single toppling outside F. As it is true for all boundary sites of F, the whole forbidden subconfiguration remains untoppled after the single toppling of the rest of the lattice. The recurrent configurations contain no forbidden subconfigurations and must topple completely on adding a particle at i_0.

The spreading of the avalanche initiated at i_0 resembles the propagation of fire from a given point. For this reason, the procedure described above is called the "burning algorithm".

7.2.3 Spanning Trees

The burning algorithm makes it possible to visualize the avalanche process and to get the other description of recurrent configurations. It is useful to introduce the index of updatings counting them from the first toppling at i_0. All unstable sites at the step n topple simultaneously producing new unstable sites that topple at the next step $n+1$. Initially, the only site i_0 is unstable which topples at the step $n = 1$. Consider an arbitrary site i. Let n be the step at which this site gets unstable. Then the site i must possess at least one neighbor toppled at the step n. Let ξ be the number of such neighbors toppled to the step n. By assumption, the height z_i obeys the inequalities

$$4 - \xi < z_i \leq 4 \,. \tag{7.19}$$

If $\xi = 1$, we mark by red the bond connecting the site i with the site toppled at the step n. If $\xi > 1$, we select the bond by which the instability (or "fire") reaches i depending on z_i. To this end, we create a list of preference by enumerating all neighboring sites of i in a lexicographic order. In the case of square lattice, the neighbor down will be numbered by 1; the left, by 2; the right, by 3; and the upper, by 4. Now, we choose from ξ candidates the site that occupies the $(z + \xi - 4)$–th position in the list of preference. For instance, if two neighboring sites of i, upper and lower, topple to the step n and $z_i = 4$, we mark by red the upper bond numbered by 4 as it is the second in the list of preference.

The given algorithm makes it possible to avoid ambiguity in the choice of the red bonds and to construct a unique portrait of the recurrent configurations.

As all sites of the recurrent configuration must topple, the graph G formed by red bonds covers all sites of the lattice. A graph like that is called the *spanning graph*. Each site of the lattice gets burnable only once during the fire, therefore G does not contain closed loops and is a *spanning tree*.

Thus, starting with the definition of forbidden subconfigurations, we get the one–to–one correspondence between recurrent sandpile configurations on the lattice and spanning trees on the same lattice. The spanning tree is a strongly correlated object because the possibility of local changes in the branch structure depends on the global structure of the tree. This explains the way in which an uncorrelated collection of heights becomes correlated in the recurrent state in the course of relaxation.

Moreover, the tree analogy provides a useful representation for determination of statistical properties of the sandpile model in the critical state.

Firstly, the number of recurrent configurations can be obtained immediately from the Kirchhoff theorem [195]. This famous combinatorical theorem

Fig. 7.3 A spanning tree on the lattice of linear size $L = 8$.

expresses the number of spanning trees of a given lattice, N_T, via the determinant of the Laplacian matrix defined on this lattice:

$$N_T = \det(\Delta) \; . \tag{7.20}$$

Due to the equivalence between recurrent states and trees established above, we obtain Eq.(7.15) again.

Secondly, correlations between different parts of the spanning trees reflecting the correlations in the sandpile can be obtained relatively easily.

For example, let us evaluate the correlations between two leaves of the tree. We consider the leaf as a site connected to the rest of the graph by the single vertical bond directed upward from the given site. To find the correlations between leaves, we fix two of them at sites i and i' separated by a distance $\mathbf{R}$. The bonds connecting i and i' with their lower, left and right neighbors j_1, j_2, j_3 and j_1', j_2', j_3' are not occupied by branches of the tree. So, we can remove them from the lattice. The diagonal elements of the Laplacian matrix Δ count the number of bonds incident to a given site, whereas the nondiagonal ones correspond to bonds themselves. Then, the Laplacian of the modified lattice Δ' will differ from Δ by elements

$$\Delta'_{ij_1} = \Delta'_{ij_2} = \Delta'_{ij_3} = 0$$

$$\Delta'_{j_1 j_1} = \Delta'_{j_2 j_2} = \Delta'_{j_3 j_3} = 3$$

$$\Delta'_{ii} = 1$$

and by similar elements with indices i', j'_1, j'_2, j'_3. The probability of finding both leaves at i and i' follows directly from Eq.(7.20):

$$\mathrm{Prob}(i, i') = \frac{\det\Delta'}{\det\Delta} \tag{7.21}$$

It is convenient to introduce the defect matrix $\delta = \Delta' - \Delta$ and the matrix $G = \Delta^{-1}$, the well known two–dimensional lattice Green function given in the limit $L \to \infty$ by

$$G(\mathbf{r}, \mathbf{r}) - G(\mathbf{r}, \mathbf{r}') = \tag{7.22}$$

$$= \frac{1}{2(2\pi)^2} \int\int_0^{2\pi} d\alpha d\beta \frac{1 - \cos(x - x')\alpha \cos(y - y')\beta}{2 - \cos\alpha - \cos\beta} ,$$

where $\mathbf{r}$ and $\mathbf{r}'$ are the radius-vectors of points i and i' with coordinates (x, y) and (x', y'). Making use of these matrices, we obtain from Eq.(7.21)

$$\mathrm{Prob}(i, i') = \det(I + G\delta) , \tag{7.23}$$

where I is the identity matrix. Crossing out those rows and columns of δ containing only zero elements, we obtain $\mathrm{Prob}(i, i')$ in the form of the determinant of an 8×8 matrix.

In principle, the evaluation of the determinant in Eq.(7.23) is straightforward, but tedious. The leading asymptotics, however, can be derived from the following arguments. The Green function $G(\mathbf{r}, \mathbf{r}')$ depends only on the difference of coordinates $\mathbf{r} - \mathbf{r}' = \mathbf{R}$ and behaves as $\ln(R)$ for large R. Matrix elements G_{ij} enter into the expansion of the determinant in combinations corresponding to variations of the first and second coordinates, that is, $G''(R) \sim R^{-2}$. Furthermore, due to the symmetry of the matrix, the polynomials corresponding to $G''(R)$ are combined with their transposes. Keeping only leading terms, we get

$$\mathrm{Prob}(i,i') - \mathrm{Prob}(i)\mathrm{Prob}(i') \sim R^{-4} \tag{7.24}$$

where $\mathrm{Prob}(i)$ is the density of vertical leaves on the tree to be found in the next subsection.

Thus, by inspection of the actual correlations between different pieces of the spanning trees representing the recurrent states of the sandpile model, the hypothesis about the critical power–law behavior appears to be fulfilled.

7.2.4 Height Probabilities

We conclude this section by considering the question that emerges probably first when we formulate the sandpile model. What are the fractional numbers of sites having heights 1, 2, 3, 4 in the critical state? If different states are not correlated, each height appears clearly with the probability 1/4. When the sandpile tends to the critical state, the height probabilities $P(1)$, $P(2)$, $P(3)$, $P(4)$ approach stationary nontrivial values due to strong correlations between sites in the recurrent state.

To find the probabilities of heights, we have to translate the problem into the language of spanning trees. It is convenient, however, to introduce at first auxiliary quantities instead of $P(z)$.

For a given lattice site i, the set of recurrent configurations can be divided into four subsets s_1, s_2, s_3, s_4. These are defined as follows. A configuration C belongs to the subset s_1 if it remains allowed after all substitutions $z_i = 1,2,3,4$ at the site i. It belongs to the subset s_2 if it remains allowed for $z_i = 2,3,4$ but becomes forbidden for $z_i = 1$ and belongs to the subset s_3 if it remains allowed for $z_i = 3,4$ and becomes forbidden for $z_i = 1,2$. The subset s_4 contains configurations allowed only for $z_i = 4$. All admitted substitutions at a given site correspond to equal numbers of configurations. Therefore, the height probabilities $P(1)$, $P(2)$, $P(3)$, $P(4)$ can be written in the form

$$P(1) = \frac{N_1}{4N_R}\,, \tag{7.25}$$

$$P(2) = P_1 + \frac{N_2}{3N_R} , \tag{7.26}$$

$$P(3) = P_2 + \frac{N_3}{2N_R} , \tag{7.27}$$

$$P(4) = P_3 + \frac{N_4}{N_R} , \tag{7.28}$$

where N_R is given by Eq.(7.15) and N_ν is the number of recurrent configurations in the subsets $s_\nu, \nu = 1, 2, 3, 4$.

Secondly, we introduce the ordering of the trees with respect to the additional site i_0 called the *root* of the tree and introduced above in connection with the burning algorithm. We shall say that a site i is the *predecessor* of a site j if the unique path from i to the site i_0 along the given tree goes through j.

Let us now turn to the description of the subset s_1. If a configuration C is allowed for $z_i = 1$, it remains allowed after the substitutions $z_i = 2, 3, 4$. Due to Eq.(7.25), N_1 is equal to the number of recurrent configurations with $z_i = 1$ multiplied by 4. Let us fix $z_i = 1$. For every recurrent configuration, there exists a burning procedure which burns the site i after all its nearest neighbors j_1, j_2, j_3, j_4. Constructing a tree in accordance with the burning algorithm, we find that the site i has no predecessors among its neighbors.

To describe s_2, we use the definition again. The substitution $z_i = 1$ converts an arbitrary configuration C of this subset into a forbidden one C'. This means that a forbidden subconfiguration (FSC) appears which contains the site i with $z_i = 1$, one of sites j_1, j_2, j_3, j_4, say j_1 with $z_{j_1} \geq 1$, and some k connected sites ($k \geq 0$) including none of sites j_2, j_3, j_4 because if one of j_2, j_3, j_4 also belongs to FSC, then the configuration C' remains forbidden after the substitution $z_i = 2$.

Let $S(C)$ be the FSC resulting from the substitution $z_i = 1$ in C. We construct a modified lattice in the following way. We delete the boundary bonds connecting the sites of $S(C)$ to the rest of the lattice with the exception of the only bond connecting the site i with one of the sites j_2, j_3, j_4 (j_2 for definiteness). For each bond deleted, we also decrease the maximum height at the two end sites of the bond by 1. In this way, we obtain a new toppling rule matrix $\Delta'(S)$ which depends on the form of a given FSC. For each recurrent configuration C with $z_i = 2$ a burning procedure exists which does

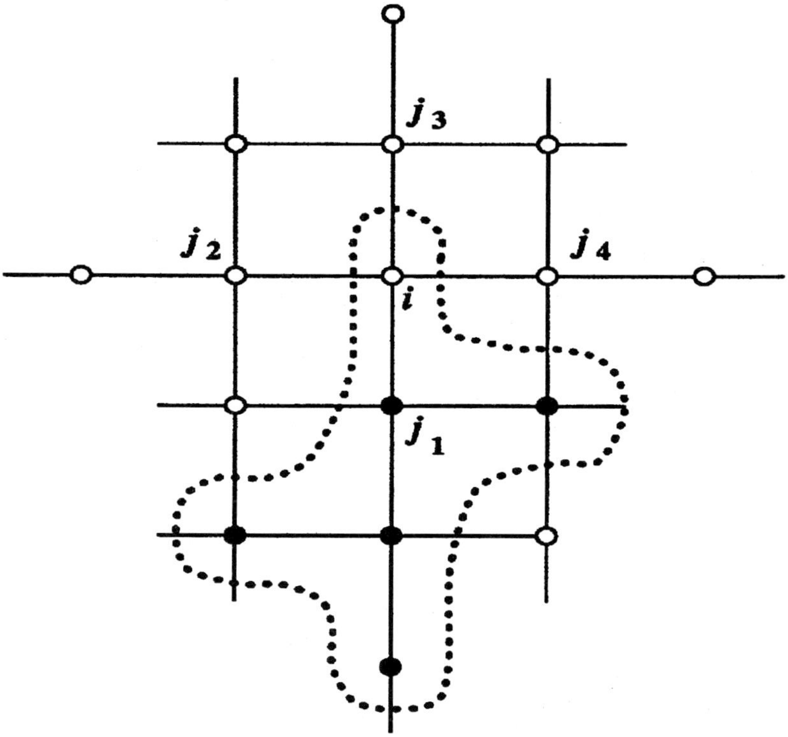

Fig. 7.4 A configuration of the class s_2. The dotted line confines the forbidden configuration resulting after substitution $z_i = 1$. Black sites together with the site i belong to the subtree T_1. None of the sites represented by open circles is a predecessor of i.

not depend on the presence of deleted bonds. Therefore, the set of recurrent configurations on the modified lattice is in one–to–one correspondence with the set of configurations s_2 that generates S by the substitution $z_i = 1$.

Constructing a tree by the burning algorithm on the modified lattice, we obtain a spanning tree T satisfying the following conditions:

1. T contains bonds $j_1 i$ and $i j_2$;

2. Deletion of the bond $i j_2$ divides T into two subtrees T_1 and T_2 such that the sites i and j_1 belong to T_1 and the sites i_0, j_2, j_3, j_4 belong to T_2;

3. The bonds $i j_3$ and $i j_4$ are always absent among the bonds of T.

The rules (1) – (3) imply that the site j_1 is the nearest predecessor of i and none of the sites j_2, j_3, j_4 are predecessors of i.

The descriptions of s_3 and s_4 are similar to those of s_1 and s_2.

Summarizing, we can formulate the rule for determination of the number of configurations N_ν in the subsets $s_\nu, \nu = 1, 2, 3, 4$ in the following form:

$$N_\nu = T(\nu - 1) , \tag{7.29}$$

where $T(n)$ is the number of trees for which the site i has n predecessors. The height probabilities $P(1)$, $P(2)$, $P(3)$, $P(4)$ then follow directly from Eq.(7.15) and Eqs.(7.25) – (7.28).

For example, let us find $P(1)$. According to the rule (7.29), N_1 is equal to the number of trees having a leaf at i. Four possible positions of the leaf are equivalent, so $P(1)$ is equal to the density of vertical leaves Prob(i) entering into Eq. (7.24). Prob(i) is given by Eq.(7.21) with the 4×4 matrix Δ' defined for a single leaf at the site i. Taking rows and columns in the natural order i, j_1, j_2, j_3 and evaluating the integrals in Eq.(7.23) we can write the non–zero part of the symmetric matrix G

$$\begin{pmatrix} G_{ii} & G_{ij_1} & G_{ij_2} & G_{ij_3} \\ \ldots & G_{j_1j_1} & G_{j_1j_2} & G_{j_1j_3} \\ \ldots & \ldots & G_{j_2j_2} & G_{j_2j_3} \\ \ldots & \ldots & \ldots & G_{j_3j_3} \end{pmatrix}$$

in the explicit form

$$\begin{pmatrix} G(0,0) & G(0,0) - \frac{1}{4} & G(0,0) - \frac{1}{4} & G(0,0) - \frac{1}{4} \\ \ldots & G(0,0) & G(0,0) - \frac{1}{\pi} & G(0,0) - \frac{1}{\pi} \\ \ldots & \ldots & G(0,0) & G(0,0) - 1 + \frac{2}{\pi} \\ \ldots & \ldots & \ldots & G(0,0) \end{pmatrix} .$$

The defect matrix δ is

$$\begin{pmatrix} -3 & 1 & 1 & 1 \\ 1 & -1 & 0 & 0 \\ 1 & 0 & -1 & 0 \\ 1 & 0 & 0 & -1 \end{pmatrix}$$

Inserting these matrices into the expression

$$\text{Prob}(i) = \det(I + G\delta) \tag{7.30}$$

quite similar to Eq.(7.23), one obtains (Majumdar and Dhar [350])

$$P(1) = \text{Prob}(i) = \frac{2}{\pi^2}\left(1 - \frac{2}{\pi}\right) . \tag{7.31}$$

The evaluation of remaining probabilities $P(2)$, $P(3)$, $P(4)$ is based also on the rule (7.29). One, two or three predecessors of a given site can appear in these cases, which requires a more cumbersome analysis of the spanning tree structure. Nevertheless, this problem is also solvable [430] and the final result can be expressed in a closed form. The numerical values of the obtained probabilities are

$$P(1) = 0.073\ldots ; \qquad P(2) = 0.173\ldots ;$$

$$P(3) = 0.306\ldots ; \qquad P(4) = 0.446\ldots \tag{7.32}$$

The constructive description of the recurrent configurations, the evaluation of correlation functions and height probabilities gives, in principle, a complete information about the structural properties of the stationary critical state. In the next section, we will consider dynamical aspects of the sandpile model associated with the statistics of avalanches.

7.3 Statistics of Avalanches

7.3.1 Waves of Topplings

The study of avalanches needs an extension of the tree representation. To this end, we will consider an avalanche process in more detail. The commutativity property (7.10) admits an arbitrary order of topplings of nonstable sites during an avalanche. We choose a particular but completely natural order amongst these. Namely, let us add a particle to the site i having the height 4 in an allowed configuration C. We topple it once and then topple

The trees of the second class contain the bond $i_0 i$. On deleting the bond $i_0 i$ from a tree of this class, a subtree T_i gets disconnected. Considering the site i as a root of T_i we obtain a two rooted situation where a spanning tree on the lattice consists of two disconnected clusters T_i and T_{i_0}.

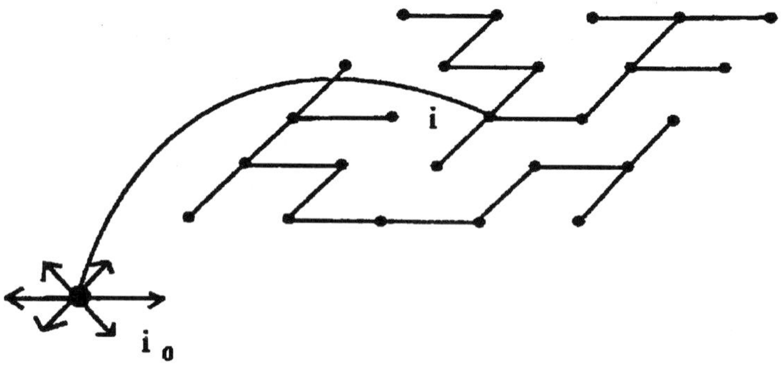

Fig. 7.5 A two-rooted spanning tree representing a certain wave in an avalanche started at the site i_0. The auxiliary lattice is obtained by adding the bond $i_0 i$. Arrows at i_0 denote bonds going to all boundary sites.

The one-rooted trees correspond to recurrent configurations. The configurations corresponding to the two-rooted trees are, in general, forbidden with respect to the sandpile dynamics on the original lattice. The exceptions are the last waves because in this case, the root i lies on the boundary of the cluster and the inequality (7.17) does no longer hold.

Thus, in addition to the correspondence between the recurrent states and one-rooted spanning trees, we get the one-to-one correspondence between all waves of topplings and all two-rooted spanning trees. The avalanche is displayed now as a collection of successive two-rooted trees and the problem of critical exponents mentioned in the Introduction gets reduced to finding the probability distributions for subtrees of this collection.

7.3.2 Waves and Green Functions

The graph representation of waves enables us to link the toppling process and the lattice Green function G_{ij}. For this purpose, we shall prove the following proposition.

Theorem: For an arbitrary connected graph Γ with a fixed vertex i_0,

$$G_{ij} = \frac{N^{(ij)}}{N} , \tag{7.33}$$

where $N^{(ij)}$ is the number of two–rooted spanning trees having the roots i_0 and i, such that both vertices i and j belong to the same one–rooted subtree; N is the total number of spanning trees on Γ.

Proof: Let Δ be a symmetric Laplacian matrix of the graph Γ. The Kirchhoff's matrix theorem reads.

$$N = \det\Delta^{(i_o)} , \tag{7.34}$$

where the matrix $\Delta^{(i_o)}$ is obtained from Δ by deleting the column and row corresponding to the root i_0 to create the open boundary conditions at i_0. By the Kirchhoff's formula for resistance, the number $N^{(i)(j)}$ of two-component spanning trees having the vertices i and j in *different* components is

$$N^{(i)(j)} = \det\Delta^{(i)(j)} , \tag{7.35}$$

where the matrix $\Delta^{(i)(j)}$ is obtained from $\Delta^{(i_o)}$ by deleting the columns and rows corresponding to the vertices i and j.

Instead of deleting elements of $\Delta^{(i_o)}$, we can add ϵ to the elements $\Delta^{(i_o)}_{ii}$ and $\Delta^{(i_o)}_{jj}$ and $-\epsilon$ to the elements $\Delta^{(i_o)}_{ij}$ and $\Delta^{(i_o)}_{ji}$, thus obtaining the new matrix $\Delta^{(i_o)}_{\epsilon}$. Then

$$N^{(i)(j)} = \lim_{\epsilon\to\infty} \frac{1}{\epsilon}\det\Delta^{(i_o)}_{\epsilon} \tag{7.36}$$

and we can evaluate the ratio $(N^{(i)(j)}/N)$ by the formula

$$\frac{\det\Delta^{(i_o)}_{\epsilon}}{\det\Delta^{(i_o)}} = \det(I + G\delta) , \tag{7.37}$$

where $\delta = \Delta_\epsilon^{i_0} - \Delta^{i_0}$ and $G = (\Delta^{i_0})^{-1}$ is the solution of the Poisson equation with the boundary conditions $G_{i_0 k} = 0$ for all k. Direct evaluation in Eq.(7.37) leads to

$$\frac{N^{(i)(j)}}{N} = G_{ii} + G_{jj} - G_{ij} - G_{ji} \; . \tag{7.38}$$

Putting $i = i_0$ we also have

$$\frac{N^{(i_0)(j)}}{N} = G_{jj} \tag{7.39}$$

for all $j \neq i_0$.

The number $N^{(i_0)(j)}$ is the sum of two parts

$$N^{(i_0)(j)} = N^{(i_0 i)(j)} + N^{(i_0)(ij)} \; , \tag{7.40}$$

where the notation (ij) means that both vertices belongs to one component. Analogously, we have

$$N^{(i)(j)} = N^{(i_0 i)(j)} + N^{(i)(i_0 j)} \; . \tag{7.41}$$

Since, by definition, $N^{(i_0)(ij)} = N^{(ij)}$, the statement of the theorem is a simple consequence of the linearity of the equations (7.38) – (7.41).

Due to the relationship between two–rooted trees and waves, we conclude that NG_{ij} is the number of waves initiated at the site i and involving the site j.

The derived result is in agreement with the observation by Dhar [114] that G_{ij} is the expected number n_{ij} of topplings at the site j due to the avalanche caused by adding a particle at i. Indeed, as each wave corresponds to exactly one toppling of all its sites, n_{ij} coincides with the expected number of waves involving the site j.

7.3.3 Critical Exponents for Waves

Now, we can collect the observations obtained above to find the critical exponents corresponding to the wave distributions of the sandpile model on the two-dimensional lattice.

We shall start with the statistics of the last waves because the exact result by Dhar, Manna and Majumdar [115] is known here. We have seen above that the last wave corresponds to the rooted subtree having the root at the boundary. The alternative way to get the tree representation of the last wave is to cut a bond and disconnect a branch from the one-rooted tree representing the recurrent configuration appearing after the last wave. The probability $P(s)$ that on deleting of a randomly chosen bond exactly s sites get disconnected from the tree varies as $s^{-11/8}$ for large s [115]. The derivation of this result is based on scaling arguments and the fact that the chemical path on a spanning tree has the fractal dimension 5/4. The chemical path l between two sites separated by the Euclidean distance r is the unique path connecting them along the edges of the tree. For the given fractal dimensionality $r \sim l^{4/5}$. Due to scaling arguments, the perimeter of the disconnected cluster l is distributed as

$$\mathrm{Prob}(l \geq l_0) \sim l_0^{-3/5} . \tag{7.42}$$

Assuming that the cluster is compact, that its area behaves as $\sim r^2$, one gets

$$\mathrm{Prob}(s \geq s_0) \sim s_0^{-3/5}, \tag{7.43}$$

which leads to $D(s) \sim s^{-11/8}$.

Using this result we can derive the asymptotics of the probability distribution for sizes of waves. To this end, we consider the dynamics of the avalanche as a sequence of waves without reference to particular avalanches which every wave belongs to. In other words, given a sequence of avalanches consisting of waves, we ignore the pauses between avalanches and investigate properties of the collection of waves. An arbitrary wave of this collection is called the wave of general form.

To find the size distribution of general waves, we consider all possible waves that belong to the avalanches starting at a fixed point i deeply inside the lattice. All these waves are in one–to–one correspondence with the recurrent configurations of the sandpile model on the lattice with the additional bond connecting the sites i_0 and i. As all recurrent configurations have the same probability, all general waves are also equally likely.

A general wave differs from the last one by the location of the root. The position of the root in the general wave is distributed uniformly over the whole area in contrast to the boundary location for the last wave. Therefore, given a wave of the area s, an additional factor s/l appears in the probability distribution. Remembering that

$$l \sim s^{5/8} \tag{7.44}$$

we get

$$\text{Prob}(s_{general} = s) \sim \frac{s}{l}\text{Prob}(s_{last} = s) \sim \frac{1}{s} \, . \tag{7.45}$$

Naturally, the obtained asymptotics must be consistent with the known asymptotics of the Green function $G(r) - G(0) \sim \ln r$. According to the graph interpretation, the Green function $G(r)$ is an expected number of rooted subtrees of two–rooted trees which contain the root and the point separated from the root by the distance r within one component. The relative number of such subtrees is proportional to the integral of $\text{Prob}(s_{general} = s)$ over all areas exceeding s, that is,

$$G(r) = \int_s^{\infty} \frac{1}{s'} ds' \, . \tag{7.46}$$

The lower limit of the integral corresponds to the leading asymptotics of the Green function.

The critical exponents of the general wave are not related directly to exponents of the avalanche distributions because considering the general wave we lose the information on a concrete avalanche which it belongs to. Fortunately, there exists a situation where all avalanches consist of one and only one wave and the avalanche distribution coincides with the distribution of general waves. Consider the sandpile model with open boundary conditions, for which the sinks are situated at all boundary sites of the lattice. In this case, avalanches starting at the boundary consist of the only wave because the second toppling is impossible due to loss of particles.

The asymptotic form of the boundary Green function in the continuum limit is

$$G(\mathbf{r}) \sim \log|\mathbf{r}-\mathbf{a}| - \log|\mathbf{r}+\mathbf{a}| \sim \frac{(\mathbf{a},\mathbf{r})}{\mathbf{r}^2}\,, \tag{7.47}$$

where $\mathbf{a}$ is a unit vector perpendicular to the boundary. Correspondingly, the probability that the front of the avalanche exceeds r is

$$\mathrm{Prob}(r' > r) \sim \frac{1}{r}\,, \tag{7.48}$$

which leads, after differentiation, to the radius distribution $D(r) \sim 1/r^2$. Using the relations $s \sim r^2$ and $D(s)ds \sim D(r)dr$, we get the sought probability distribution

$$D(s) \sim \frac{1}{s^{3/2}}\,. \tag{7.49}$$

Thus, the exponent τ in Eq.(7.6) is equal to (3/2) for the boundary avalanches. The relationship between the exponents τ and α can be found from the following arguments. By the construction of trees, the avalanche process follows the branch structure of the tree. Then, the duration of avalanche t varies as the chemical distance of the tree $l \sim r^{5/4}$. This implies that

$$5\tau = 8\alpha - 3\,. \tag{7.50}$$

We obtain $\alpha = 9/5$ for the boundary avalanches.

The above arguments can be easily generalized to boundaries forming an arbitrary angle γ. It is known from the theory of complex variables that the Green function of the Laplacian in the region bounded by an angle α has the form

$$G(x,y) = \frac{1}{2\pi}\mathrm{Im}(z^{-\pi/\gamma})\,, \tag{7.51}$$

where $z = x + iy$, (x,y) are the Cartesian coordinates. Then, the function $G(r)$ decays as

$$G(r) \sim r^{-\pi/\gamma} \tag{7.52}$$

for any direction apart from arms of the angle. This leads to the distribution

$$D(r) \sim r^{-1-\pi/\gamma} . \tag{7.53}$$

Again using the relations $s \sim r^2$ and $D(s)ds \sim D(r)dr$, we get the asymptotics

$$D(s) \sim s^{-1-\pi/2\gamma} , \tag{7.54}$$

which corresponds to $\tau = 1 + \pi/2\gamma$ and $\alpha = 1 + 4\pi/5\alpha$.

The analytical results can be verified numerically by Monte Carlo simulations. Table 7.1 shows the numerical results for the lattices with sizes up to 100 sites having angles $\gamma = \pi/2$, π, $3\pi/2$, 2π obtained with statistics up to a million avalanches.

α	$\pi/2$	π	$3\pi/2$	2π
τ	1.9	1.51	1.32	1.21
exact	2	3/2	4/3	5/4

Tab. 7.1 Angular critical exponents for multiples of $\pi/2$.

The angle 2π is of special interest. In this case, avalanches start at the top of a cut of the plane. The geometry of avalanches closely resembles the one occurring deep inside the lattice. So, one can expect that the critical exponents in both cases are in close agreement. Indeed, the difference between numerical estimations by Manna [352] $\tau = 1.22$ and $\alpha = 1.38$ and the boundary exponents near the cut $\tau = 1.25$ and $\alpha = 1.40$ is not more than 3 per cent.

7.3.4 Critical Exponents for Avalanches

Self–similarity of avalanches implies self–similaryity of their parts. Therefore, one can expect that the size difference between successive waves $\Delta s = s_k - s_{k+1}$ obeys also a power law

$$\Delta s \sim s^{\beta} . \tag{7.55}$$

The exponent β, if exists, can be related to τ by a scaling relation. Let n denote the number of waves in an avalanche which coincides with the number of topplings at the site i. Equation (7.55) can be rewritten in the differential form $ds/dn \sim s^\beta$ or, equivalently,

$$dn \sim \frac{1}{s^\beta} ds \ . \tag{7.56}$$

The wave of size s belongs to an avalanche of size $S \geq s$ which has the probability $P(S \geq s) \sim s^{1-\tau}$. Then, the distribution of waves belonging to diverse avalanches is

$$D(s) \sim \frac{1}{s^{\beta+\tau-1}} \ . \tag{7.57}$$

Comparing Eq.(7.57) with Eq.(7.45), we obtain the scaling relation

$$\beta + \tau = 2 \tag{7.58}$$

Majumdar and Dhar [351] introduced an exponent y assuming that n scales with the size of an avalanche as $n \sim s^{y/2}$. To be consistent, the exponents β and y must be related as

$$2\beta + y = 2 \ . \tag{7.59}$$

Grassberger and Manna [176] introduced clusters of sites A_n toppled $\geq n$ times, $n \geq 1$, during an avalanche. If waves of a given avalanche obey the relations $F_1 \supset F_2 \supset ... \supset F_n$ strictly, the structure of waves coincides completely with that of clusters A. At the same time, Dhar and Manna [116] who investigated inverse waves registered situations when the wave F_k overlaps the preceding one F_{k-1}. They argued that these events are nevertheless relatively rare and on the average the last waves scale as the clusters of maximal topplings. Simulations show generally that the distributions of waves F and clusters A follow the same asymptotic law. Taking into account these observations, we neglect the overlapping of waves and deal only with the decrease of waves.

The above construction allows us to determine β from scaling arguments. To this end, we have to link the decrease in the size of waves Δs with the spanning tree characteristics.

Let $T_i(F_k)$ be the subtree with the root i corresponding to the wave F_k. As all sites involved into F_k topple exactly once, all internal sites of F_k remain unchanged. The wave F_{k+1} following F_k will repeat its order of topplings until the relaxation process reaches the boundary of F_k. Accordingly, the subtree $T_i(F_{k+1})$ that represents F_{k+1} will coincide with $T_i(F_k)$ as long as its sites have no predecessors among the boundary sites of F_k. Denote by B_j a set of sites of $T_i(F_k)$ having a boundary site j as a predecessor. Actually, B_j is a branch of $T_i(F_k)$ attached to the subtree at the point j. If the site j gets stable with respect to the next wave F_{k+1}, all sites of B_j get stable too as the toppling process penetrates into B_j via the point j. As a result, the sites of B_j as well as the site j itself contribute to Δs. Generally, Δs consists of all boundary sites $j_1, j_2, \ldots$ of the wave F_k getting stable with respect to F_{k+1} and of sites of all sets B_{j_1}, $B_{j_2}, \ldots$ having $j_1, j_2, \ldots$ as predecessors.

In Fig. 7.6 we show a typical form of the set contributing to Δs. The external contour Γ represents the boundary sites of the wave F_k, and the loops γ_i correspond to the sets B_i. By construction, two main quantities determine Δs: the length of the contour Γ and the area of loops γ.

Denoting by R a linear extent of the wave F_k, we can estimate the length of the contour Γ as $R^{5/4}$ since Γ is a chemical path on the dual spanning tree. Then, the contribution from Γ gives

$$\Delta s \sim R^{5/4} \sim s^{5/8} , \tag{7.60}$$

which implies 5/8 for the exponent β. We shall see, however, that the leading contribution comes from the second quantity determined by the interior of loops γ.

Consider a single loop γ. It is characterized by the distance l between points x and y where it is attached to the contour Γ and the linear extent r (see Fig. 7.6). The cluster surrounded by γ is a subtree having a fixed root at one of the two boundary sites, say x. According to (7.45), the trees of linear extent r are distributed as $D(r) \sim 1/r$. The root can occupy any of r^2 positions inside γ. Therefore, subtrees with a fixed root are distributed as $1/r^3$. Let us consider a circle C of radius l having the center at point x. The average number of intersections between C and Γ is of order $l^{1/4}$ due to fractal dimensions of the chemical path. The point y can occupy any of

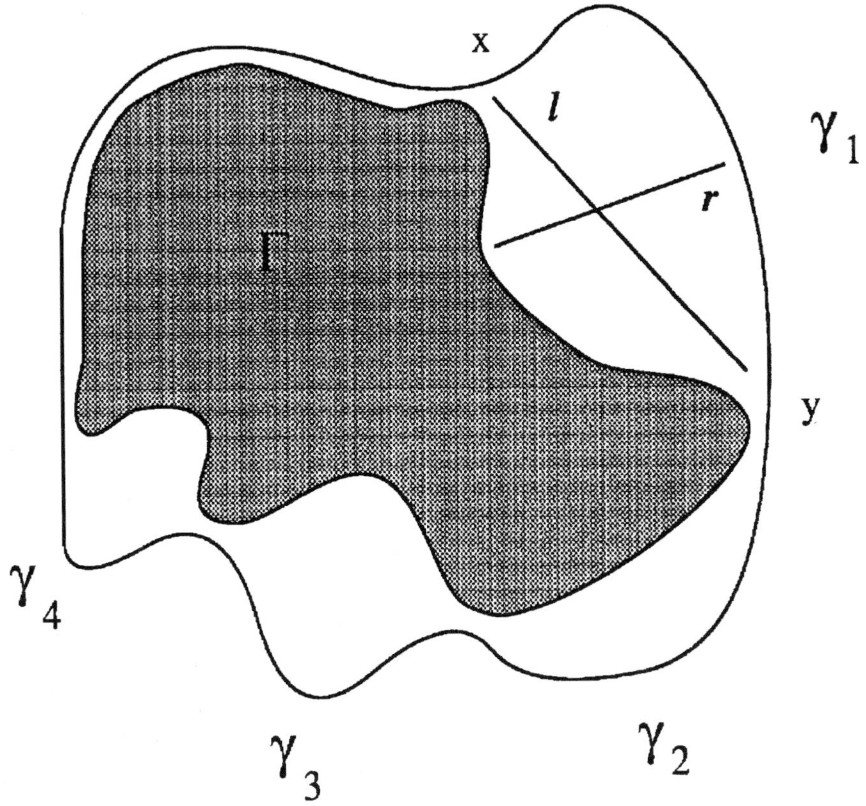

Fig. 7.6 Subsets of the lattice sites contributing to Δs. The external closed curve shows a wave. The hatched region is the next wave. $\gamma_1, \gamma_2, \gamma_3, \gamma_4$ are loops. The loop γ_1 of linear extent r is attached to the external contour Γ at points x and y separated by distance l.

l points of C with equal probability. Thus, we obtain the asymptotic joint distribution of loops

$$D_\gamma(l, r) \sim \frac{l^{1/4}}{r^3 l} \,. \tag{7.61}$$

The maximal extent of both r and l is of order R. The minimal extent of r is of order l whereas l is bounded from below by the lattice spacing. Integrating over r and l, we obtain the contribution to Δs from the single loop γ

$$\Delta_\gamma s \sim \int_1^R \int_l^R r^2 D_\gamma(l,r) dr dl \sim R^{1/4} . \tag{7.62}$$

The number of loops is proportional to the length of Γ, that is, $R^{5/4}$. Then, the total Δs is

$$\Delta s \sim R^{3/2} \sim s^{3/4} . \tag{7.63}$$

Comparing Eq.(7.63) with Eq.(7.55) and using Eq.(7.58) we finally get $\beta = 3/4$ and $\tau = 5/4$.

The obtained exponents complete, in principle, the list of known results for the two–dimensional sandpile model. However, it should be noticed that the derivation of the bulk exponent τ is based just on heuristic scaling arguments rather than on a rigorous analytical theory. The construction of a theory of that sort and a full mathematical description of the avalanche dynamics is a problem of great importance due to its applicability to a variety of physical processes exhibiting self–organized criticality.

8 Aggregation and Structure Formation in Reaction–Diffusion Processes in Chemical Systems

V. N. Kuzovkov

8.1 Introduction

8.1.1 Macroscopic Description

The most simple method of description of chemical reactions consists in the macroscopic description by applying the methods of formal chemical kinetics (Benson [30], Frank-Kamenetski [155], Ebeling [125], Polak [427]). However, the macroscopic method of description does not take into account structural properties of the reacting systems. The only parameter describing the process is assumed to consist in the concentration of the different reacting species, $n_i(t)$. It is assumed (Zhabotinsky [586]) that such a method of description is appropriate for a well–mixed homogeneous chemically reacting system.

In the considered case, the reaction may be described by a system of ordinary differential equations

$$\frac{dn_i(t)}{dt} = F_i(n_1, \ldots, n_f) \qquad i = 1, \ldots, f \tag{8.1}$$

where f denotes the number of components. The, in general, non–linear functions F_i are determined based on the mass action law (Benson [30], Zhabotinsky [586]).

In the most realistic models of physico–chemical processes, the process is described by applying the concept of mono– and bimolecular reactions (Zhabotinsky [586]). For such reactions, the functions F_i represent bilinear functions of the concentrations. For example, reactions of the type

$A + B \to C$ (or $A + B \to 0$) may be described by kinetic equations of the form

$$\frac{dn_\nu(t)}{dt} = -Kn_A(t)n_B(t), \quad \nu = A, B\,. \tag{8.2}$$

Here K is the rate constant of the bimolecular reaction.

A particular application of such type of reactions are processes of annihilation of pairs AB of Frenkel defects in solids described by the reaction $A + B \to 0$.

If the concentrations of both species have the same values, Eq.(8.2) has the simple solution

$$n(t) = \frac{n(0)}{1 + n(0)Kt}\,. \tag{8.3}$$

Asymptotically, for large times a decrease of the concentrations is found governed by the law $n(t) \propto t^{-\alpha}$ with the classical value of the exponent α equal to one.

From the point of view of the theory of selforganisation (Ebeling [125], Polak [427], Zhabotinsky [586], Nicolis and Prigogine [393]), elementary reactions are oversimplified. They cannnot give a description of processes of formation of spatio–temporal structures: the number of nonlinear equations and the degree of nonlinearity is not sufficient for such purposes. Popular models, used widely in the theory of selforganisation processes, consider, as a rule, sufficiently complex chains of reactions including trimolecular and higher–order reaction stages. In this way, a sufficient degree of nonlinearity is reached. On the other hand, as mentioned in Ref. [586], the use of trimolecular and higher order reactions may be considered as a compact description of a system of reactions consisting of mono– and bimolecular reactions, only, but with a large number of degrees of freedom.

The investigations of elementary reactions carried out in the last twenty years allow to correct the pessimistic judgement with respect to elementary reactions. In macroscopic homogeneous systems, structure formation processes have been observed for very different elementary reactions. The result of structural changes is quite significant: the value of the coefficient α describing the asymptotic behaviour of the chemical reaction is changed. Moreover, this parameter becomes dependent on the spatial dimensionality of the system. A strong analogy with critical phenomena, studied widely in physics, is found.

Before we are going over to an outline of the new results, we consider the prehistory of the question and different methods of description of the kinetics.

8.1.2 Mesoscopic Description

In the theory of selforganizing chemical systems, diffusion processes play a significant role (Ebeling [125], Polak [427], Zhabotinsky [586], Nicolis and Prigogine [393]). Applying a structureless method, diffusion processes are described by a generalization of the system of equations (8.1) taking into account spatial inhomogeneities. Instead of equations for the macroscopic concentrations $n_i(t)$, equations for the local concentrations $C_i(t)$ are formulated. In this way, the system of ordinary differential equations (8.1) is transformed into a system of partial differential equations

$$\frac{\partial C_i(\mathbf{r},t)}{\partial t} = D_i \nabla^2 C_i(\mathbf{r},t) + F_i(C_1,\dots,C_f) \,, \tag{8.4}$$

with a function $F_i(C_1, C_2, \dots, C_f)$ being of the same form as in Eq.(8.1). For the elementary reaction $A + B \to 0$, considered earlier, this system of equations gets the form

$$\frac{\partial C_A(\mathbf{r},t)}{\partial t} = D\nabla^2 C_A(\mathbf{r},t) - KC_A(\mathbf{r},t)C_B(\mathbf{r},t) \,, \tag{8.5}$$

$$\frac{\partial C_B(\mathbf{r},t)}{\partial t} = D\nabla^2 C_B(\mathbf{r},t) - KC_A(\mathbf{r},t)C_B(\mathbf{r},t) \,. \tag{8.6}$$

Hereby the simplifying assumption of equal values of the coefficients of diffusion $D_A = D_B$ was made.

From the point of view of the theory of selforganizing systems, the degree of non–linearity in Eqs.(8.5) and (8.6) in insufficient. Diffusion processes will lead in the course of time to a vanishing of the spatial inhomogeneities in the considered reaction volume. This process may be characterized by a diffusion length $\xi_D(t) = \sqrt{Dt}$. By applying this definition, the local excess of the concentration in a d–dimensional spatial system will occupy a volume $V_d = \xi_D(t)^d$ more or less homogeneously.

8.1.3 Microscopic Description

The foundations of the microscopic description of the kinetics of elementary reactions were developed in the pioneering paper by Smoluchowski (1917) [520]. Smoluchowski, for the first time, gave a microscopic interpretation of the phenomenological rate constant K of the chemical reaction for the case of a diffusion limited binary reactions. He developed the method of determination of this quantity. From a present day interpretation, Smoluchowski's method has to be considered as being of intuitive character. The solution of the many particle problem is replaced by the solution of a certain two–body problem, which allows an analytic solution. It is, however, of interest to note that more recent sufficiently fundamental analyses (Waite [562, 563], Leibfried [303]) have shown the principal validity of Smoluchowski's theory. In these more fundamental analyses, the kinetics of bimolecular reactions was analyzed from the point of view of a many–particle system. However, the solution was carried out, again, by some approximate method giving support to the intuitive solution as developed by Smoluchowski.

For a demonstration of the method, we use here the simplest model of an elementary reaction step, the *black sphere.* According to this model, pairs AB do not exist if the distance between the molecules is less than r_0 (r_0 is the reaction radius).

In the reaction $A + B \to 0$, controlled by diffusion processes, an elementary reaction step (pair annihilation) occurs if the distance between both components becomes equal or less than r_0. In this way, in the microscopic theory the first structure parameter appears having a dimension of a length.

According to Smoluchowski, the kinetic equation (8.2) retains its form. The *reaction rate constant* K becomes, however, in general a function of time $K = K(t)$. The value of K is determined by the correlational flux through the surface of the sphere of radius r_0,

$$K(t) = \gamma_d r_0^{d-1} \left|\mathbf{j}(r_0, t)\right| . \tag{8.7}$$

For the determination of this flux Smoluchowski's equation

$$\frac{\partial Y(r,t)}{\partial t} = \nabla \mathbf{j}(r,t), \quad \mathbf{j}(r,t) = D \nabla Y(r,t) \tag{8.8}$$

has to be solved.

Here γ_d is the surface area of a surface of unit radius ($\gamma_d = 2, 2\pi, 4\pi$ for $d = 1, 2, 3$, respectively). $D = D_A + D_B$ is the effective diffusion coefficient.

Eq.(8.8) is equivalent to the following model problem. Let us assume that in the origin of the coordinate system a trap is located. The trap is able to absorb an unlimited number of particles. At a moment of time $t = 0$, the space is filled with non–interacting particles. The mobility of the particles is determined by their diffusion coefficient D, the concentration in the initial moment of time equals $Y(r, 0) = 1$. Moreover, it is assumed that the concentration at large distances, $Y(\infty, t)$, is sustained at the same level. The problem is to find the flux via the surface enclosing the trap.

Taking into account the normalization condition for the concentration of the moving particles, the function $Y(r, t)$ is more natural to be interpreted as some correlation function. It describes the probability to find a particle at the distance r from the trap at time t.

Eq. (8.8) has to be solved, consequently, by applying the following boundary condition:

(a) Smoluchowski's condition

$$Y(r_0, t) = 0 \tag{8.9}$$

corresponding to the chosen reaction model;

(b) The boundary condition

$$\lim_{r \to \infty} Y(r, t) = 1 \tag{8.10}$$

representing the natural condition of decreasing correlation for large distances.

It is easy to prove that Eq.(8.8) describes some effective two–body problem. The connection of this problem with the original problem is not so easy to find out. One may say that Smoluchowski's many–body problem is replaced by an ensemble of two–body problems allowing an analytic solution.

A characteristic property of Smoluchowski's equation is the existence of a stationary solution for the case of three spatial dimensions $d = 3$. This solution has got the form

$$Y(r, \infty) = Y_0(r) = 1 - \frac{r_0}{r} \,. \tag{8.11}$$

From this solution the rate constant K for the diffusion limited case (Smoluchowski's constant) may be determined as

$$K(\infty) = K_0 = 4\pi D r_0 \,. \tag{8.12}$$

The solution shows that the reaction constant is proportional to the diffusion coefficient D of the particles and to a length parameter, the radius of the reaction r_0. In this way, Smoluchowski's method gives not only the verification of the existence of a reaction constant K (at least, asymptotically, for large times). It allows, moreover, a determination of this constant for simple models of chemical reactions.

The existence of a stationary solution of Eq.(8.11) indicates that in three spatial dimensions the diffusion length $\xi_D = \sqrt{Dt}$, introduced earlier, does not play any role in the asymptotic region. However, this length scale is of importance in the intermediate stage, when

$$K(t) = K_0 \left(1 + \frac{r_0}{\sqrt{\pi}\xi_D(t)} \right) \tag{8.13}$$

holds.

If systems with lower spatial dimensionality are considered, then the situation becomes quite different. Stationary solutions of the diffusion equation do not exist any more. In the two–dimensional case ($d = 2$) (Goesele [175], Kotomin and Kuzovkov [263]), we may write asymptotically (for large times)

$$K(t) \approx \frac{2\pi D}{\ln\left(c\xi_D(t)/r_0\right)} \,. \tag{8.14}$$

Here the parameter c is a constant. In the one–dimensional case ($d = 1$), we have

$$K(t) = \frac{2D}{\sqrt{\pi}\xi_D(t)} \,. \tag{8.15}$$

In other words, for large values of time the rate constant becomes equal to zero ($K(\infty) = 0$). It follows that the description of the kinetics of chemical reactions, in terms of the formal kinetics discussed in section 8.1.1, is not verified by the microscopic theory.

In the microscopic theory, the rate constant of the chemical reaction is interpreted for the first time as some structural characteristic of the system of particles, which is determined by the competition between typical length and time scales. For neutral particles interactions over large distances, due to the existence of some interaction potential, do not occur. In this case, two length scales exist. One of these length scales is determined by the radius of the reaction, r_0, it is a property of an elementary reaction step. In the three–dimensional case, it determines the spatial distribution of particles around the trap (Eq.(8.11)).

In reactions proceeding in lower spatial dimensions, in the course of evolution of the correlation function $Y(r,t)$ a new length scale $\xi_D(t)$ evolves, determined by diffusion processes. In these cases, the correlation function equals zero ($Y(r,t) \cong 0$) in the vicinity of the trap (for $r < \xi_D(t)$). In this region of space, the trap effectively absorbed all particles, while beyond this region the particles do not feel the influence of the trap ($Y(r,t) \cong 1$ for $r > r_0$).

If one substitutes the expressions for $K(t)$ (Eqs.(8.13)–(8.15)) into Eq.(8.2), then for equal concentrations, again, an asymptotic expression for the decrease of the concentration with time of the form $n(t) \propto t^{-\alpha}$ is found. For spatial dimensions $d = 2$ and $d = 3$, the value of the exponent α coincides with the classical result $\alpha = 1$. The logarithmic decrease of $K(t)$ for $d = 2$ (Eq.(8.14)) is too slight to change its value. In the one–dimensional case, however, a discontinuous change of α is found. α becomes equal to $\alpha = 1/2$, it means that the reaction proceeds slowlier compared with the cases $d = 2$ and $d = 3$.

The values of the exponent α obtained in the sketched way can be denoted as classical ones, however, they are derived here in the framework of a microscopic theory.

8.1.4 Account of Interactions Between the Particles

The generalization of the outlined results to cases when interactions between the particles have to be taken into account is relatively easy to carry out (Debye [110], Waite [564], Eyring et al. [138], Gardiner [166]). In the expression for the density of correlational fluxes $\mathbf{j}(r,t)$ (Eq.(8.8)), an additional term has to be added accounting for the directed motion due to the interaction force

$$\mathbf{j}(r,t) = D\left(\nabla Y(r,t) + \beta Y(r,t)\nabla U_{AB}(r)\right) . \tag{8.16}$$

In this equation, the parameter β is defined via $\beta = 1/k_B T$, $U_{AB}(r)$ is the interaction potential between a pair of particles A and B.

If we restrict ourselves to the more simple three–dimensional case ($d = 3$), then a detailed analysis verifies the existence of a stationary value of the reaction rate constant, K_0. In analogy to the results of Smoluchowski, this quantity may be written in the form

$$K_0 = 4\pi D R_{eff} . \tag{8.17}$$

R_{eff} is the effective radius of the reaction. It is a function of the reaction radius, r_0, and of a new spatial length scale, R, reflecting the range of interaction between the particles. For Coulomb interactions, it can be identified with the so–called Onsager radius [110] $R = e^2/\varepsilon k_B T$. It is equal to the distance where the Coulomb energy $e^2/\varepsilon R$ equals the energy of thermal motion $k_B T$. If the inequality $R \ll r_0$ holds, we have to set $R_{eff} = r_0$, in the opposite case $R_{eff} = R$. In this way, the reaction rate constant remains a structural characteristic of the system. Asymptotically, it is determined by the largest of the length scales characterizing the reaction or the interaction between the particles.

In low–dimensional systems, in addition to the both mentioned finite length scales R and r_0, a third time–dependent diffusion length $\xi_D(t)$ appears, again.

In a critical reconsideration of Eq.(8.16), a number of limitations can be mentioned implicitely contained in Smoluchowski's method. The main limitation is that the many–body problem is reduced to an ensemble of two–body problems. For two–body problems it is a characteristic feature that the diffusion coefficients of the reacting species enter the theory only as a sum, i.e., as $D = D_A + D_B$. It is assumed, moreover, that for the description of the interactions it is sufficient to use a pair potential $U_{AB}(r)$. For the original many–body problem we have to expect that the kinetics may be dependent also on other parameters of the system, in particular, on the interaction potentials of particles of the same species $U_{AA}(r)$ and $U_{AB}(r)$, on the partial diffusion coefficients, for example, in form of a ratio $\kappa = D_A/D$. The way of incorporating of such additional factors is not straightforward once one remains in the framework of Smoluchowski's approach.

First attempts to go beyond the framework established by Smoluchowski [303, 562, 563] gave his method, one one side, a more sound basis. It was shown that Smoluchowski's equations are obtained from the correct relations, having the form of an infinite chain of coupled equations, as the result of some approximation. The approximation applied may be denoted as linearization, since Smoluchowski's equation are linear functions of the correlation functions $Y(r,t)$. The task to reformulate the mathematical method and the basic ideas of the macroscopic theory of chemical kinetics became particularly up–to–date after the publication of the fundamental papers by Zeldovich and Ovchinnikov [405, 580].

8.1.5 Fluctuational Kinetics

In these papers [405, 580] for the first time the attention was drawn to the necessity of an appropriate account of concentration fluctuations in the kinetic description of bimolecular processes. The method of formal chemical kinetics describes, as outlined above, homogeneous macroscopic systems by macroparameters, the concentrations of the different species, $n_i(t)$. The condition of macroscopic homogeneity, however, does not contradict the existence of a certain spectrum of fluctuations in the many–body system. It is known, for example, that ideal systems are characterized by a Poisson spectrum of fluctuations. If in an arbitrary volume V the average number of particles is given by $\langle N_i \rangle = n_i V$ then the dispersion of the number of particles is determined by

$$\left\langle (N_i - \langle N_i \rangle)^2 \right\rangle = \langle N_i \rangle \ . \tag{8.18}$$

This natural inhomogeneity in the distribution of particles in space may have significant consequences with respect to the kinetics of chemical reactions.

For an illustration we restrict ourselves here, again, to the case of a diffusion controlled reaction of the type $A + B \to 0$. If one assumes that the elementary step in the reaction proceeds sufficiently fast, then the rate of the reaction is determined by the diffusional transport of the particles to each other, i.e., by diffusion processes. It has been pointed out already that diffusion leads to a decrease of local inhomogeneities in the system. At a time t, the homogenization takes place in a part of space characterized by a length scale $\xi_D(t)$. At any arbitrary moment of time, the distribution of particles in the volume considered may be illustrated by dividing the whole volume into compartments with a linear size $\xi_D(t)$. Inside each of the compartments, the concentration may be considered as the same. In the course of homogenization, the reactions inside the compartments will be completed. As the result, inside one of such compartments, all particles of one of the considered species, A or B, disappear, if the number of one of the species at the beginning of the process was larger the other one.

It follows that for large times, some block structure should be formed in the system. In each of such blocks, particles only of one of the components exist. The survival of species in one of such blocks is determined hereby by the initial distribution of particles. Once such a stage is reached, the further reaction may occur only near the boundaries of the compartments. This restriction has to lead to an effective decrease of the reaction. It remains to give a quantitative estimate of the discussed effect.

The reactions in systems with such a block structure are determined by the size of the blocks, i.e., by the value of $\xi_D(t)$. Consequently, one has to expect that the details of the kinetics of the reaction, i.e., any microscopic parameters of the theory like the radius of the reaction, r_0, are not significant for the determination of the asymptotic law of the reaction. For estimates of the course of the reaction, therefore, mesoscopic methods may be applied.

We start with Eqs.(8.5) and (8.6) and assume that the macroscopic homogeneity is conserved. However, local changes in the concentration due to thermal fluctuations are allowed. For the average concentrations, we have

$$\overline{C_A(\mathbf{r},t)} = \overline{C_B(\mathbf{r},t)} = n(t) \ . \tag{8.19}$$

Further, as new variables we introduce the difference

$$q(\mathbf{r},t) = C_A(\mathbf{r},t) - C_B(\mathbf{r},t) \tag{8.20}$$

and the sum

$$s(\mathbf{r},t) = C_A(\mathbf{r},t) + C_B(\mathbf{r},t) \tag{8.21}$$

of the local concentrations. In terms of these new variables, the system of Eqs.(8.5) and (8.6) gets the following form:

(a) The equation for the difference of the concentrations

$$\frac{\partial q(\mathbf{r},t)}{\partial t} = D\nabla^2 q(\mathbf{r},t)\,, \tag{8.22}$$

is a linear diffusion type equation. The general solution of this equation may be expressed via the Green's function $G(\mathbf{r},\mathbf{r}',t)$ of Eq.(8.22) as

$$q(\mathbf{r},t) = \int G(\mathbf{r},\mathbf{r}';t) q_0(\mathbf{r}')d\mathbf{r}'\,. \tag{8.23}$$

(b) The equation for the sum of the concentrations $s(\mathbf{r},t)$ is non–linear, i.e.,

$$\frac{\partial s(\mathbf{r},t)}{\partial t} = D\nabla^2 s(\mathbf{r},t) + \frac{K}{2}\left(q(\mathbf{r},t)^2 - s(\mathbf{r},t)^2\right)\,. \tag{8.24}$$

Its solution is, in general, a complicated task.

However, in the limit of an infinitely fast reaction ($K \to \infty$) it is possible to obtain a solution of the degenerate equation in the form

$$s(\mathbf{r},t) = |q(\mathbf{r},t)|\,. \tag{8.25}$$

In this case, the non–linear problem is reduced to a linear problem of diffusion type, to the solution of Eq.(8.22).

The solution of Eq.(8.25) allows a simple interpretation. The whole volume of the chemical system is divided into different blocks, characterized by different values of the quantity $q(\mathbf{r},t)$. In the regions, where $q > 0$ holds, the relations $C_A = q$ and $C_B = 0$ hold. Vice versa, for $q < 0$ we have $C_A = 0$ and $C_B = q$. Moreover, since

$$C_A(\mathbf{r},t) = \frac{1}{2}\left(q(\mathbf{r},t) + |q(\mathbf{r},t)|\right)\,, \tag{8.26}$$

holds we have by averaging Eq.(8.26) with Eq.(8.19) and $\overline{q(\mathbf{r},t)} = 0$

$$n(t) = \frac{1}{2}\overline{|q(\mathbf{r},t)|}\,. \tag{8.27}$$

The Langevin–like source term

$$\delta\varphi(\mathbf{r},t) = \varphi_A(\mathbf{r},t) - \varphi_B(\mathbf{r},t) \tag{8.35}$$

obeys the condition $\langle\delta\varphi(\mathbf{r},t)\rangle = 0$, it is characterized by Poisson statistical properties

$$\big\langle\delta\varphi(\mathbf{r},t)\delta\varphi(\mathbf{r}',t')\big\rangle = 2p\delta(\mathbf{r}-\mathbf{r}')\delta(t-t') \ . \tag{8.36}$$

Here p is a measure of the rate of formation of new particles of types A or B per unit volume, $\delta(x)$ is Dirac's deltafunction.

An analysis of Eq.(8.34) with the initial condition $q(\mathbf{r},t) = 0$ (homogeneous system with absence of excess particles) shows that at arbitrary moments of time the fluctuational spectrum of the quantity $q(\mathbf{r},t)$ is not a Gaussian distribution (8.29). Consequently, if we switch on at some moment of time a source of particles, we get a totally different slowlier kinetic law of the reaction, not identical with Eq.(8.30).

Of principal interest is also the question concerning the law of growth of the concentration $n(t)$ for an unlimited in time process of creation of particles. It has been observed that the saturation of the concentration (a stationary state) is found only in three spatial dimensions. In lower dimensional cases, a stationary solution does not exist. A block structure is formed in the volume with respect to the distribution of reacting species. This block structure is characterized by a diffusion length $\xi_D(t)$, it corresponds to segregation of particles of different types. In two–dimensional systems, the concentration weakly increases with time according to a logarithmic law, in the one–dimensional case the deviation from a stationary behaviour is relatively insignificant and can be described by a power law.

As a certain peak of the development of the method, which is based on the application of the difference in the concentrations, the papers [524, 525, 526] may be considered. In these investigations, to equations (8.34) of the mesoscopic approach terms are added describing the directed motion in a potential. These terms are non–local in their nature. Therefore, in [524, 525, 526] a number of additional approximations has been made. It is assumed that the interaction of the particles is of pseudo–Coulomb type, i.e., that the interaction has got the form $U_{\lambda\mu}(r) = \pm e_\lambda e_\mu u(r)$. The *charges* have the values $|e_\lambda| = e$, the function $u(r)$ is short–ranged and integrable. Assuming as earlier that the diffusion length is the determining spatial length scale, in [524, 525, 526] a non–linear equation for $q(\mathbf{r}, t)$ is obtained in the form

$$\frac{\partial q(\mathbf{r},t)}{\partial t} = \nabla \left(D + w\,|q(\mathbf{r},t)|\right) \nabla q(\mathbf{r},t) + \delta\varphi(\mathbf{r},t)\,. \tag{8.37}$$

It follows that the effects of interaction of the particles are characterized by one parameter w, only. This parameter w is greater than zero for systems where particles of different type attract and particles of the same species repulse each other. An attraction of species of the same type is characterized by $w < 0$.

In [524] it was shown that for values of the parameter w less than zero Eq.(8.37) looses its meaning. In such cases, the solution of the equation has singularities, corresponding to a collaps, i.e., to an irreversible agglomeration of particles of the same type. In such a situation, it becomes incorrect to assume that the reactions are determined by the diffusion length, only. In contrast, for such situations, the details of the distribution of the particles at small distances become dominating. Such details may be described adequately by microscopic approaches, only.

For pseudo–Coulomb systems ($w > 0$), such problems do not occur explicitely. It has been shown that in reactions without creation of new particles the reaction law is identical with Eq.(8.30), it does not depend at all on the value of the parameter w. This result can be understood easily. In the considered reactions, similar to the case of reactions of neutral particles, a block structure is formed, characterized by some diffusion length. In each of the blocks, particles of only one kind are concentrated preferably. The interaction between these particles adds an insignificant structural detail: particles of the same type repulse and will not be found near to each other.

In reactions with creation of new particles, the interaction leads to some effect, however, in the one–dimensional case, only. Instead of a value $\alpha = 1/4$ in the power law for the macroscopic concentration $n(t) \propto t^{\alpha}$ a value $\alpha = 1/5$ is obtained [524].

8.2 Microscopic Theory of the Reaction $A + B \to 0$

8.2.1 Reduced Description of the Fluctuational Spectrum of the System

Parallel to the description of the fluctuational kinetics in mesoscopic terms, also the mathematical formalism of the microscopic theory has been developed further. In a series of papers [262, 274, 275, 276, 277, 278, 279, 280, 281], the basic principles were laid. A systematic outline of this approach is given also in the reviews [263, 282].

From a formal point of view, a chemical spatially homogeneous system is equivalent to other homogeneous many–particles systems like gases and liquids. All these systems may be described by the same sets of structural characteristics. It is hereby not essential that in chemical systems the number of particles may change and that non–equilibrium processes occur. Such factors determine only the form of the equations of the theory, the basic structural characteristics are conserved.

Many–particle systems may be characterized, in general, by a developed spectrum of fluctuations. The correct description of such spectra may require the determination of an infinite number of different functions. It is therefore reasonable to start with the most simple structural characteristics. These are the macroscopic concentrations, i.e., the densities of particles $n_\nu(t)$. The quantity $n_\nu(t)$ is a measure of the average distance between the particles.

As a next step in the description, pair correlation functions $F_{\lambda\mu}(r,t)$ may be introduced, known in the statistical physics of dense gases and liquids as radial distribution functions [16, 589]. The correlation functions allow a simple physical interpretation. Assume in the origin of the coordinate system a particle of the type λ is located. The expression

$$C_\mu^{(\lambda)}(r,t) = n_\mu F_{\lambda\mu}(r,t) \tag{8.38}$$

determines the average density of particles of species μ at time t at a distance r from the origin. For large distances, the role of the central particle should decrease (decrease of the correlation) and the quantity $C_\mu^{(\lambda)}(r,t)$ coincides with the macroscopic concentration $n_\mu(t)$. The boundary conditions of decrease of the correlation for the pair correlation function is given thus by

$$\lim_{r \to \infty} F_{\lambda\mu}(r, t) = 1 \,. \tag{8.39}$$

Note that in statistical physics commonly not the function $F_{\lambda\mu}(r, t)$ is denoted as the correlation function but the combination $F_{\lambda\mu}(r, t) - 1$. In chemical kinetics, however, it is more convenient to use the definition characterized by the asymptotic behavior as given by Eq.(8.39).

The knowledge of the pair correlation functions allows the determination of the dispersion of the number of particles in an arbitrary volume V [16, 589]. It can be shown [263] that for an average value of the number of particles of type ν in the volume V, given by $\langle N_\nu \rangle = n_\nu(t) V$, the relative dispersion is determined by

$$\frac{\left\langle (N_\nu - \langle N_\nu \rangle)^2 \right\rangle}{\langle N_\nu \rangle} = 1 + \frac{n_\nu(t)}{V} \int\int_V \left(F_{\nu\nu}(|\mathbf{r} - \mathbf{r}'| - 1\right) d\mathbf{r} d\mathbf{r}' \tag{8.40}$$

If correlations in the spatial distributions of the particles of component ν do not exist, then the relation $F_{\nu\nu}(r, t) = 1$ holds. In this case, the right hand side of Eq.(8.40) equals one and Eq.(8.40) is transformed into Eq.(8.18), i.e., a Poissonian spectrum of fluctuations is found. Obviously, any deviation of the correlation functions from the value one result in deviations of the spectrum of fluctuations and in a Non–Poissonian behavior.

In a next stage, triple and higher order correlation functions may be introduced. These functions do not allow such a simple physical interpretation. In principle, correlation functions of any order represent independent structural characteristics of a many–body system.

For many–particle systems of any kind it is found [16, 589] that the equations determining the structural characteristics consist of a coupled infinite chain of equations, the BBGKY equations. This is valid also for chemically reacting systems [263, 282]. The accurate equation for the macroscopic concentration $n_\nu(t)$ contains in the right hand side (via the reaction rate constants) the pair correlation functions. The pair correlation functions are determined via triple correlations etc. By this reason, the system of equations can be solved only involving certain approximations.

For the physics of many–particle systems, a reduced description of the spectrum of fluctuations is usually applied by cutting the infinite set of equations by some appropriate approximation [16, 589]. This is also the case in the description of the kinetics of bimolecular reactions. The reduced description is introduced by applying only the simplest structural characteristics like the macroscopic concentration $n_\nu(t)$ and the pair correlation functions $F_{\lambda\mu}(r,t)$ [263, 282]. Hereby Kirkwood's approximation is used [257], where the triple correlations are expressed via the pair correlation functions.

The equations of chemical kinetics represent, in this way, an analogon of the BGY–equations, well–known in the statistical physics of dense gases and liquids. Naturally, the equations of chemical kinetics differ in their form from the BGY–equations, they describe principally different many–particle systems. It is interesting to note that in statistical physics also a number of other integral equations are used. However, their extension to problems of chemical kinetics is a difficult task [263, 282].

In the case of an elementary bimolecular reaction $A+B \to 0$, the general expressions may be written in more detail. In the case of equal concentrations of the reacting species, the reaction is characterized by only one macroscopic parameter, $n_\nu(t) = n(t)$. The analysis may be carried out in a similar way also without assuming equality of the concentrations. There exist three independent pair correlation functions. For their specification we use the following abbreviations. The correlation function of the same species we denote by $X_\nu(r,t) = F_{\nu\nu}(r,t)$, the correlation function of different species by $Y(r,t) = F_{AB}(r,t) = F_{BA}(r,t)$.

The pair correlation functions give not a comprehensive but sufficiently accurate picture of the mutual position of the particles. For an illustration, we consider the spatial distribution and the pair correlation functions in the so–called tunneling recombination [491, 492]. Note that the region of applicability of the microscopic theory is not restricted to diffusion controlled reactions. Of significant interest are also the description of non–moving particles [318, 491, 492]. Tunneling recombination processes are characterized by the rate $\sigma(r)$ of recombination of pairs, AB, at a distance r of the pair. This function is given by $\sigma(r) = \sigma_0 \exp(-r/r_0)$, where σ_0 and r_0 are constants. A comprehensive analysis of the reaction $A + B \to 0$ showed [263, 282] that a number of properties of the reaction are not connected with the degree of mobility of the particles. This conclusion is valid also with respect to the block structure discussed earlier.

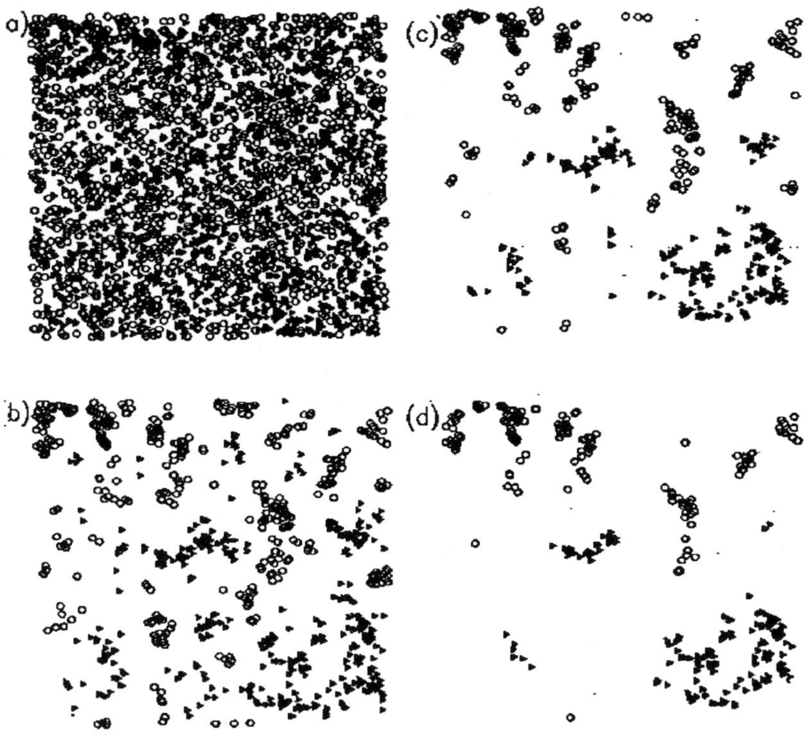

Fig. 8.1 Spatial distribution of A and B particles during a reaction process on a $1000 \cdot 1000$ square lattice (with periodic boundary conditions) after starting with 10^4 particles of each kind. The distributions are shown for *(a)* $t = 10$, *(b)* $t = 10^5$, *(c)* $t = 10^{10}$, and *(d)* $t = 10^{15}$.

On Fig. 8.1 the spatial distribution of particles is shown for reactions proceeding in two spatial dimensions [492]. The result was obtained by computer modelling of this process. The system was normalized in such a way that the relations $\sigma_0 = 1$ and $r_0 = 5$ hold. It can be seen that in the course of the reaction agglomerates of particles of the same species survive. At any arbitrary moment of time the system may be divided into sufficiently large compartments in such a way that inside one of such blocks only particles of one of the species are found. The linear sizes of the compartments grow with time. Similar agglomerates are found in computer simulations [540] for diffusion controlled reactions. In these cases, the distributions are more smooth due to diffusion processes.

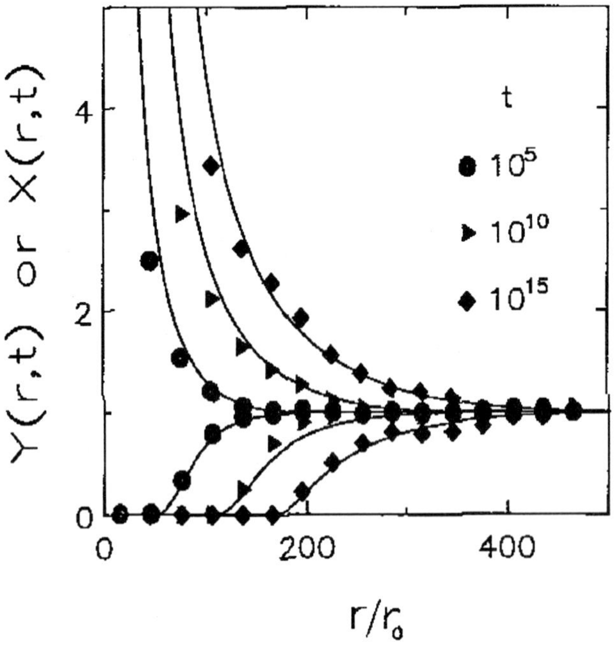

Fig. 8.2 Correlation functions for like (X) and unlike (Y) particles, calculated numerically with $d = 2$ (solid lines). The symbols mark the results from numerical simulations: each value is an average over 20 realization of the process, smoothed out over an interval $6r_0$.

On Fig. 8.2 the pair correlation functions are shown corresponding to the distribution of particles represented in Fig. 8.1. The results are compared with the results of a numerical solution of the equations of the microscopic theory [491, 492]. Since for large times the number of particles in the system decreases, averaging procedures were carried out over the different computer realizations and in spatial regions of the sizes $\Delta r = 6r_0$. The comparison of these two approaches (microscopic theory and computer simulation) shows that the reduced description of the fluctuational spectrum of the system does not lead to considerable mistakes in the determination of the correlation functions. The deviations in the curves describing the $n(t)$ kinetics are negligible as well [318, 491, 492]. In other words, for a bimolecular reaction $A + B \to 0$ the accuracy of the superposition approximation by Kirkwood [257] is sufficiently high. This conclusion holds also for diffusion controlled reactions [263, 282].

In the behavior of the correlation functions, a simple dependence can be obtained. The existence of a length scale $\xi(t)$ is found. This quantity may be denoted as the correlation length of the system. For $r < \xi(t)$, the correlation length of particles of different types $Y(r,t)$ is nearly equal to zero. Pairs AB does not coexist practically at such distances. At the same scale, the correlation function of particles of the same type $X(r,t)$ significantly exceeds their asymptotic value. This property is connected with the agglomeration of species of the same type. Note the singularity in the value of the correlation function $X(r,t)$ for small distances. It is an indication of the high probability to observe a particle A (or B) in the immediate vicinity of a particle of the same type. The aggregates of the particles in the absence of diffusional smoothening are very compact.

As shown in [491, 492], in the case of tunneling recombination of immobile particles the correlation length is equal to the tunneling length $\xi(t) = r_0 \ln(\sigma_0 t)$. It grows unlimited with time. The law of decrease of the concentration may be written in the form

$$n(t) \propto \xi(t)^{-d/2} , \tag{8.41}$$

similar to Eq.(8.31) and with the same value of the exponent α equal to $\alpha = d/2$.

8.2.2 Kinetic Equations for Diffusion Controlled Reactions

The system of kinetic equations for the description of diffusion controlled reactions of the type $A+B \to 0$ for the case of equal concentrations $n_\nu(t) = n(t)$ has the form

$$\frac{dn(t)}{dt} = -K(t)n(t)^2 . \tag{8.42}$$

The equation for the macroscopic concentrations has to be supplemented by the determination of the rate constant of the reaction $K(t)$, i.e.,

$$K(t) = \gamma_d r_0^{d-1} \left|\mathbf{j}(r_0,t)\right| . \tag{8.43}$$

For the calculation of the rate constant one has to add a system of equations for the pair correlation functions

$$\frac{\partial Y(r,t)}{\partial t} = \nabla \mathbf{j}(r,t) - 2n(t)K(t)Y(r,t)J_d(X) \;, \tag{8.44}$$

$$\mathbf{j}(r,t) = D\nabla Y(r,t) \;, \tag{8.45}$$

$$\frac{\partial X_\nu(r,t)}{\partial t} = \nabla \mathbf{j}_\nu(r,t) - 2n(t)K(t)X_\nu(r,t)J_d(Y) \;, \tag{8.46}$$

$$\mathbf{j}_\nu(r,t) = 2D_\nu \nabla X_\nu(r,t) \;. \tag{8.47}$$

Here the notations $D = D_A + D_B$, $X(r,t) = (X_A(r,t) + X_B(r,t))/2$ are used. The functionals $J_d(Z)$ will be given after introduction of dimensionless variables. These functionals occur only in terms arising from the superposition approximation due to Kirkwood [257].

The system of equations for the dynamics of correlations has to be supplemented by the boundary conditions. In addition to the natural condition of weakening of the correlations at infinite distances (8.39), Smoluchowski's condition (8.9) has to be fulfilled. This condition is connected with an additional one, the absence of pairs of particles at distances less than r_0 (i.e., $Y(r < r_0) = 0$). These conditions are consequences of the black sphere model [263, 282] applied in the derivation of the basic equations. The particles are considered here as point masses, their finite size is more naturally accounted for by the form of the interaction potential. In this way, we have as an additional condition the absence of a correlational flux at the origin of the system of coordinates

$$\lim_{r\to 0} \gamma_d r^{d-1} \left|\mathbf{j}_\nu(r,t)\right| = 0 \;. \tag{8.48}$$

The equations describing the dynamics of correlations are solved with the standard assumption that in the initial state correlations are equal to zero

$$X_\nu(r,0) = Y(r,0) = 1 \;. \tag{8.49}$$

In other words, only those spatial correlations of the particles are investigated, which occur in the course of the chemical reaction.

If one compares the equations developed here with Smoluchowski's expressions, then it is easy to establish the interrelations between both theoretical approaches. Smoluchowski's equation follows as a consequence by a linearization of the basic equations for the dynamics of the correlations, i.e., by neglection of non–linear terms in the equations containing the functionals of the type $J_d(Z)$. In the course of linearization, superficial equations for the correlation functions of particles of the same type may be omitted. Their solution does not have any influence on the value of the reaction rate constant. Moreover, the initial condition (8.49) for the function $X_\nu(r,t)$ coincides with the stationary solution of the linearized equations. Equation (8.44) for the correlation function of particles of different types goes over immediately into Eq.(8.8).

It has to be noted that the procedure applied for the linearization of the dynamics of the correlations cannot be given any additional support by ideas like the smallness of the concentrations (for dilute systems) since, from a mathematical point of view, the non–linear terms in the equations for the dynamics of the correlations represent either attractors or repellers of the correlations.

For diffusion controlled reactions it is reasonable to go over to new dimensionless variables, applying two natural parameters: the radius of the reaction r_0 and the effective diffusion coefficient $D = D_A + D_B$. The reduced variables are given by

$$r' = r/r_0 \,, \quad t' = Dt/r_0^2 \,, \tag{8.50}$$

$$n'(t') = \gamma_d r_0^d n(t) \,, \quad D'_\nu = D_\nu/D \,. \tag{8.51}$$

In the new variables the equations for the macroscopic concentrations are not changed but the expression for the rate constant of the chemical reaction is simplified

$$\frac{dn(t)}{dt} = -K(t)n(t)^2 \,, \quad K(t) = \left. \frac{\partial Y(r,t)}{\partial r} \right|_{r=1} \,. \tag{8.52}$$

The equations for the dynamics of the correlations (8.44)–(8.47) also conserve the form. Some of the parameters have to be replaced by

$$D = 1 \,, \quad D_A = \kappa \,, \quad D_B = 1 - \kappa \,, \tag{8.53}$$

where $\kappa = D_A/D$ is a asymmetry parameter obeying the condition $(0 \leq \kappa \leq 1)$. For an illustration, we consider only two limiting cases of symmetric $(\kappa = 1/2; D_A = D_B)$ and asymmetric $(\kappa = 0, D_A = 0)$ diffusion.

The functionals $J_d(Z)$ are determined by the following equations, they depend on the the dimensionality of the space where the reaction takes place [263],[282]

$$J_1(Z) = \frac{1}{2}\left(Z(r+1,t) + Z(|r-1|,t)\right) - 1 , \tag{8.54}$$

$$J_2(Z) = \frac{1}{\pi}\int_{-1}^{1} \frac{(Z(r',t)-1)\,d\zeta}{\sqrt{1-\zeta^2}}, \quad r' = \sqrt{1+r^2+2r\zeta} , \tag{8.55}$$

$$J_3(Z) = \frac{1}{2r}\int_{|r-1|}^{r+1} \left(Z(r',t)-1\right) r'dr' . \tag{8.56}$$

It is easy to verify that the correlation functions in the functionals $J_d(Z)$ are found only in form of differences $Z(r,t) - 1$. These differences characterize the deviation of the fluctuational spectrum of the system from a Poisson law. Sources of correlations are found in the system of kinetic equations (8.44)–(8.47), if such deviations from the Poisson spectrum (8.49) occur. The mechanism of formation of Non–Poisson behaviour in the bimolecular reaction $A + B \rightarrow 0$, considered here, is contained in the dynamics of the correlations and not artificially introduced from outside. This is due to the supplemental condition of absence of pairs, AB, at distances less than the radius of the reaction (*black sphere model*). In new dimensionless variables, this equation gets the form $Y(r \leq 1, t) = 0$. In this way, the functional $J_d(Y)$ becomes different from zero (negative) already at the initial moment of time. Since this functional is contained in the equation for the correlation function of particles of the same type (8.46), a source of Non–Poisson fluctuations for them occurs. Non–Poissonian fluctuations in the location of particles of the same type are the origin of sources of correlations in eq (8.44).

In this way, elementary correlations in the system of particles occur immediately in the course of the reaction, they are transformed later on into more complicated correlations. As the result a spatial distribution of particles of a given type is developed.

The system of integro–differential equations (8.44), (8.46) and (8.52) is characterized by a high degree of non–linearity and complexity, comparable with the systems commonly analyzed in Synergetics. We will see that the solution of the equations governing the dynamics of the correlations for a number of systems has the form of travelling waves. Since the correlation functions obey radial symmetry, the considered waves have also the property of radial symmetry. They describe processes of segregation in the distribution of the reacting species like the formation of block structures.

8.2.3 Kinetics of Diffusion Controlled Reactions

A systematic investigation of the kinetics of diffusion controlled reactions of the type $A + B \to 0$ was carried out in [264]. After the introduction of dimensionless variables and the proper choice of the initial conditions (8.49) for the correlation functions, the solution of the system of kinetic equations depends only on the dimensionality of space, d , the asymmetry coefficient of diffusion, κ, and the initial concentration, $n(0)$. As shown [263, 282], only the first two of these parameters determine the asymptotics of the reaction. The third parameter, $n(0)$, determines only the time scale for the transition to the new fluctuational dependencies.

On Figs. 8.3–8.4, the $n(t)$–curves are shown for reactions proceeding in one, two and three spatial dimensions, respectively. The group of curves 1–3 corresponds to three different values of the initial concentration, $n(0) = 1$, $n(0) = 0.1$ and $n(0) = 0.01$, respectively. The full and dashed curves refer to the solution of the complete system of equations (8.44), (8.46) and (8.52). The solid curve refers to the case of asymmetric diffusion ($D_A = 0$), the dashed curve to the symmetric case $D_A = D_B$. For comparison, by a dotted curve the results as obtained by Smoluchowski's method Eqs.(8.8) and (8.52) are given.

As evident from the figures, the range of applicability of Smoluchowski's method is determined by the value of the initial concentration, $n(0)$. With decreasing initial concentration, the time interval, where this method is applicable, increases. For large values of the initial concentration, the range of applicability may tend to zero.

For large times (in a logarithmic scale), the kinetic curves, calculated by Smoluchowski's method and based on Kirkwood's superposition approximation, differ. They predict, thus, different asymptotic power laws of the reaction. The change of the slope is particularly expressed in low–dimensional systems.

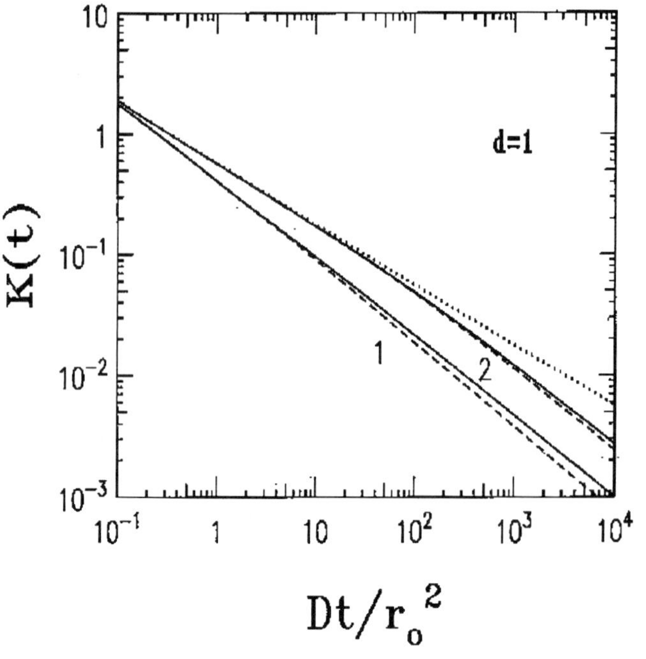

Fig. 8.7 Kinetics of the diffusion-controlled recombination $K(t)$ for $d = 1$ [264]. The solid curve shows the superposition approximation for $D_A = 0$, the dashed line shows the superposition approximation for $D_A = D_B$, and the dotted line shows the linear approximation (Smoluchowski). The initial reactant concentrations are $n(0) = 1.0, 0.1$ (curves 1 to 2, respectively).

different values of the spatial dimensionality, d, the initial concentration, $n(0)$, and the coefficient of diffusional asymmetry, κ. The notations are the same as in Figs. 8.5 – 8.6. It can be seen that in low dimensional systems the asymmetry of the diffusion is not important, the curves with different values of κ have similar values of the slope. In the three–dimensional case, the deviations are sufficiently large. Such behavior corresponds to the existence of different asymptotic values of α for symmetric and asymmetric diffusion, respectively.

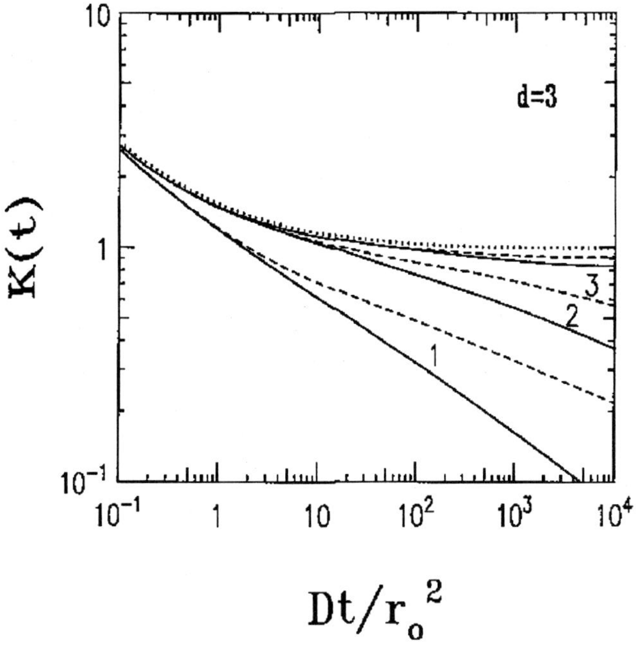

Fig. 8.8 Kinetics of the diffusion-controlled recombination $K(t)$ for $d = 3$ [264]. The solid curve shows the superposition approximation for $D_A = 0$, the dashed line shows the superposition approximation for $D_A = D_B$, and the dotted line shows the linear approximation (Smoluchowski). The initial reactant concentrations are $n(0) = 1.0, 0.1, 0.01$ (curves 1 to 3, respectively).

8.2.4 Block Structure in the Distribution of the Reacting Species

The change of the kinetic curves is a consequence of certain structural changes in the system. These structural changes may be followed by analyzing the characteristics of the pair correlation functions (see Figs. 8.9–8.10).

The behavior of the pair correlation functions shows certain similarities for different spatial dimensions. Let us consider, for example, Fig. 8.9 corresponding to the case $d = 1$. The correlation function of particles of different types $Y(r,t)$ has the characteristic shape of a smoothened step function. We have $Y(r,t) \cong 0$ for $r < \xi(t)$ and $Y(r,t) \cong 1$ for $r > \xi(t)$. Here $\xi(t)$ is some length scale increasing with time. At such length scales, particles of different types do not coexist practically. For $r > \xi(t)$, particles AB are distributed

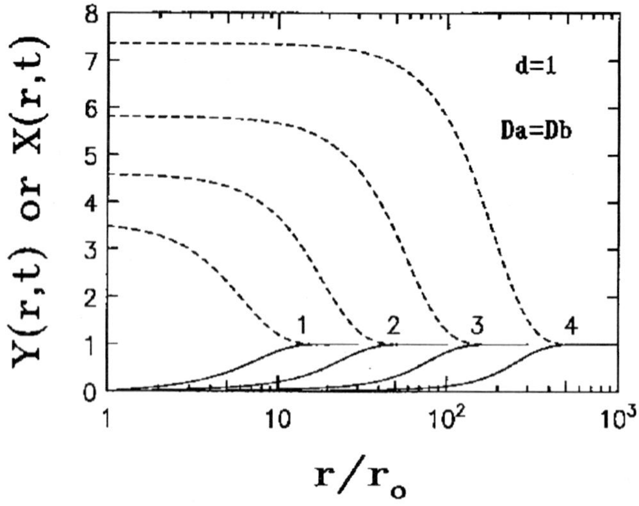

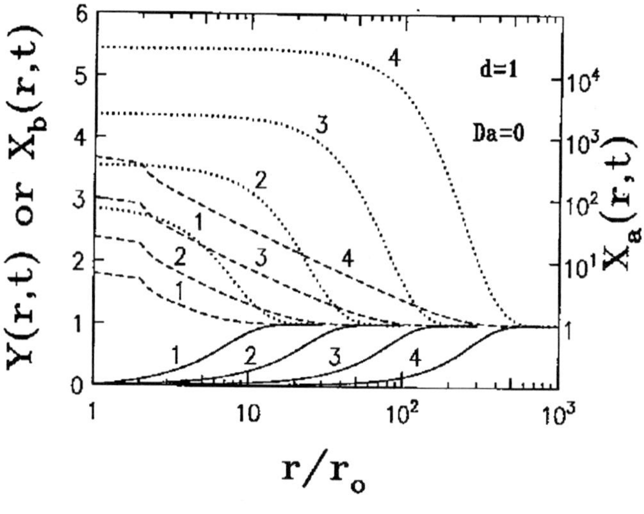

Fig. 8.9 Correlation functions of the same kind, $X_\nu(r,t)$, with $\nu = A, B$ (broken or dotted curves), and of different kind, $Y(r,t)$ (solid curves), versus the (logarithmically plotted) particle distance r for $d = 1$ and initial concentration $n(0) = 1$. Figure *(a)* shows the case $D_A = D_B$; in figure *(b)* ($D_A = 0$) the right-hand scale belong to the broken curves. The curves labelled 1-4 refer to $t = 10^1, 10^2, 10^3$, and 10^4, respectively.

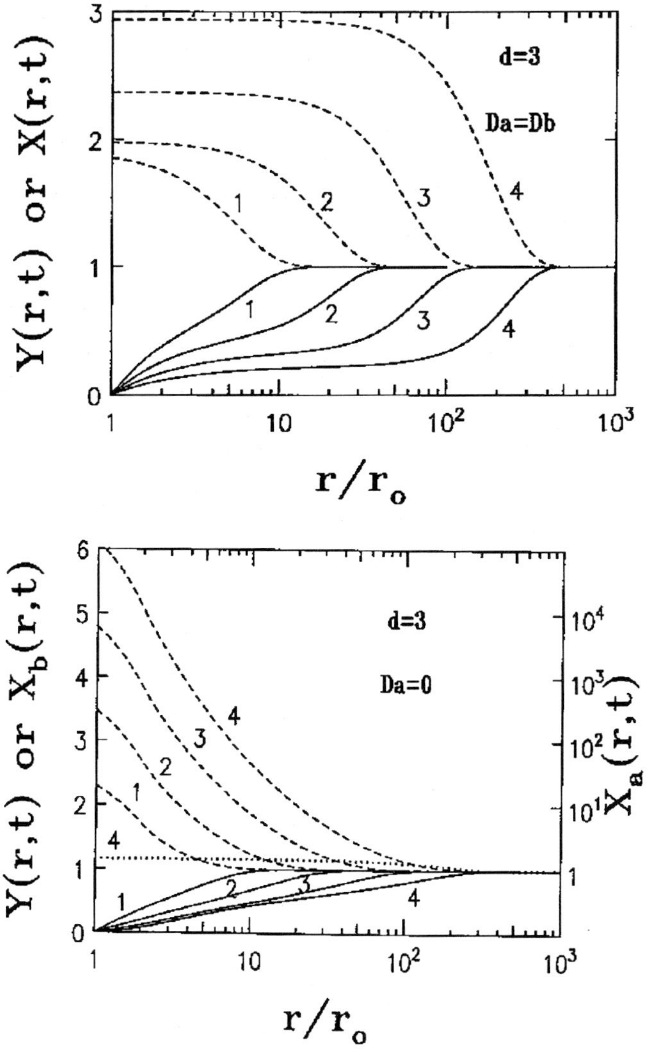

Fig. 8.10 Correlation functions of the same kind, $X_\nu(r,t)$, with $\nu = A, B$ (broken or dotted curves), and of different kind, $Y(r,t)$ (solid curves), versus the (logarithmically plotted) particle distance r for $d = 1$ and an initial concentration $n(0) = 1$. Figure *(a)* shows the case $D_A = D_B$; in figure *(b)* ($D_A = 0$) the right-hand scale belong to the broken curves. The curves labelled 1-4 refer to $t = 10^1, 10^2, 10^3$, and 10^4, respectively.

chaotically, there are no correlations between them. On the figures the r –axis is given in a logarithmic scale. It can be seen that the steps in the curves $Y(r,t)$ move by an order in r with a change of time of two orders of magnitude. It follows that $\xi(t)$ may be identified with the diffusion length $\xi_D(t)$, determined by the square root of t. In the same scale $\xi(t) = \xi_D(t)$, the correlation functions of the mobile particles (dashed curves on Figs. *(a)* and dotted curves on Figs. *(b)*) have the characteristic shape of a plateau ($X(r,t) \cong$ const) and tend rapidly to the asymptotic value for $r > \xi(t)$. It follows that the probability of detection of particles of the same type for distances $r < \xi(t)$ from the central one is considerably larger than the asymptotic value. It is constant at the same scale. In other words, the mobile particles are distributed at length scales $\xi(t)$ sufficiently smoothly. This is different from the results shown on Fig. 8.2 where the correlation functions of particles of the same type exhibited some kind of singularity at small distances. Obviously, the difference is caused by diffusional smoothening.

These features correspond to the formation of block–type structures as discussed above. In each of the blocks of size $\xi(t)$ particles of only one kind dominate ($Y(r,t) \cong 0$). The mobile particles cover the volume more or less regularly. Which of the species dominates in a given block is due to chance. In the vicinity, particles of both types may be found with the same probability. With increasing $r > \xi(t)$ the correlation function tends rapidly to the asymptotic value.

For the spatial distribution of immobile particles (the case of asymmetric diffusion; Figs. *(b)*) a different type of dependencies is found. For an illustration, the respective correlation functions are given by dashed curves on Figs. *(b)* in logarithmic coordinates. The maximum value of the correlation function of the immobile particles, $X_A(0,t)$, exceeds by some orders of magnitude the respective value of the mobile particles, $X_B(0,t)$. For an illustration and qualitative discussion, it may be assumed that the immobile particles are grouped into superparticles. Hereby the volume $V_d = \xi_D(t)^d$ is occupied by only one of such superparticles. The number of single particles, N, in such a superparticle is determined by $N = n(t)V_d = n(t)\xi_D(t)^d$. Obviously, the formation of superparticles has practically no influence on the kinetics of the reaction in low dimensional systems.

Results for the reactions proceeding in two spatial dimensions practically do not differ from the one–dimensional results. The correlation function $Y(r,t)$ also has the form of a kink, however, the step is more smoothened, in particular, in the case of asymmetric diffusion. Moreover, in the case of asymmetric diffusion the growth of the maximum of the correlation function of the mobile particles is inhibited considerably.

Even more inhibited are the fluctuational effects in the three–dimensional case, see Fig. 8.10. The kink in the function $Y(r, t)$ is smoothened already in the case of symmetric diffusion. Even for times equal to $t = 10^4$ a background value $Y(r, t) \cong 0.1$ is observed for $r < \xi_D(t)$. In the case of asymmetric diffusion this background value is even larger. This result implies that for the considered time interval segregation is inhibited. If the system is divided, again, into compartments, in each of them in addition to the main type of species also a backgroud concentration of species of the other type is found. For asymmetric diffusion, the correlation function of the mobile particles for $r < \xi_D(t)$ only slightly exceeds their asymptotic value. In terms of the model of superparticles, we can give the following interpretation: superparticles of the immobile species are formed on the background of a smoothened distribution of the mobile particles.

Part II

Selected Applications

9 Theoretical Determination of the Number of Clusters Formed in Nucleation–Growth Processes

J. Schmelzer, V. V. Slezov

9.1 Introduction

Classical nucleation theory, as worked out first by Stranski, Kaischew [228], Becker, Döring [26], Volmer [557], Frenkel [158], Zeldovich [579] and others and discussed in detail in chapter 2, is dealing primarily with the determination of the steady–state nucleation rate, J, i.e., the estimation of the number of supercritical clusters formed per unit time interval in a unit volume of a thermodynamically metastable system.

In the further development of the theory, non–steady state effects have been intensively studied connected with the finite time required for the system to reach steady–state conditions (for an overview see, for example, [187]).

There exists, however, also another reason why the determination of the steady–state nucleation rate cannot give a comprehensive information concerning the course of nucleation processes. In most practical applications, a steady–state can be established in a system only for a limited period of time. This effect is due to a depletion of the state of the system, i.e., the decrease of the number of particles which can be incorporated into the new phase (see, e.g., [39, 180, 471]).

Depletion effects lead to a non–linear feedback and a high degree of complexity of the problem to be analyzed. In particular, as the result of such depletion effects only a finite number of clusters develops in the system.

These conclusions are illustrated in Fig. 9.1. Similarly to Fig. 2.2, the evolution of the cluster size distribution $N(j,t)$ is shown here for different moments of time. However, in contrast to the calculations leading to Fig. 2.2, this time the realistic condition of conservation of the number of particles is taken into account.

The determination of the number of clusters formed in a thermodynamically metastable system in dependence on the initial supersaturation and the average size of the clusters observed initially in nucleation experiments is a task beyond the scope of classical nucleation theory and most of its modifications and generalizations. The knowledge of these characteristics is, however, of great technological importance allowing to vary the dispersity of the newly evolving phase in the ambient phase and in this way the properties of the respective materials.

In addition to classical applications like in materials science and engineering (see, e.g., [97, 225]), the knowledge of the dependence of the number of clusters (or bubbles) formed in nucleation–growth experiments on the initial supersaturation may shed some light also on such problems like the understanding of fragment or cluster size distributions in molecular or nuclear clustering processes (see, e.g., [378, 482, 518] and chapter 10) or certain stages in the evolution of the early universe (cf. [230]).

The determination of these characteristics – the number of clusters formed in nucleation–growth experiments and their average size in dependence on the initial supersaturation in the system – is the aim of the present chapter. Hereby we follow with modifications the method as outlined in a paper by Slezov, Schmelzer, Tkatch [511] for kinetic limited nucleation. However, here we will concentrate the attention to nucleation–growth processes governed by diffusion limited growth.

The results of the analysis allow to compare the outcome of nucleation – growth processes for the both considered modes of segregation giving some insight into the problem how sensitive the number of clusters formed depends on the considered mode of growth. Moreover, for a number of applications – like segregation in glass–forming melts [187], vacamcy cluster evolution under irradiation [565] or bubble formation in viscous polymeric liquids (cf. [316, 317]) – nucleation is believed to be more adequately described if the diffusional mechanism of growth is assumed. Since for diffusion limited growth the transport of segregating particles proceeds effectively over distances of the order R, where R denotes the size of the cluster, both modes of growth may be expected to lead to widely equivalent results for the immediate nucleation stage.

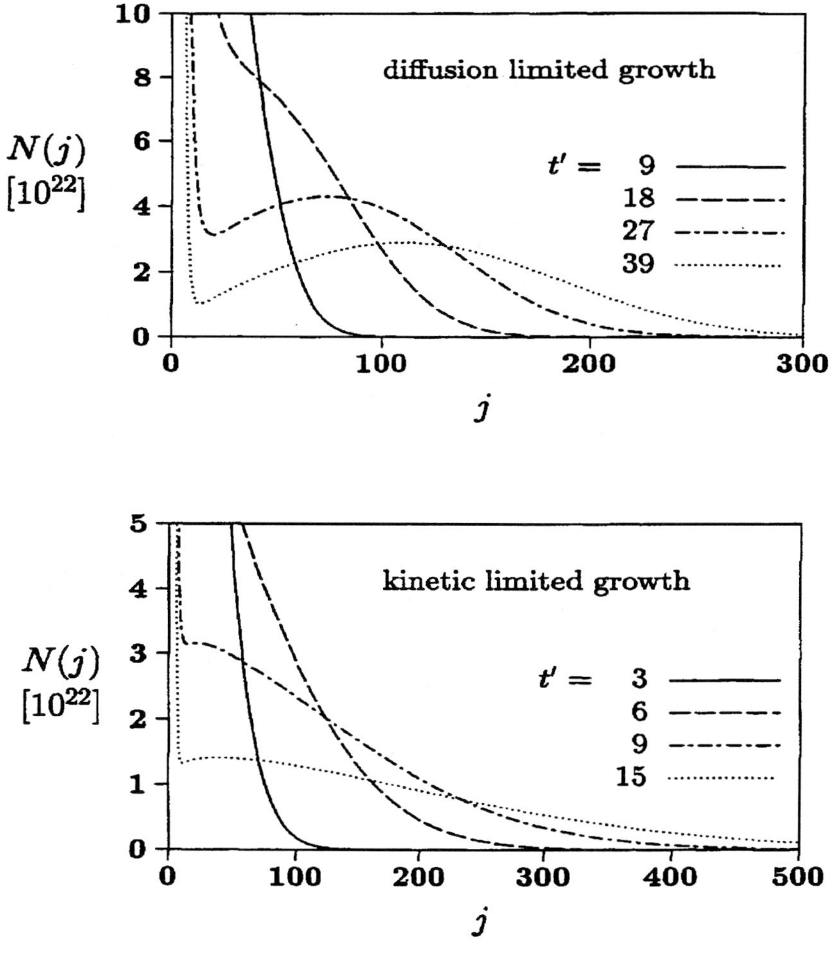

Fig. 9.1 Different stages of the time evolution of the cluster size distribution, $N(j,t)$, as obtained by the numerical solution of the set of equations (2.2) and (2.3) with the kinetic coefficients (2.79) and (2.80). In the calculations, it is supposed that the total number of monomers (free particles and particles bound to clusters of different sizes) is conserved. In contrast to the results shown in Fig. 2.2, a steady-state is not approached but an evolution process is found resulting in the formation of a finite number of clusters in the system. For further details of the calculations, the values of the parameters and the definition of the reduced time scale t', see, e.g., Schmelzer et al. [322, 483].

By the subscript β the chemical potential μ per building unit in the ambient phase is specified, while α refers to the respective value in the newly evolving phase (both taken at a temperature T and a pressure p). σ is the surface tension or interfacial specific energy.

In terms of the number of monomers in the cluster, Eq.(9.7) can be reformulated to give

$$\Delta G(j) = -j\Delta\mu + \alpha_2 j^{2/3}\,, \qquad \alpha_2 = 4\pi\sigma\left(\frac{3\omega_s}{4\pi}\right)^{2/3}\,. \tag{9.8}$$

With these notations, the critical cluster size j_c, corresponding to a maximum of the Gibbs free energy, is determined by

$$\left.\frac{\partial\Delta G(j)}{\partial j}\right|_{j=j_c} = -\Delta\mu + \frac{2}{3}\alpha_2 j_c^{-1/3} = 0\,, \qquad j_c^{1/3} = \frac{2\alpha_2}{3\Delta\mu}\,. \tag{9.9}$$

Moreover, for the second derivative of ΔG we obtain

$$\left.\frac{\partial^2\Delta G(j)}{\partial j^2}\right|_{j=j_c} = -\frac{2}{9}\alpha_2 j_c^{-4/3} = -\left(\frac{9}{8}\right)\left[\frac{(\Delta\mu)^4}{\alpha_2^3}\right]\,. \tag{9.10}$$

These expressions are needed for the subsequent derivations. In particular, the value of G at $j = j_c$ is given by

$$\Delta G(n_c) = \frac{1}{3}\alpha_2 j_c^{2/3} = \left(\frac{4}{27}\right)\frac{\alpha_2^3}{(\Delta\mu)^2}\,. \tag{9.11}$$

Moreover, the range of j–values δj in the vicinity of the extremum of ΔG, where the inequality $\Delta G(j_c) - \Delta G(j) \leq k_B T$ holds, and thermal fluctuations determine the motion of the clusters in cluster size space, can be written as

$$|j - j_c| \leq \delta j = \frac{1}{\sqrt{-\dfrac{1}{2k_B T}\left(\left.\dfrac{\partial^2\Delta G}{\partial j^2}\right|_{j=j_c}\right)}}\,. \tag{9.12}$$

As mentioned in chapter 2, for simplicity of the notations we consider segregation in a perfect solution. In this case, we have

$$\Delta\mu = k_B T \ln\left(\frac{c}{c_\infty}\right) . \tag{9.13}$$

c_∞ is the concentration of the segregating particles in the ambient phase in equilibrium with the newly evolving phase at a planar interface at the given values of pressure p and temperature T.

The time interval τ_1 after which a steady state is established in the range $0 \leq j \leq g$, $g = j_c + \delta j$ can be written according to Eq.(2.57) generally as

$$\tau_1 \cong \frac{(\delta j)^2}{w^{(+)}(j_c)} . \tag{9.14}$$

With Eqs.(9.5), (9.9), (9.10) and (9.12) we obtain

$$\tau_1 \cong \frac{6}{(\alpha_1 c_0)}\left[\frac{j_c(c_0)}{\beta}\right] , \qquad \beta = \frac{2\alpha_2}{3k_B T} . \tag{9.15}$$

After the time τ_1, steady–state conditions are established in the range of cluster sizes $j \leq g$. The expression for the steady–state nucleation rate J reads then (see, again, chapter 2)

$$J = c w^{(+)}(j_c)\Gamma_{(z)} \exp\left(-\frac{\Delta G(j_c)}{k_B T}\right) , \tag{9.16}$$

$$\Gamma_{(z)} = \sqrt{-\frac{1}{2\pi k_B T}\left(\left.\frac{\partial^2 \Delta G}{\partial j^2}\right|_{j=j_c}\right)} . \tag{9.17}$$

For diffusion limited growth Eq.(9.18) is obtained as a special case (cf. Eq.(2.84)), i. e.,

$$J = \left(\frac{4\pi}{3}\right)\left(\frac{D c^2 a_s}{j_c^{1/3}}\right)\sqrt{\frac{\alpha_2}{\pi k_B T}} \exp\left\{-\frac{\Delta G}{k_B T}\right\} . \tag{9.18}$$

The change of the supersaturation affects the value of the nucleation rate mainly via the dependence of the work of critical cluster formation on the value of the concentration. According to Eqs.(9.9), (9.11) and (9.13) we may write

$$\Delta G[j_c(c)] = \Delta G[j_c(c_0)] + \left(\frac{\partial \Delta G}{\partial \Delta\mu}\right)[\Delta\mu(c) - \Delta\mu(c_0)] \tag{9.19}$$

or

$$\frac{\Delta G[j_c(c)]}{k_B T} \cong \frac{\Delta G[j_c(c_0)]}{k_B T} - j_c(c_0)\left(1 - \frac{c}{c_0}\right) . \tag{9.20}$$

Here c_0 is the concentration in the initial state and c the actual concentration of the segregating particles in the solution.

With these dependencies, the nucleation rate can be written as

$$J(c) = J(c_0)\exp\left[-j_c(c_0)\varphi\right] , \qquad \varphi = 1 - \frac{c}{c_0} . \tag{9.21}$$

Steady–state nucleation takes place (by definition) until the concentration is diminished to such a degree that the relation

$$j_c(c_0)\varphi \cong 1 \tag{9.22}$$

holds. Typically, the critical cluster contains more than about ten particles. Therefore, the variations of the concentration in the course of steady–state nucleation are relatively small.

Above formulated dependencies give the basis for an estimate of the moment of time when dominating nucleation is replaced by a stage of dominating independent growth of the already formed supercritical clusters. The respective analysis will be carried out in the next section.

9.3 Time Interval for Steady–State Nucleation

For a derivation of conclusions concerning the number of clusters, formed by nucleation–growth processes, and its dependence on the initial supersaturation, one has to obtain an estimate of the time interval for steady–state nucleation to occur. This derivation is based on the mass balance equation

$$\Delta_0 = \Delta + \int_0^\infty f(j,t) j \, dj \; , \quad \Delta_0 = c_0 - c_\infty \; , \quad \Delta = c - c_\infty \; . \tag{9.23}$$

Here c_0 is the initial, c the actual and c_∞ the equilibrium concentration of the segregating particles in the ambient phase. In the course of the transition the actual concentration approaches the equilibrium value and we have $(\Delta(t = 0) = \Delta_0, \Delta(t \to \infty) \to 0)$.

A derivation of Eq.(9.23) with respect to time yields

$$\frac{d\Delta(t)}{dt} = - \int_0^\infty \left(\frac{\partial f(j,t)}{\partial t} \right) j \, dj = \int_0^\infty \left(\frac{\partial J(j,t)}{\partial j} \right) j \, dj \tag{9.24}$$

and after partial integration

$$\frac{d\Delta(t)}{dt} = - \int_0^\infty J(j,t) \, dj \; . \tag{9.25}$$

For the range of cluster sizes $j \leq g$ we assume that steady–state conditions are fulfilled $(J(j,t) = J(\Delta))$, while for $j > g$ the flux has to be considered, in general, as a function of j and t. It follows

$$\frac{d\Delta(t)}{dt} = - \int_0^g J(\Delta) \, dj - \int_g^\infty J(j,t) \, dj \tag{9.26}$$

$$= -J(\Delta) g - \int_g^\infty J(j,t) \, dj \; .$$

The first term on the right hand side of Eq.(9.26) accounts for the influence of the newly formed clusters of size g on the supersturation Δ while the second term reflects the influence of the clusters with sizes $j > g$.

Both terms on the right hand side of Eq.(9.26) contribute to the change in the supersaturation. Therefore, as a next step the second term in Eq.(9.26) has to be evaluated.

For $j \geq g$, the motion in cluster size space is determined by the deterministic flow term and Eqs.(9.3) and (9.4) yield

$$\frac{\partial J(j,t)}{\partial t} = w^{(+)}(j,t)\left\{\frac{1}{k_B T}\frac{\partial \Delta G(j)}{\partial j}\right\}\left(\frac{\partial J(j,t)}{\partial j}\right) \tag{9.27}$$

$$\text{for} \quad j \geq g \, . \tag{9.28}$$

Hereby the boundary conditions

$$J(g,t) = J(\Delta) \tag{9.29}$$

have to be fulfilled.

The term

$$v(j,t) = \frac{dj}{dt} = -w^{(+)}(j,t)\left\{\frac{1}{k_B T}\frac{\partial \Delta G(j)}{\partial j}\right\} \tag{9.30}$$

represents the velocity of deterministic motion in cluster size space or the deterministic growth rate (cf. Eq.(2.26)). For the considered case of diffusion limited growth, it gets the form (see Eqs.(9.5), (9.9) and (9.13))

$$\frac{dj}{dt} \cong \alpha_1 \left[c \ln\left(\frac{c}{c_\infty}\right)\right] j^{1/3} \, . \tag{9.31}$$

Hereby surface terms have been neglected in the determination of the growth equation.

The solution of Eq.(9.27) with the boundary conditions Eq.(9.29) can be written as

$$J(j,t) = J[g, t_0(t,j)] \, , \qquad j \geq g \, . \tag{9.32}$$

$t_0(t, j)$ is the characteristic curve of Eq.(9.27). It is determined by the solution of Eq.(9.31) in the form

$$j^{2/3} - g^{2/3} = \left(\frac{2\alpha_1}{3}\right) \int\limits_{t_0(t,j)}^{t} \left[c \ln\left(\frac{c}{c_\infty}\right)\right] dt \ , \qquad j \geq g \ , \tag{9.33}$$

or

$$j^{2/3} - g^{2/3} = \left(\frac{2\alpha_1}{3}\right) \left[c \ln\left(\frac{c}{c_\infty}\right)\right] [t - t_0(t, j)] \ . \tag{9.34}$$

The largest cluster formed at time t corresponds to $t_0 = 0$, while $t_0 = t$ refers to $j = g$. Thus we have

$$t_0(t, g) = t \ , \qquad t_0(t, j_{max}) = 0 \ . \tag{9.35}$$

j_{max} is the number of segregating particles in the largest cluster formed till time t in the system.

Since for a given moment of time the flux J is different from zero in the range of cluster sizes $j \leq j_{max}$, only, we may write (cf. Eq.(9.26))

$$\int\limits_g^\infty J(j, t) dj = \int\limits_g^{j_{max}} J(j, t) dj \ . \tag{9.36}$$

By changing the variable of integration (cf. Eq.(9.32)) we have further

$$\begin{aligned} \int\limits_g^{j_{max}} J(j, t) dj &= \int\limits_t^0 J[g, t_0(t, j)] \frac{dj}{dt_0} dt_0 \\ &= \int\limits_0^t J[g, t_0(t, j)] \left[\frac{dj(t - t_0)}{dt}\right] dt_0 \ . \end{aligned} \tag{9.37}$$

Here $v(j, t - t_0) = [dj(t - t_0)/dt]$ is the growth rate of a cluster of size j at a moment of time $(t - t_0)$. In the derivation, Eq.(9.34) was employed.

A substitution of these results into the balance equation Eq.(9.26) yields

$$\frac{d\Delta(t)}{dt} = -J(j_c)g - \int_0^t J[g, t_0(j,t)] \left[\frac{dj(t - t_0)}{dt}\right] dt_0 \, . \tag{9.38}$$

Instead of the parameter Δ we may also use the quantity φ (cf. Eq.(9.21)) as a measure of the relative deviation of the concentration from its initial value. In terms of φ, Eq.(9.38) reads

$$\frac{d\varphi(t)}{dt} = \frac{J(\varphi)g}{c_0} + \int_0^t \frac{J[g, t_0(j,t)]}{c_0} \left[\frac{dj(t - t_0)}{dt}\right] dt_0 \, . \tag{9.39}$$

A partial integration of the second term on the right hand side of Eq.(9.39) yields

$$\frac{d\varphi}{dt} = \frac{J[j_c(c_0)]g}{c_0} \left\{1 + \frac{2\alpha_1}{3g^{2/3}} \left[c_0 \ln\left(\frac{c_0}{c_\infty}\right)\right] t\right\}^{3/2} \, . \tag{9.40}$$

In this expression terms proportional to $[J(c_0)/c_0]^2$ have been neglected.

An integration of this expression with the initial condition $\varphi(0) = 0$ results in

$$\varphi(t) = \frac{2A}{5B}\left[(1 + Bt)^{5/2} - 1\right] \tag{9.41}$$

$$A = \frac{J[j_c(c_0)]g}{c_0} \, , \qquad B = \frac{2\alpha_1}{3g^{2/3}} \left[c_0 \ln\left(\frac{c_0}{c_\infty}\right)\right] . \tag{9.42}$$

Taking into account the condition Eq.(9.22), we find as an estimate for the time of steady–state nucleation

$$\tau_N \cong \left(\frac{5}{2}\right)^{2/5} \left[\frac{c_0}{J[j_c(c_0)]}\right]^{2/5} \frac{1}{\left[\frac{2j_c(c_0)}{3}(\alpha_1 c_0) \ln\left(\frac{c_0}{c_\infty}\right)\right]^{3/5}} \, . \tag{9.43}$$

With Eqs.(9.5), (9.9), (9.10), (9.13), (9.16), (9.17) and (9.18) this relation is transformed to

$$\tau_N \cong \frac{a_s^2}{D(\omega_s c_0)\beta^{4/5}} \exp\left[\frac{2}{5}\left(\frac{\Delta G[j_c(0)]}{k_B T}\right)\right] , \tag{9.44}$$

$$\beta = \frac{2\alpha_2}{3k_B T} = \frac{8\pi}{3}\left(\frac{\sigma a_s^2}{k_B T}\right) . \tag{9.45}$$

Eqs.(9.15) and (9.44) yield

$$\frac{\tau_N}{\tau_1} \cong \frac{\beta^{1/5}}{2j_c(c_0)} \exp\left(\frac{2}{5}\frac{\Delta G}{k_B T}\right) \gg 1 . \tag{9.46}$$

It follows that the time interval of steady–state nucleation exceeds by far the non–steady state time–lag.

9.4 Number of Clusters and Their Average Sizes

Once the time interval of intensive steady–state nucleation is estimated, immediately the number of supercritical clusters may be determined via

$$N = \int_0^{\tau_N} J(\Delta)\, dt , \qquad J(\Delta) = J(\Delta_0) \exp(-j_c(c_0)\varphi) . \tag{9.47}$$

Rewriting Eq.(9.47) in the form

$$N = \int_0^{\tau_N} J(\Delta_0) \left\{1 + \frac{[J(\Delta) - J(\Delta_0)]}{J(\Delta_0)}\right\} dt \tag{9.48}$$

we get

$$N = J(\Delta_0)\tau_N \left[1 + \frac{1}{\tau_N}\int_0^{\tau_N} \frac{[J(\Delta) - J(\Delta_0)]}{J(\Delta_0)} dt\right] . \tag{9.49}$$

For the second term on the right hand side of Eq.(9.49) we have with Eqs.(9.21) and (9.22)

$$\frac{1}{\tau_N}\left|\int_0^{\tau_N}\frac{[J(\Delta)-J(\Delta_0)]}{J(\Delta_0)}dt\right| \le \frac{1}{\tau_N}\left|\int_0^{\tau_N}\frac{[J(\tilde{\Delta})-J(\Delta_0)]}{J(\Delta_0)}dt\right| \tag{9.50}$$
$$= |\exp(-j_c(c_0)\varphi(\tau_N)) - 1| < \frac{2}{3} .$$

We may write thus in a good approximation

$$N \cong J(\Delta_0)\tau_N . \tag{9.51}$$

We find with Eqs.(9.18) and (9.44)

$$N \cong \frac{c_0}{n_c^{1/3}\beta^{3/10}}\exp\left[-\frac{3}{5}\frac{\Delta G}{k_B T}\right] . \tag{9.52}$$

As evident from Eq.(9.52), the number of clusters formed in the system is determined mainly by the initial supersaturation and the value of the interfacial specific energy (via the parameters α_2 or β, respectively, cf. Eqs.(9.8) and (9.44)), it does not depend on the diffusion coefficient D.

The stage of dominating nucleation is followed, in general, by a stage of dominating independent growth of the supercritical clusters present in the system [39, 471, 511, 516]. In this stage, the supercritical clusters grow at the expense of the excess monomers and the supersaturation tends to zero $[\Delta(t) = (c - c_\infty) \to 0]$. After this process is finished the transformation has reached a certain degree of completion (cf. [39]) and goes over into a third stage of competitive growth or Ostwald ripening (cf. [310, 471, 508] and chapters 2 and 13).

In nucleation–growth experiments commonly states are observed initially corresponding to the end of the stage of independent growth. In the calculation of the initial average sizes of the clusters observed experimentally one has to take into account, therefore, the deterministic growth of the supercritical clusters. This growth is stopped after the supersaturation reaches values near zero (for a more precise formulation see, e.g., [471, 544]).

Since we have determined already the number of clusters, N, formed in the course of nucleation, by a purely thermodynamic argumentation it is possible to give estimates also of the average size, $\langle n\rangle$, or the average radius, $\langle R\rangle$, of the clusters and its dependence on the initial supersaturation. Approximately, we may write the mass balance equation for the final state of independent growth in the form (for more details see, again, [471, 544])

$$(c_0 - c_\infty) = \langle j\rangle N = \frac{4\pi\langle R\rangle^3}{3\omega_s}N\,, \tag{9.53}$$

which is equivalent to

$$\langle R\rangle^3 = \left(\frac{3\omega_s c_\infty}{4\pi}\right)\left(\frac{1}{N}\right)\left[\left(\frac{c_0}{c_\infty}\right) - 1\right]\,. \tag{9.54}$$

A substitution of the expression for N into Eq.(9.54) yields, finally,

$$\langle R\rangle \cong a_s \exp\left[\frac{1}{5}\frac{\Delta G}{k_B T}\right]\,. \tag{9.55}$$

9.5 Time Interval of Independent Growth

Neglecting surface effects, the growth equation may be written in the form

$$\frac{dR}{dt} = \frac{(D\omega_s)}{R}c\left[\ln\left(\frac{c}{c_\infty}\right)\right]\,. \tag{9.56}$$

From the mass balance equation we have further, assuming that all clusters are nearly of the same size,

$$c = c_0 - NR^3\left(\frac{4\pi}{3\omega_s}\right)\,. \tag{9.57}$$

A substitution into Eq.(9.56) yields (see also [508])

$$\frac{dR}{dt} = \frac{D\omega_s}{R}\left[c_0 - \left(\frac{4\pi N}{3\omega_s}\right)R^3\right]\left[\ln\left(\frac{c}{c_\infty}\right)\right]\,. \tag{9.58}$$

As an estimate for the duration τ_{gr} of the stage of independent growth we obtain thus

$$\tau_{gr} \cong \frac{\langle R \rangle^2}{2D\omega_s c_0} \qquad (9.59)$$

resulting with Eq.(9.55) in

$$\tau_{gr} \cong \frac{a_s^2}{2D(\omega_s c_0)} \exp\left[\frac{2}{5}\frac{\Delta G}{k_B T}\right] . \qquad (9.60)$$

The ratio (τ_{gr}/τ_N) can be written as

$$\left(\frac{\tau_{gr}}{\tau_N}\right) = \frac{\beta^{4/5}}{2} . \qquad (9.61)$$

The parameter β may be written as (cf. Eqs.(9.9) and (9.45))

$$\beta = n_c^{1/3} \ln\left(\frac{c_0}{c_\infty}\right) . \qquad (9.62)$$

It turns out that τ_{gr} is considerably larger than τ_N but of the same order of magnitude. It follows as a consequence that in the range of the initial parameter values, for which intensive nucleation may take place, the inequality $\tau_{gr} \gg \tau_N$ holds. In this way the time τ_{compl}

$$\tau_{compl} = \tau_1 + \tau_N + \tau_{gr} \qquad (9.63)$$

required to reach the stage of competitive growth is determined, in general, mainly by the value of τ_{gr}.

9.6 Results of Computer Calculations

In experiments and also in numerical studies of segregation processes the initial supersaturation is most easily changed by varying the initial concentration c_0. To study the dependence of the number of clusters, N, their average size, $\langle R \rangle$, the time–lag τ_1, the time interval of active nucleation, τ_N, and the time interval of independent growth τ_{gr} on the initial supersaturation we introduce the notations

$$y = \left(\frac{c_0}{c_\infty}\right) , \qquad \chi = \ln\left(\frac{c_0}{c_\infty}\right) \tag{9.64}$$

as different possible measures of the initial supersaturation in the system. Eqs.(9.15), (9.44), (9.52), (9.55) and (9.60) can be reformulated then to give

$$\tau_1 = \frac{6\beta^2}{(\alpha_1 c_\infty)} \frac{1}{y[\ln y]^3} , \tag{9.65}$$

$$\tau_N \cong \frac{a_s^2}{D(\omega_s c_\infty)\beta^{4/5}} \left(\frac{1}{y}\right) \exp\left[\frac{1}{5}\left(\frac{\beta^3}{[\ln y]^2}\right)\right] , \tag{9.66}$$

$$\tau_{gr} \cong \frac{a_s^2}{2D(\omega_s c_\infty)} \left(\frac{1}{y}\right) \exp\left[\frac{1}{5}\left(\frac{\beta^3}{[\ln y]^2}\right)\right] , \tag{9.67}$$

$$N \cong \left(\frac{c_\infty}{\beta^{13/10}}\right) [y \ln y] \exp\left[-\frac{3}{10}\left(\frac{\beta^3}{[\ln y]^2}\right)\right] , \tag{9.68}$$

$$\langle R \rangle \cong a_s \exp\left[\frac{1}{10}\left(\frac{\beta^3}{[\ln y]^2}\right)\right] . \tag{9.69}$$

It is easily seen that τ_1, τ_N and τ_{gr} decrease with increasing value of the initial supersaturation y.

Moreover, a derivation of Eqs.(9.68) and (9.69) with respect to y yields

$$\frac{d \ln N}{d \ln y} = \left\{1 + \frac{1}{(\ln y)}\left[1 + \frac{3\beta^3}{5(\ln y)^2}\right]\right\} , \tag{9.70}$$

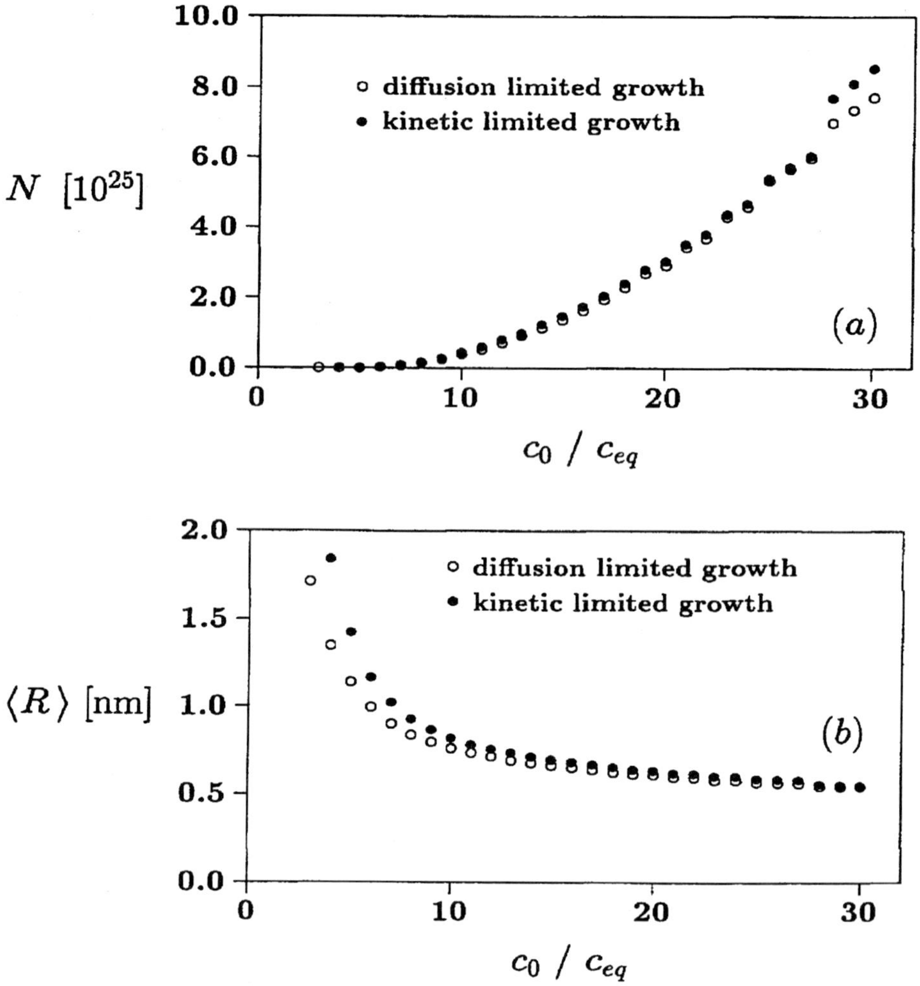

Fig. 9.2 Number of clusters, N, formed in nucleation processes *(a)*, and the average size, $\langle R\rangle$, of the clusters in the final stage of dominating independent growth *(b)* as a function of the ratio $y = (c_0/c_\infty)$. In agreement with the analytical results derived here N increases and $\langle R\rangle$ decreases with increasing initial supersaturation (for the details see Ludwig, Schmelzer [322]).

$$\frac{d\ln\langle R\rangle}{d\ln y} = -\frac{\beta^3}{5[\ln y]^3} \,. \tag{9.71}$$

It is evident that the inequalities

$$\frac{d\ln N}{d\ln y} > 0\,, \qquad \frac{d\ln\langle R\rangle}{d\ln y} < 0 \tag{9.72}$$

are fulfilled.

For comparison, in Fig.9.2 the number of clusters, N, and the average size of the clusters, $\langle R\rangle$, are shown as functions of $y = (c_0/c_\infty)$. The curves are obtained based on the numerical solution of the set of kinetic equations underlying classical nucleation theory resulting in the continuums approximation in Eq.(9.1) (for the details of the calculations see [322, 483]).

In agreement with the analytical results, derived here, it is found that

- the number of clusters increases with increasing supersaturation;
- the average cluster radius decreases with increasing supersaturation, the rate of decrease is diminished with increasing values of χ.

Moreover, the curve can be described by the analytical expressions even quantitatively well.

9.7 Discussion

In 1926 von Weimarn [566], based on the analysis of experimental results, argued that the average size of the clusters, $\langle R\rangle$, formed in nucleation processes may be expressed by a power law of the form ($\langle R\rangle \propto y^{-q_1}$; q_1=constant> 0). Since power laws play an outstanding role in physics [567], in general, and in the theory of phase transformation processes, in particular, in a recent paper [322] the problem was analyzed whether, indeed, the average cluster size and the number of clusters obey dependencies of such type or, more accurately, of the form

$$\langle R\rangle \propto \chi^{-q_1}\,, \qquad N \propto \chi^{q_2} \tag{9.73}$$

with constant positive values of q_1 and q_2.

Based on the numerical solution of the kinetic equations of classical nucleation theory it was shown that fits of the form as given by Eqs.(9.73) cannot be applied for the whole range of initial supersaturations. Taking into account the results of the theoretical analysis outlined here this finding is not a surprise since the expressions Eq.(9.52) and Eq.(9.55) for N and $\langle R \rangle$ cannot be written in such a simple form.

Eqs.(9.52) and (9.55), and in particular the equivalent expressions as given by Eqs.(9.70) and (9.71), show some similarity with Kashchiev's nucleation theorem [236, 409]. It is believed that mentioned relations may play in future a similarly important role. In particular, the may be used to test different approaches for the determination of the work of critical cluster formation, a problem which is discussed inztensively till now [481, 273].

Finally, we would like to mention that an even more detailed description of the first stages of first–order phase transitions may be given. The respective method and results are outlined in a recent paper by Slezov and Schmelzer [516]. In addition to the topics discussed here, detailed expressions for the evolution of the cluster size distribution function and the flux in cluster size space are given for the first stages of the nucleation – growth process. Moreover, the analysis is carried out both for diffusion and kinetic limited nucleation – growth processes as well as for phase separation in non–ideal solutions. This way, an immediate comparison with experiment for various real systems has become possible.

10 Self–Organization in Multifragmentation in Molecular and Nuclear Collision Processes

J. Schmelzer, G. Röpke

10.1 Introduction

Multifragmentation processes in nuclear or molecular collisions are widely studied in order to obtain some insight into the properties of the nuclei and nuclear matter, respectively, atomic or molecular clusters and the interaction forces between the basic building units (see, e.g., [392, 489]). Hereby different approaches in the theoretical interpretation of the results have been developed.

Based on the pioneering work of Finn et al. [149] a large amount of work has been carried out on the investigation of multifragmentation in nuclear collision processes. In particular, the attention has been directed to the theoretical interpretation of fragment size distributions of the form $N(A) \propto A^{-\tau}$ and the derivation of conclusions concerning the equation of state of nuclear matter from such dependencies (for recent overviews see [145, 378]). $N(A)$ is the number of fragments consisting of A nucleons.

Already Finn et al. [149] proposed an interpretation of the observed dependencies based on the assumption of a liquid–gas transition of nuclear matter by clustering processes in an expanding hot nuclear system. These clustering processes were proposed to be described by Fisher's droplet [150] or percolation models [529]. Both mentioned approaches lead to similar predictions for the shape of the cluster size distribution function in the vicinity of the respective critical points.

Percolation models in the vicinity of the critical percolation point result in expressions $N(j) \propto j^{-\tau}$ for the cluster size distribution with $\tau \cong 2.2$ [113, 133, 529]. Similarly, in the vicinity of the liquid–gas critical point

Fisher's droplet model leads to distributions of the type $N(j) \propto j^{-\tau}$ with τ having values in the range $2 < \tau < 3$ (or more precisely $2 < \tau < 2.5$ [255]). These predictions are in reasonable agreement with the experimental results obtained by Finn et al. [149]. Here j is the number of particles in a percolation cluster, respectively, a drop.

As it turns out the coincidence in the results of these at a first glance seemingly different approaches is not accidental [98, 239, 411, 529]. Mapping the partition function of the system appropriately onto the percolation model [98, 239], the percolation critical points come into coincidence with the thermal critical points. A physical interpretation of this procedure was given by Pan and das Gupta [411]. They took into account that two neighboring nucleons can be considered as bound only if their relative velocity is within some upper limit determined by the binding energy of the two nucleons.

However, similar dependencies but with varying values of τ, at part considerably beyond the mentioned interval, were found later for very different values of the incident energy. These results made the interpretation in terms of critical point properties somewhat vague [541, 412].

To overcome such difficulties an apparent or effective value of τ was introduced denoted by τ_{app}. It was suggested that in the metastable (or unstable) initial states the evolving cluster size distributions are generally of the form $N(A) \propto A^{-\tau_{app}}$. Near the critical point τ_{app} is expected to coincide with the parameter τ occuring in Fisher's model [412, 428]. Moreover, it was observed that τ_{app} has a more or less well–expressed minimum as a function of the incident energy [306, 412, 428].

This approach sketched here briefly implies that Fisher's droplet model, leading to droplet size distributions $N(j) \propto j^{-\tau}$ in the immediate vicinity of the critical point (see [150, 152]), is applied for the whole region of unstable respectively metastable initial states, where phase transformations are, in general, of first order and cannot be interpreted in terms of critical point properties.

Nevertheless, till now, in the analysis of experimental results on fragmentation processes Fisher's droplet model – in this generalized interpretation – is widely used [378], as it seems not being aware of the limits of applicability of such type of model for interpreting real cluster size distributions evolving in first–order phase transformation processes.

More explicitely, it will be shown in the subsequent analysis that Fisher's droplet model cannot give an adequate description of the cluster size distributions developing in the course of time, except for thermdynamically stable states and for states in the immediate vicinity of the critical point. This statement refers, in particular, also to clustering processes occuring as the result of nuclear collisions in an expanding hot nuclear system.

Starting from initial states at a sufficiently large distance from the critical point, phase formation processes are of first order and proceed, in general, via nucleation–growth or spinodal decomposition. While spinodal decomposition type models have been applied already widely to an interpretation of nuclear fragmentation [95, 96, 203, 413, 420], this is so far not the case for nucleation – growth models [106]. Moreover, it has been shown that the nucleation–growth model works well even in the region of thermodynamically unstable initial states, in particular, when the interactions between the ambient phase particles are short–ranged [39]. In this way, it is highly desirable to check whether nucleation–growth models can be applied with success to model fragment size distributions evolving in nuclear multifragmentation processes.

Nuclear collision processes are, at present, intensively studied theoretically by different microscopic approaches like quantum molecular dynamics (QMD) [4] or by numerical calculations based on the Boltzmann–Uehling–Uhlenbeck (BUU) or similar equations [32, 64, 538]. Taking into account that in these approaches a comprehensive description of cluster formation from first principles is not incorporated up to now, the formulation of an alternative model – the discussed here nucleation–growth model – may give an additional tool for the understanding of the complex phenomena occuring in nuclear collision processes. Moreover, in developing such a semi–microscopic approach to the description of nuclear multifragmentation we – following Pethick et al. [420] – may consider the results of QMD–, BUU– or similar calculations as "experimental results" to be explained in physical terms.

In addition, in the mesoscopic approach considered the input parameters of the theory are thermodynamic and kinetic properties of nuclear matter. By this reason, a comparison of experimental and theoretical results may give thus immediately an answer whether the assumptions made concerning the properties of nuclear matter are correct or not.

By applying a nucleation–growth model, it assumed from the very beginning that multifragmentation can be understood as the result of clustering processes in a thermodynamically unstable nucleonic system. A comprehensive discussion of the variety of other possible mechanisms of multifragmentation resulting eventually in similar dependencies is out of the scope of the present analysis (cf., e.g., [3, 572]). For comparison, only two alternative approaches are discussed here briefly.

The first one is a classical fragmentation study (see also [487]). In this approach, the energy transferred to the nucleonic or molecular ensemble results in an immediate fragmentation into a more or less large number of fragments. In analyzing this approach we concentrate the attention on a relatively new aspect of the role of dissipative forces in fragmentation processes. Though we use a simplified one–dimensional model, it is believed that the respective study is of general heuristic value for the understanding of the outcome of collision processes, in general.

As an illustration of a second widely applied method, we give a sketch of basic features of statistical fragmentation studies. As an example, the fragmentation of a charged liquid drop is analyzed (cf. also [505]).

In the third and most widely discussed here approach, we assume that the collision energy transfers the system initially into a highly excited gas of nucleons, atoms, or molecules. In the further expansion of the hot system of particles the system cools down and fragments are formed by condensation processes in the expanding system. In this approach, multifragmentation is understood as a kind of selforganization process in a non–equilibrium system.

10.2 Multifragmentation and Dissipation: A Numerical Simulation

10.2.1 Description of the Model

Chain models are widely used in physics as a means of study of different physical phenomena both for the understanding of qualitative aspects as well as for quantitative estimates. One particular example in this respect is the widely employed Frenkel–Kontorova model [157] allowing to explain a broad spectrum of phenomena like charge–density waves, domain boundaries, magnetic structures, crack formation in overlayers (see [370]).

As another example, the investigations of Fermi, Pasta, Ulam [146] can be mentioned, where by means of chain models the approach to equilibrium was attempted to be studied. The respective investigations raised a large amount of discussions concerning basic assumptions of statistical physics.

From another point of view, the intuition of physicists concerning the problem what is going on in collisions of particles, is trained to a large extent by simplified model experiments (e.g., collisions of billard balls, interaction via Coulomb or gravitational forces), where the motion of the particles is determined by conservative forces.

In a recent paper by Du, Li and Kadanoff [578] it was shown by considering a one–dimensional system of particles that quite unexpected effects may occur if some dissipation of energy takes place in the collision processes. The model was motivated originally by studies of flow behavior of granular materials [221].

Another area of physics, where such effects are believed to be of considerable significance, is the description of collision processes of heavy ions. Attempts to describe theoretically the interaction and fission of two nuclei are shown to be more successful assuming a dissipation of energy [162, 392].

Above mentioned developments were the motivation to examine the collision between two nuclei using the simple model of two linearly arranged chains of particles colliding with each other. In the collision processes momentum is conserved but dissipation of energy is allowed.

More definitely, we consider two one–dimensional chains of identical particles (representing in the idealized model two nuclei). It is assumed that the particles interact only with the nearest neighbors. These interactions are realized by a conservative force depending on the distance between two particles and a dissipative force depending on the relative velocity between them. The dissipative force is assumed to be of importance also only at distances where an interaction between the particles by conservative forces is effective. As it will be shown, by taking into account dissipation it is possible to get a variety of fragmentation channels in contrast to the case when only elastic interactions are taken into account.

10.2.2 Description of the Interactions Between the Particles

We consider a system of linearly arranged identical particles. Between two of these particles an effective conservative force acts, used for modelling the interactions between nucleons [392]. Denoting by $r = x_{n+1} - x_n$ the distance between two particles we have

$$F_p(r) = \frac{a}{r^{12}} - b\exp(-cr) \; . \qquad (10.1)$$

Since we are interested in qualitative results we choose the constants a, b and c in such a way that the force becomes equal to zero at a distance $r_0 = 1$.

In our simulation program, we used the following constants

$$a = 5 \; , \qquad b = 100 \; , \qquad c = \ln\left(\frac{b}{a}\right) = \ln(20) \; . \qquad (10.2)$$

In addition to the conservative force $\vec{F_p}$ depending on the distance r only, we introduced a force $\vec{F_d}$, which depends on the relative velocity between two neighboring particles with the velocities $\vec{v_1}$ and $\vec{v_2}$ as

$$\vec{F_d}(\vec{v_2} - \vec{v_1}) = R(\vec{v_2} - \vec{v_1}) \; . \qquad (10.3)$$

Here R is the dissipation–factor (coefficient of friction), it is a measure of the strength of the dissipative force, i.e., it determines how much kinetic energy is dissipated in the oscillations inside the chains. The direction of $\vec{F_d}$ is chosen in such a way that it reduces the kinetic energy of the relative motion. Further, the dissipative force is assumed to act only between nearest neighbors when they are located within the distance $2r_0$.

10.2.3 Simulation

To start the simulation, we define a number N_1 of particles separated by the rest–distance r_0 from each other. To all particles the same initial value of the velocity v is assigned to. The second chain, consisting of N_2 particles and formed in the same way, is initially at rest ($v = 0$). In the mentioned way, we modelled the initial conditions for nuclear collision processes: two nuclei in their ground states one moving into the direction of the other without internal energy of oscillations.

As the next step, we start the simulation by solving Newton's equations of motion for each of the $N_1 + N_2$ particles. We have chosen the numerical implementation in the simple form

$$s(t_0 + \Delta t) = s(t_0) + v(t_0)\Delta t + \frac{a}{2}(\Delta t)^2 \, ,$$
$$v(t_0 + \Delta t) = v(t_0) + a\Delta t \, . \tag{10.4}$$

To get meaningful results, it is important that for the time step Δt a sufficiently small value is taken. For the simulations we applied the following values of Δt and the additional parameters

$$\Delta t \leq 0.001 \, , \qquad R = 0 \ldots 6 \, , \qquad v(t_0) = 0 \ldots 8 \, . \tag{10.5}$$

10.2.4 Results

By chosing different values for N_1, N_2, v and R we got a variety of different fragmentation channels with respect to the number and the size of the fragment chains. Some of these are shown on Fig. 10.1 for the example $N_1 = 3$ and $N_2 = 10$. In the parameter space of v and R, one can hereby distinguish between three main regions referring to different reaction channels (see Fig. 10.2).

At first there exists a region in parameter space, in which both chains fuse to form a chain of 13 particles. Such processes of fusion occur for high values of R and relatively small values of v. In these cases, the friction decreases the kinetic energy of the first chain very fast so that not enough energy remains to separate a part from the chain, again.

The other limiting region consists of the area of relatively high velovities v and of low values of the dissipation parameter R. Here one finds a behavior similar to the case of elastic collisions, when the same number of particles, which has arrived, leaves the other side of the chain after the collision process.

Between these limiting regions in parameter space, values for R and v exist, resulting in different other fragmentation channels: two chains with different sizes, compared with the states before the collision process, are found as well as fragmentation processes into more than two parts. One particular interesting example consists in a collision process, where in addition to two large fragments a small one is formed between them. Such processes are

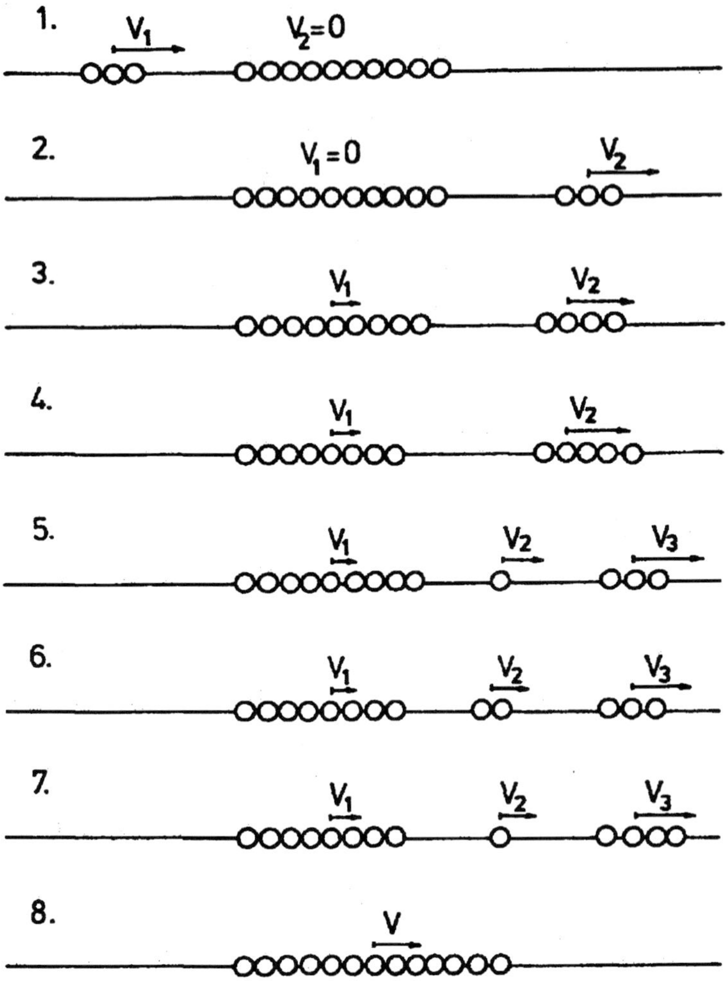

Fig. 10.1 Observed fragmentation channels in the collision of $N_1 = 3$ and $N_2 = 10$ particles: *[1]*: 2 chains before the collision, *[2]*: 2 chains after an elastic collision, *[3–4]*: 2 chains after a collision with some dissipation of energy ($v_1 < v_2$), *[5–7]*: 3 chains after a collision with a higher value of the dissipated energy ($v_1 < v_2 < v_3$), *[8]*: One chain formed as the result of the collision, the ratio of the dissipated energy is higher compared with the previous cases.

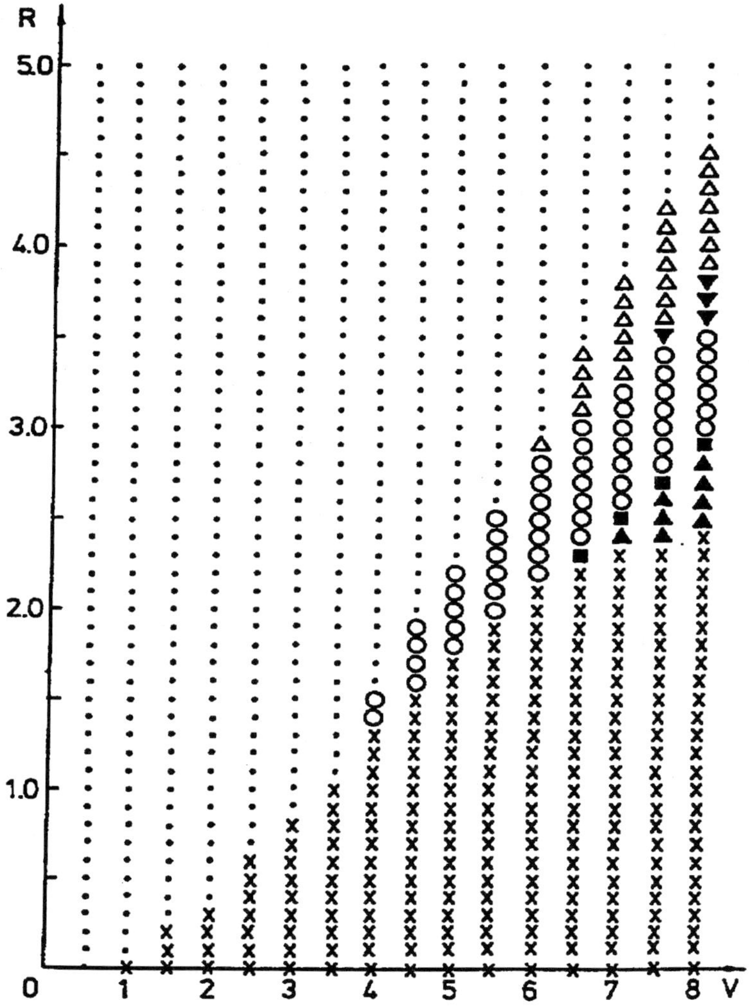

Fig. 10.2 Fragmentation behavior in the parameter space initial velocity v versus coefficient of friction R for $N_1 = 3$ (N_1 – number of particles in the first chain), $N_2 = 10$ (N_2 – number of particles in the second chain). The following fragmentation channels are found: (●) : chain of 13 particles, (×) : 2 chains of 10/3 particles, (○) : 2 chains of 9/4 particles, (△) : 2 chains of 8/5 particles, (▲) : 3 chains of 8/2/3 particles, (▼) : 3 chains of 8/1/4 particles, (■) : 3 chains of 9/1/3 particles.

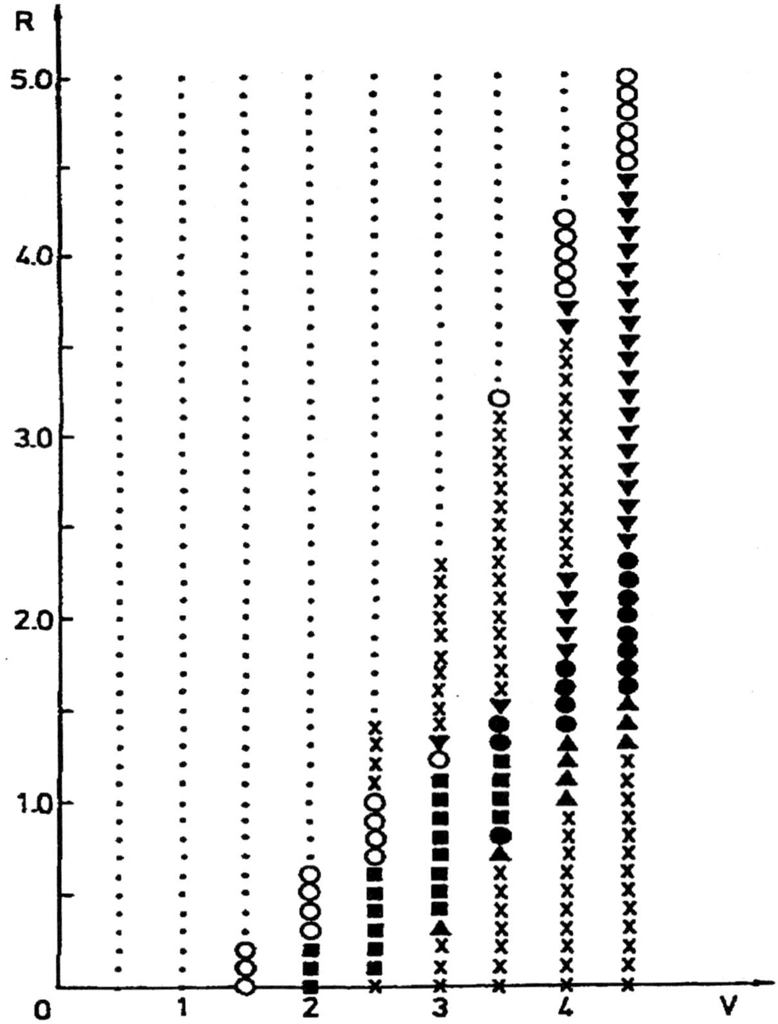

Fig. 10.3 Fragmentation behavior in the parameter space initial velocity v versus coefficient of friction R for $N_1 = 7$ (N_1 – number of particles in the first chain), $N_2 = 10$ (N_2 – number of particles in the second chain). The following fragmentation channels are found: (•) : chain of 17 particles, (×) : 2 chains of 10/7 particles, (○) : 2 chains of 9/8 particles, (■) : 3 chains of 9/2/6 particles, (▲) : 3 chains of 10/1/6 particles, (▼) : 3 chains of 9/1/7 particles, (●) : 4 chains of 9/1/1/6 particles.

intensively studied in nuclear collisions where they are denoted as neck–emissions [192].

We find also a small range of v–values in the case of totally elastic collisions ($R = 0$), where fragmentation into more than two parts and also the formation of only one chain take place. Although in this case any kind of friction due to relative motion is excluded, the kinetic energy dissipates, again, this time into the internal energy of oscillations of the chains.

Similar results as discussed in detail above, we got also by examining a collision of two chains with $N_1 = 7$ and $N_2 = 10$ (cf. Fig. 10.3). In this case, the respective regions for the different fragmentation channels are larger and moved to lower values of R and v. The reason is that the kinetic energy of 7 particles is higher than that of 3 particles having the same velocity.

10.2.5 Discussion

The present investigation shows that the consideration of dissipative forces in a very simple model of the collision of one–dimensional chains allows to reproduce different results of the experimental research with methods of classical physics. It is important to point out that the use of only conservative potentials in this model would not allow to obtain the variety of fragmentation channels as discussed above.

10.3 Statistical Approaches to Fragmentation

10.3.1 Description of the Model

In the following considerations we are going to analyze basic features of statistical fragmentation models and some consequences. The advantage of such models is that only relatively few informations are required concerning the process considered (for example, conservation of energy and mass). By this reason, the theory is of very general nature and widely applicable to different fragmentation processes (see also [93, 135, 179, 364, 505]). However, such kind of theories fail if the dynamics of the process determines the final fragment size distributions.

We analyze here the following model: We assume that we have a compact object consisting initially of N_0 distinguishable primary particles (in the considered application, these are the nucleons). These primary particles may form in the course of the fragmentation process different fragment size distributions.

As usual we will denote by j the number of primary particles in a cluster, j_{max} is the number of primary particles in the largest cluster in the considered distribution, N_j is the number of clusters consisting of j particles.

By definition we have

$$N_0 \geq j_{max} \geq 1 \tag{10.6}$$

and

$$N_0 = \sum_{j=1}^{j_{max}} j N_j \,. \tag{10.7}$$

As a first step in the subsequent analysis, we would like to determine the number $\Omega(N_1, N_2, N_3, \ldots, N_{j_{max}})$ of the possible realizations of a given distribution $(N_1, N_2, N_3, \ldots, N_{j_{max}})$ by the N_0 particles present in the system. We obtain [93, 505]

$$\Omega(N_1, N_2, N_3, \ldots, N_{j_{max}}) = \frac{N_0!}{\prod_{j=1}^{j_{max}} N_j!(j!)^{N_j}} \,. \tag{10.8}$$

In the further analysis, consequences from the basic equation (10.8) are derived.

10.3.2 Application

10.3.2.1 General Method

For the application of Eq.(10.8) to the interpretation of experimental results we postulate in agreement with the general principles of statistical physics [295] that the probability of a given distribution is proportional to the number of realizations Ω of the respective state (see also [179]).

In order to determine the distribution which is realized in a given process with the highest probability one has to determine therefore the maximum of Ω, respectively, $\ln \Omega$. Hereby the constraints (for example, conservation of mass) have to be taken into account. If one takes into account conservation of mass as given by Eq.(10.7) only, then one gets

$$\frac{\partial}{\partial N_i} \left\{ \ln(\Omega) + \lambda \left(\sum_{i=1}^{i_{max}} i N_i - N_0 \right) \right\} = 0 \tag{10.9}$$

or

$$\frac{\partial \ln(\Omega)}{\partial N_i} + \lambda i = 0 \; . \tag{10.10}$$

One obtains

$$N_i = \frac{\exp(\lambda i)}{i!} \; . \tag{10.11}$$

The Lagrange parameter λ will be determined such as to fulfil the constraints. We get after some straightforward transformations

$$\lambda = \ln \left\{ \ln(N_0) - \ln \left[\ln(N_0) - \ln \ln(N_0) \right] \right\} \; . \tag{10.12}$$

The total number of clusters in the distribution can be estimated then as

$$\sum_{j=1}^{j_{max}} N_j = \sum_{j=1}^{j_{max}} \frac{X^j}{j!} \cong \exp(X) = \frac{N_0}{\ln(N_0) - \ln \ln(N_0)} \; , \tag{10.13}$$

$$X = \exp(\lambda) \; . \tag{10.14}$$

The probability of occurence of a cluster of size j in the distribution is given by

$$P(j) = \frac{N_j}{\sum\limits_{j=1}^{j_{max}} N_j} = \exp(-X) \frac{X^j}{j!} \; . \tag{10.15}$$

In application of the equations derived in the present section one has to have in mind that they gain in accuracy with increasing values of N_0. For small values of the total number of primary particles, it is preferable to apply the discrete dependence as given by Eq.(10.8).

10.3.2.2 Conservation of Mass and Energy

So far in the analysis, only the constraint of mass conservation was taken into account. We would like to demonstrate now what kind of changes occur if in addition also the condition of energy conservation is considered as it is commonly the case in physical applications.

We assume that as the result of heavy–ion collisions a compound nucleus is formed as an intermediate object characterized by an excitation energy E_{ex}. In the course of the fragmentation only such fragment size distributions may be formed where the total energy is equal or less than the energy of the compound nucleus.

The energy of a nucleus is approximated here according to the Bethe – Weizsäcker relation by a sum of surface, volume and Coulomb energy terms (see also [357]). We have generally for a system in the ground state

$$E_{nucleus} = C_O A^{2/3} + C_C Z^2 A^{-1/3} - C_V A \,, \tag{10.16}$$

$$C_O = 18.34\,\text{MeV} \,, \quad C_C = 0.71\,\text{MeV} \,, \quad C_V = 15.85\,\text{MeV} \,. \tag{10.17}$$

For a further simplification of the analysis we set, moreover, $Z = A/2$ resulting in

$$E_{nucleus} = C_O A^{2/3} + \frac{C_C}{4} A^{5/3} - C_V A \,. \tag{10.18}$$

The energy of the excited compound nucleus can be written then as

$$E_{comp} = C_O N_0^{2/3} + \frac{C_C}{4} N_0^{5/3} - C_V N_0 + E_{ex} \,. \tag{10.19}$$

Similarly we have for any possible fragment distribution in the final state

$$E_{distr} = \sum_{j=1}^{j_{max}} N_j \left(C_O j^{2/3} + \frac{C_C}{4} j^{5/3} - C_V j \right) \,. \tag{10.20}$$

In general, in order to obtain the total energy of the final state the quantity E_{distr} has to be supplemented by the kinetic energy E_{kin} of the fragments.

Now, again the problem of a constraint extremum has to be solved where the constraints are given both by mass and energy conservation. In this case, two Lagrange parameters have to be determined.

However, in applying the condition of energy conservation, for example, in the form $E_{distr} = E_{comp}$ (in such a formulation the kinetic energy of the fragments is assumed to be negligible) one finds that generally for the majority of values of the energy of the initial state (compound nucleus) no fragment size distribution exists having the same energy. Mathematically this fact is expressed by the absence of solutions for the Lagrange parameters.

The boundary condition corresponding to energy conservation has to be formulated therefore in a somewhat different way. We will use here the constraint

$$E_{distr} \leq E_{comp} . \tag{10.21}$$

However, applying such a boundary condition the determination of the extremum of the number of configurations becomes a considerably more complicated task. In the further analysis we will apply therefore numerical methods of solution.

10.3.2.3 Results of Numerical Computations

In order to solve above formulated problem, a computer program was developed. This program solves the following tasks:

- Determination of the whole set of possible distributions which are allowed by the condition of conservation of mass;
- Selection of those distributions which fulfil the condition $E_{distr} \leq E_{comp}$;
- Determination of the number of realizations Ω of these distributions;
- Specification of the distribution with the highest value of $\Omega = \Omega_{max}$;
- Printout of Ω_{max} and the distribution corresponding to it.

In the first and second steps of the calculations, the quantities N_0 and j_{max} have to be specified as parameters. Generally one should set $N_0 = j_{max}$. Moroever, in the second step one has also to assign definite values to the parameter E_{ex}.

The number of distributions N_Σ increases considerably with an increase of the number of primary particles in the system N_0. An example is given in Table 10.1 for illustration.

Particle number N_0	Number of Distributions N_Σ	Particle number N_0	Number of Distributions N_Σ
1	1	16	231
2	2	17	297
3	3	18	385
4	5	19	490
5	7	20	627
6	11	21	792
7	15	22	1002
8	22	23	1255
9	30	24	1575
10	42	25	1958
11	56	26	2436
12	77	27	3010
13	101	28	3718
14	135	29	4565
15	176	30	5604

Tab. 10.1 Number of fragment distributions in dependence on the number of primary particles N_0.

For relatively small values of the number of primary particles N_0, the number of distributions is relatively small. By this reason, for a given value of N_0 several distributions may exist with similar values of $\Omega \cong \Omega_{max}$. Such effects do not occur for higher values of N_0.

To check the computer program, as a second application we calculate the most probable fragment size distribution taking $N_0 = 100$ and use as constraints only the condition of conservation of mass. In this case, a direct comparison with analytical results derived above is possible. The results obtained are summarized in Table 10.2. Evidently, they are in a qualitatively good agreement (remember that the analytical expressions become correct in the limit of large numbers N_0 which is not reached here).

i	1	2	3	4	5	6	7	8
N_i (num. values)	4	6	7	6	4	2	1	0
N_i (anal. results)	3.39	5.74	6.48	5.48	3.72	2.10	1.01	0.43

Tab. 10.2 Fragment size distribution obtained for $N_0 = 100$ by computer calculations (numerical values) and according to the analytical expression Eq.(10.11).

As a next step, we take, now, energy conservation into account, in addition, in the form as given by Eq.(10.21). The results concerning the fragment size distributions formed in the system for different values of the excitation energy are summarized in Table 10.3. A schematic representation of the results is given in Fig. 10.4.

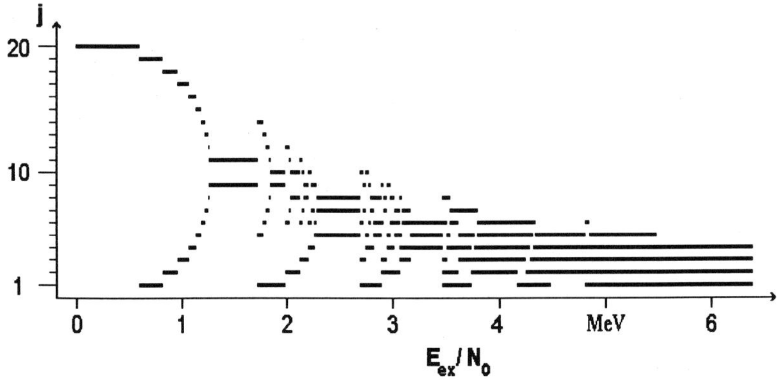

Fig. 10.4 Most probable distributions Ω_{max} in dependence on the excitation energy per nucleon E_{ex}/N_0. Hereby the number of primary particles was set equal to $N_0 = j_{max} = 20$.

In the calculations, represented in Fig. 10.4 and Table 10.3, the total number of particles in the system was set equal to $N_0 = j_{max} = 20$. The distributions of fragments, corresponding to the highest value of $\Omega = \Omega_{max}$, are represented in Table 10.3. In Fig. 10.4, it is shown graphically whether frag-

0.60 - 0.81	1																		1
0.82 - 0.95		1																1	
0.96 - 1.06			1														1		
1.07 - 1.13				1												1			
1.14 - 1.18					1										1				
1.19 - 1.22						1								1					
1.23 - 1.25							1						1						
1.26 - 1.26								1				1							
1.27 - 1.72									1		1								
1.73 - 1.77	1				1									1					
1.78 - 1.80	1					1							1						
1.81 - 1.83	1						1					1							
1.84 - 1.84	1							1			1								
1.85 - 1.98	1								1	1									
1.99 - 2.02		1				1						1							
2.03 - 2.03		1					1				1								
2.04 - 2.12		1						1		1									
2.13 - 2.14			1			1					1								
2.15 - 2.16			1				1			1									
2.17 - 2.20			1					1	1										
2.21 - 2.23				1		1				1									
2.24 - 2.26				1			1		1										
2.27 - 2.28					1	1			1										
2.29 - 2.69					1		1	1											
2.70 - 2.72	1		1			1				1									
2.73 - 2.74	1		1				1		1										
2.75 - 2.77	1			1	1					1									
2.78 - 2.79	1			1		1			1										
2.80 - 2.82	1			1			1	1											
2.83 - 2.89	1				1	1		1											
2.90 - 2.91		1	1			1			1										
2.92 - 2.94		1	1				1	1											
2.95 - 2.97		1		1	1				1										
2.98 - 3.01		1		1		1		1											
3.02 - 3.06		1			1	1	1												
3.07 - 3.08			1	1	1			1											
3.09 - 3.16			1	1		1	1												
3.17 - 3.46				1	2	1													
3.47 - 3.50	1	1	1			1		1											
3.51 - 3.53	1	1		1	1			1											
3.54 - 3.61	1	1		1		1	1												
3.62 - 3.73	1		1	1	1		1												
3.74 - 3.76		1	2		1		1												
3.77 - 3.79		1	1	2			1												
3.80 - 4.16		1	1	1	1	1													
4.17 - 4.24	2		1	1	1	1													
4.25 - 4.28	1	2		1	1	1													
4.29 - 4.29	1	1	2		1	1													
4.30 - 4.33	1	1	1	2		1													
4.34 - 4.47	1	1	1	1	2														
4.48 - 4.80		2	1	2	1														
4.81 - 4.84	2	1	2	1		1													
4.85 - 4.96	2	1	1	2	1														
4.97 - 5.41	1	2	2	1	1														
5.42 - 5.48	2	3	1	1	1														
5.49 - 6.04	2	2	2	2															
6.05 - INF	2	4	2	1															

Tab. 10.3 Most probable fragment size distributions in dependence on the excitation energy of the compund nucleus (left column) determined by a computer program as characterized in the text. In the range of energies $0 \leq E_{ex}/N_0 \leq 0.60$ MeV no fragmentation occurs. The total number of primary particles equals $N_0 = 20$.

ments of a given size exist or not, i.e., all fragments in the most probable distributions with multiplicities $N_i > 0$ are indicated. Hereby in Fig. 10.4 only the mere existence of such fragments is expressed but not the number of the respective fragments.

It turns out that only for excitation energies exceeding $E_{ex}/N_0 = 0.60$ MeV a fragmentation process occurs. Hereby first evaporation – like processes are observed going over continuouly into symmetric binary fission. This process is repeated then at a higher level, leading to three fragments in the most probable fragmentation channel etc. With increasing values of energy the constraint of conservation of energy as applied here looses its importance for the outcome of the fragmentation process. Asymptotically such distributions are attained for large values of the excitation energy, which result from the contraint extremum procedure taking into account only conservation of mass.

Table 10.3 and Fig. 10.4 show only such distributions which correspond to the highest numbe of possible realizations, Ω_{max}, of the fragment size distributions. However, in particular for small numbers of particles in the system, there exist also a variety of different distributions with similar values of the number of realizations, Ω.

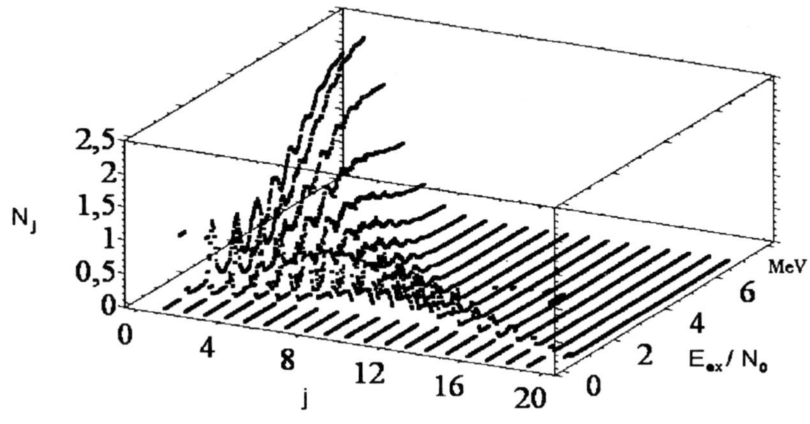

Fig. 10.5 Distribution obtained by averaging over all possible fragmentation channels. Again, the number of particles in the system equals $N_0 = j_{max} = 20$.

By this reason, we calculated in addition to the most probable distributions also those distributions which can be obtained by averaging over all possible

Obviously, the classical "equilibrium distribution" (supplemented by an additional term ($k_BT\tau \ln j$)) and Fisher's distribution (10.24) obtained by methods of equilibrium statistical physics in the framework of a droplet model approach are widely equivalent.

10.4.2 Limits of Applicability

In the part of the parameter space, where the homogeneous initial phase is stable from a thermodynamic point of view ($\Delta\mu < 0$ or $y < 1$), the change of the characteristic thermodynamic function in cluster formation $\Delta G(j)$ is a monotonicly increasing function and $N^{(e)}(j)$ (or $N^{(F)}(j)$ in Fisher's model) are monotonicly decreasing functions of j. Eqs.(10.24) and (10.33) are reduced then to Boltzmann–type expressions for the probability of heterophase fluctuations in a thermodynamic equilibrium state. However, for thermodynamically metastable or unstable states (determined by $\Delta\mu > 0$ or $y > 1$) both distributions do not describe real cluster size distributions (cf. also [299]).

With respect to Fisher's approach, which is based on the calculation of the partition function of the system, this conclusion follows immediately, for example, from a remark made by Fisher in his article ("... *to describe the equilibrium properties of a physical system we need only to calculate the partition function and use the formalism of statistical mechanics ...*" ([150], p. 250)). The extension of the results, obtained by methods of equilibrium statistical mechanics, into thermodynamic non–equilibrium states is, consequently, in general incorrect.

Fisher's method, as desired by him, allows to establish the boundaries of thermodynamically stable states and their properties, but cannot give us an adequate information concerning droplet size distributions of real systems in thermodynamically unstable regions (again, except, may be, in the immediate vicinity of the critical point). Indeed, assuming that the conditions of constancy of external pressure ($p > p_{eq}$, where p_{eq} is the equilibrium pressure for bulk liquid–gas coexistence at a temperature T) and temperature in the system can be sustained at any moment of time, the thermodynamic equilibrium state is that of a bulk liquid.

The same conclusions – inadequateness of the model to describe real cluster size distributions – can be drawn also with respect to the "equilibrium distributions" of classical nucleation theory. As it follows from the derivation sketched above, this distribution refers to a very artificial system (Szilard's model) not realized in nature (cf. chapter 2). Moreover, the stationary state considered in classical nucleation theory is not an equilibrium but a non–equilibrium steady state. Therefore, the applied method of minimization of the Gibbs free energy for the determination of the distribution function (see, e.g, [158]) lacks a thermodynamic foundation. It follows, consequently, also from a thermodynamic consideration that distributions of the type as expressed by Eqs.(10.24) and (10.33) do not reflect, in thermodynamic unstable initial states, cluster size distributions which may be established in real systems by nucleation and growth processes.

With respect to the determination of the evaporation or dissolution rates the failure of the so–called "equilibrium distribution" to describe real cluster size distributions is not crucial. It can be shown (see [510], chapter 2) that similar dependencies may be derived in an alternative way and applied for the determination of the kinetic coefficients but without assigning the meaning of a cluster size distribution in a real system to them.

Summarizing, we have to conclude that the classical "equilibrium distribution", respectively, Fisher's droplet model do not describe real cluster size distributions evolving in thermodynamically unstable states except, may be, in the immediate vicinity of the critical point. Therefore, they cannot be applied, in principle, for the interpretation of nuclear fragmentation processes taking into account the wide range of excitation energies for which distributions of the type $N(A) \propto A^{-\tau}$ occur and the interval of parameter values for τ observed experimentally. Provided, the physical nature of fragmentation can, indeed, be understood in terms of nucleation and growth, other approaches for the determination of the cluster size distributions have to be used. Two of such possible approaches will be analyzed in the next sections.

10.4.3 An Alternative Approach: The Steady–State Cluster Size Distribution of Classical Nucleation Theory

The classical theory of nucleation is based on the consideration of this process as a sequence of binary reactions. The basic equations used for a kinetic description of nucleation and growth are given by (cf. chapter 2)

$$\frac{\partial N(j,t)}{\partial t} = J(j-1,t) - J(j,t) \qquad \text{for} \qquad j \geq 2 \tag{10.34}$$

with

$$J(j,t) = w^{(+)}(j,t)N(j,t) - w^{(-)}(j+1,t)N(j+1,t) \ . \tag{10.35}$$

Here $w^{(+)}(j,t)$ represents the average number of monomers incorporated into a cluster of size j per unit time, while the kinetic coefficients $w^{(-)}(j,t)$ describe similarly the rate of emission processes.

Assuming that the number of monomers per unit volume is fixed (the supersaturation is kept constant) and that in addition the relation $N(j,t) = 0$ for $j \gg j_c$ holds, in the course of time a steady–state cluster size distribution $N^{(s)}(j)$ is established in the system.

An expression applicable in the whole range of cluster sizes was proposed by Binder and Stauffer ([39], see also [511, 543] and chapter 2). It reads (cf. Eq.(2.51))

$$\left(\frac{N^{(s)}(j)}{N^{(e)}(j)}\right) \cong \left(\frac{1}{2}\right) \operatorname{erfc}\left\{\Gamma_{(z)}\sqrt{\pi}(j - j_c)\right\} \ , \tag{10.36}$$

$$\operatorname{erfc}(\xi) = \frac{2}{\sqrt{\pi}} \int_{\xi}^{\infty} \exp(-z^2)\, dz \ . \tag{10.37}$$

With Eq.(10.38)

$$\Gamma_{(z)} = \left\{ -\frac{1}{2\pi k_B T} \left(\left.\frac{\partial^2 \Delta G}{\partial j^2}\right|_{j=j_c} \right) \right\}^{1/2} = \frac{1}{j_c}\sqrt{\frac{\Delta G(j_c)}{3\pi k_B T}} \tag{10.38}$$

this relation is equivalent to

$$\left(\frac{N^{(s)}(j)}{N^{(e)}(j)}\right) \cong \left(\frac{1}{2}\right) \operatorname{erfc}\left\{ \sqrt{\frac{\Delta G(j_c)}{3k_B T}} \left(\frac{j}{j_c} - 1\right) \right\} \ . \tag{10.39}$$

The steady–state and "equilibrium" cluster size distributions, according to above given equations, are presented qualitatively in Fig. 10.6. It is seen that the ratio $(N^{(s)}(j)/N^{(e)}(j))$ tends to zero for large values of j $(j > j_c)$.

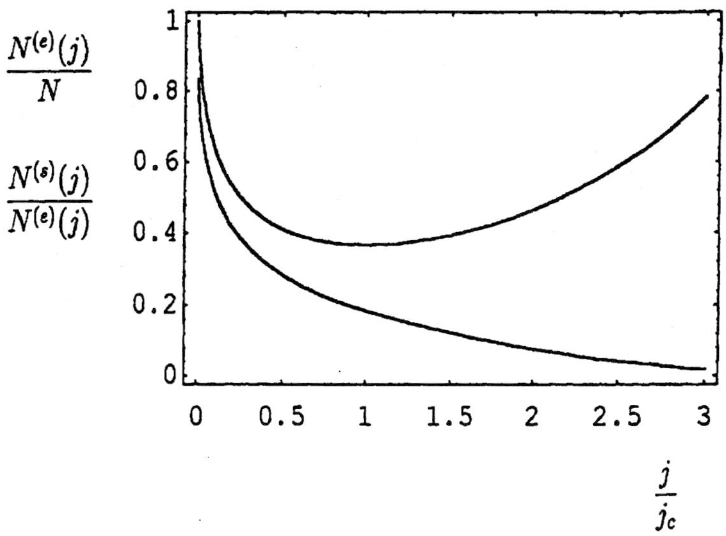

Fig. 10.6 Cluster size distribution in relative coordinates (j/j_c) according to Fisher's droplet model, respectively, the widely equivalent "equilibrium" distribution $N^{(e)}(j)$ of classical nucleation theory for thermodynamically unstable states (according to Eq.(10.28), upper curve). As the second curve the steady–state cluster size distribution is shown (Eq.(10.39), lower curve). $(\Delta G(j_c)/k_B T)$ was chosen equal to one for convenience.

Obviously, the steady–state distribution is to be preferred from a physical point of view as compared with the so–called "equilibrium distribution" or the distribution derived by Fisher. Its derivation does not involve the extension of the methods of equilibrium statistical physics to non–equilibrium states or thermodynamically not well–founded argumentations. Moreover, not only the ratio $(N^{(s)}(j)/N^{(e)}(j))$ tends to zero for large values of j but also the function $N^{(s)}(j)$ itself behaves in such a physically reasonable (for clustering processes) way. Indeed, for large values of j $(j \gg j_c)$ or, equivalently, large values of ξ we may write with Eqs.(10.26), (10.27) and (10.36)

$$\lim_{j\to\infty} N^{(s)}(j) = \frac{1}{\sqrt{\pi}} \lim_{j\to\infty} \left\{ N^{(e)}(j) \int_{\xi}^{\infty} \exp(-z^2)\, dz \right\} , \qquad (10.40)$$

$$\xi = \sqrt{\pi}\Gamma_{(z)}(j - j_c) , \qquad (10.41)$$

resulting with the approximation (valid for large positive values of ξ)

$$\int_{\xi}^{\infty} \exp(-z^2)\, dz \leq \frac{1}{\xi} \int_{\xi}^{\infty} z \exp(-z^2)\, dz \tag{10.42}$$
$$\cong \frac{1}{2\sqrt{\pi}\Gamma_{(z)} j} \exp\left(-\pi \Gamma_{(z)}^2 j^2\right)$$

in

$$\lim_{j\to\infty} N^{(s)}(j) \propto \lim_{j\to\infty} \left\{ \left(\frac{1}{j}\right) \cdot \exp\left[\left(\frac{\Delta\mu}{k_B T}\right) j\right] \exp\left(-\pi \Gamma_{(z)}^2 j^2\right)\right\} = 0\,. \tag{10.43}$$

This limit is reached, in a good approximation, already at a value of j equal to (see [511])

$$j = j_c + \Delta j\,, \qquad \Delta j = \frac{1}{\sqrt{\pi}\Gamma_{(z)}}\,. \tag{10.44}$$

Therefore, the given dependencies for $N^{(s)}(j)$ or even more sophisticated time–dependent expressions (e.g., [94, 234]) modelling in addition the approach of the distribution to the steady state could be used for an interpretation of experimental results on clustering processes, e.g., in supercooled gases or expanding nucleonic systems [85, 112]. In particular, it turns out also from the outlined here considerations that for cluster sizes $j < j_c$ the "equilibrium distribution" of classical nucleation theory, respectively, Fisher's droplet model are retained in the steady–state distribution with an accuracy of a factor of the order (1/2).

However, in most practical applications, including nucleation and growth occuring possibly as the result of heavy ion collisions, the density of monomers in the system is not kept constant but decreases with time due to aggregation and growth. It follows that the main boundary condition – constancy of the number of free monomers per unit volume – applied for the determination of the steady state distributions is not fulfilled or only approximately fulfilled for relatively short time intervals at the beginning of this process.

To establish correct time dependencies for cluster (or droplet) size distributions in such non–stationary non–equilibrium processes, one has to rely on the numerical solution of the set of kinetic equations for boundary conditions appropriate for the considered applications. In particular, one has to take into account that the total number of monomers – including free particles as well as monomers aggregated in clusters (droplets) of different sizes – is kept constant, i.e., the equation

$$N(1,t) + \sum_{j=2}^{\infty} jN(j,t) \equiv N = \text{constant} \tag{10.45}$$

has to be fulfilled. The results obtained from such an approach will be discussed in the next section.

10.4.4 Numerical Solution of the Kinetic Equations Describing Nucleation and Growth

In accordance with the classical theory of nucleation, we assume that growth and decay processes of clusters proceed only via aggregation or emission of monomers. The clusters are assumed to be of spherical size and are characterized by the number of monomers j, contained in them, or by a radius R_j. By $N(j,t)$ the number of cluster consisting of j monomers is denoted.

The time evolution of the cluster size distribution function, $N(j,t)$, is governed under these assumptions by the set of equations (10.34) and (10.35). Specific features of the mechanism of growth enter the description via the determination of the kinetic parameters $w^{(+)}$ and $w^{(-)}$.

Two mechanisms of growth, which are of importance in a number of applications, are the diffusion and kinetic limited growth modes. For diffusion–limited growth we have (for the limiting case of a perfect solution or a perfect gas; see [247, 510])

$$w^{(+)}(j,t) = 4\pi R_j D c \ , \tag{10.46}$$

$$w^{(-)}(j,t) = 4\pi R_j D c_{R_j} \ , \tag{10.47}$$

while for kinetic-limited growth

$$w^{(+)}(j,t) = 4\pi R_j^2 D \frac{c}{l_0} \ , \tag{10.48}$$

$$w^{(-)}(j,t) = 4\pi R_j^2 D \frac{c_{R_j}}{l_0} \tag{10.49}$$

holds.

Hereby c is the concentration of the segregating particles far away from the growing or dissolving clusters, c_{R_j} specifies the equilibrium concentration of segregating particles in the vicinity of a cluster of size R_j. Its value is given by (e.g. [187])

$$c_{R_j} = c_{eq} \left[\exp\left(\frac{2\sigma}{c_\alpha k_B T R_j} \right) \right] \ . \tag{10.50}$$

c_{eq} is the equilibrium concentration (or density) of the segregating component in the ambient phase in contact with the newly evolving phase at a planar interface.

Denoting by v_α the volume and by R_1 the radius of a monomeric ambient phase particle we may write further

$$v_\alpha = \frac{1}{c_\alpha} = \left(\frac{4\pi}{3} \right) R_1^3 \ . \tag{10.51}$$

Taking into account this relation, the radius R_j of a cluster of size j may be expressed as

$$R_j = \left(\frac{3 v_\alpha}{4\pi} \right)^{1/3} j^{1/3} \ . \tag{10.52}$$

For the numerical calculations we will use further a dimensionless time scale defined by

$$t' = 4\pi D c_{eq} \left(\frac{3v_\alpha}{4\pi}\right)^{1/3} t \,. \tag{10.53}$$

The parameter l_0, introduced with Eqs.(10.48) and (10.49), we identify in the calculations with R_1.

A substitution of Eqs.(10.46)–(10.53) into the basic equations (10.34) and (10.35) yields

$$\begin{aligned}\frac{\partial N(j,t')}{\partial t'} &= \left(\frac{c}{c_{eq}}\right)(j-1)^{1/3} N(j-1,t') \\ &+ j^{1/3} \exp\left(\frac{2\sigma}{c_\alpha k_B T R_{j+1}}\right) N(j+1,t') \\ &- \left[\left(\frac{c}{c_{eq}}\right) + \left(\frac{j-1}{j}\right)^{1/3} \exp\left(\frac{2\sigma}{c_\alpha k_B T R_j}\right)\right] j^{1/3} N(j,t')\end{aligned} \tag{10.54}$$

for diffusion limited growth and

$$\begin{aligned}\frac{\partial N(j,t')}{\partial t'} &= \left(\frac{c}{c_{eq}}\right)(j-1)^{2/3} N(j-1,t') \\ &+ j^{2/3} \exp\left(\frac{2\sigma}{c_\alpha k_B T R_{j+1}}\right) N(j+1,t') \\ &- \left[\left(\frac{c}{c_{eq}}\right) + \left(\frac{j-1}{j}\right)^{1/3} \exp\left(\frac{2\sigma}{c_\alpha k_B T R_j}\right)\right] j^{2/3} N(j,t')\end{aligned} \tag{10.55}$$

for kinetic limited growth.

For the numerical solutions of Eqs.(10.54) and (10.55), Euler's Polygon method is used (cf. [247]), i.e., the change of the number of clusters consisting of j monomers is calculated by

$$N(j, t' + \Delta t') = N(j,t') + \frac{\partial N(j,t')}{\partial t'} \Delta t' \,. \tag{10.56}$$

In the computations, we chose the following values of the parameters

$$l_0 = R_1 \,, \qquad \frac{2\sigma}{c_\alpha k_B T} = 0.8\,\mathrm{nm} \,, \qquad c_{eq} = 4.2 \cdot 10^{26}\,\mathrm{m}^{-3} \,. \qquad (10.57)$$

In the determination of the kinetic coefficients $w^{(-)}$ the work of formation $\Delta G(j)$ of clusters of size j is applied in the form as given by Eq.(10.27), i.e., as determined in classical nucleation theory. This choice has been made at part to demonstrate, as will be evident from the numerical results outlined subsequently, that the possibility of explanation of droplet size distributions of the form $N(j) \propto j^{-\tau}$ (or $N(A) \propto A^{-\tau}$ in nuclear fragmentation) based on a nucleation–growth model is not connected with the term $(k_B T \tau \ln j)$ additionally introduced by Fisher. This approach was used already in chapters 2 and 9 in the preparation of some of the figures.

Based on the solution of the set of kinetic equations, in Fig. 2.2 the evolution of the cluster size distribution is shown for different moments of time first for the case that the number of monomers in the system (respectively, the supersaturation) is artificially kept constant. In the course of time, a steady–state cluster size distribution is established in the system well–aproximated by the analytical expressions given in the preceding section.

As seen in Fig. 9.1, if the artificial condition of constancy of the supersaturation is omitted then the time evolution of the distribution behaves quite differently. Starting with an initial distribution consisting of monomers only, in the first stage of the development a monotonicly decreasing distribution of clusters is established rapidly. This distribution is transformed in the further evolution into a bimodal curve characterized by a peak for small cluster sizes (monomers, dimers, trimers, etc.) and a second peak representing clusters of near–critical and supercritical sizes capable to a further deterministic growth.

An inspection of Fig. 9.1 shows that monotonicly decreasing dependencies $N(j,t)$ vs j are found only in the first stages of the nucleation–growth process. In connection with the discussion of nuclear multifragmentation hereby the question is of particular interest, whether these dependencies may be approximated by curves of the form

$$N(j) \propto j^{-\tau} \,. \qquad (10.58)$$

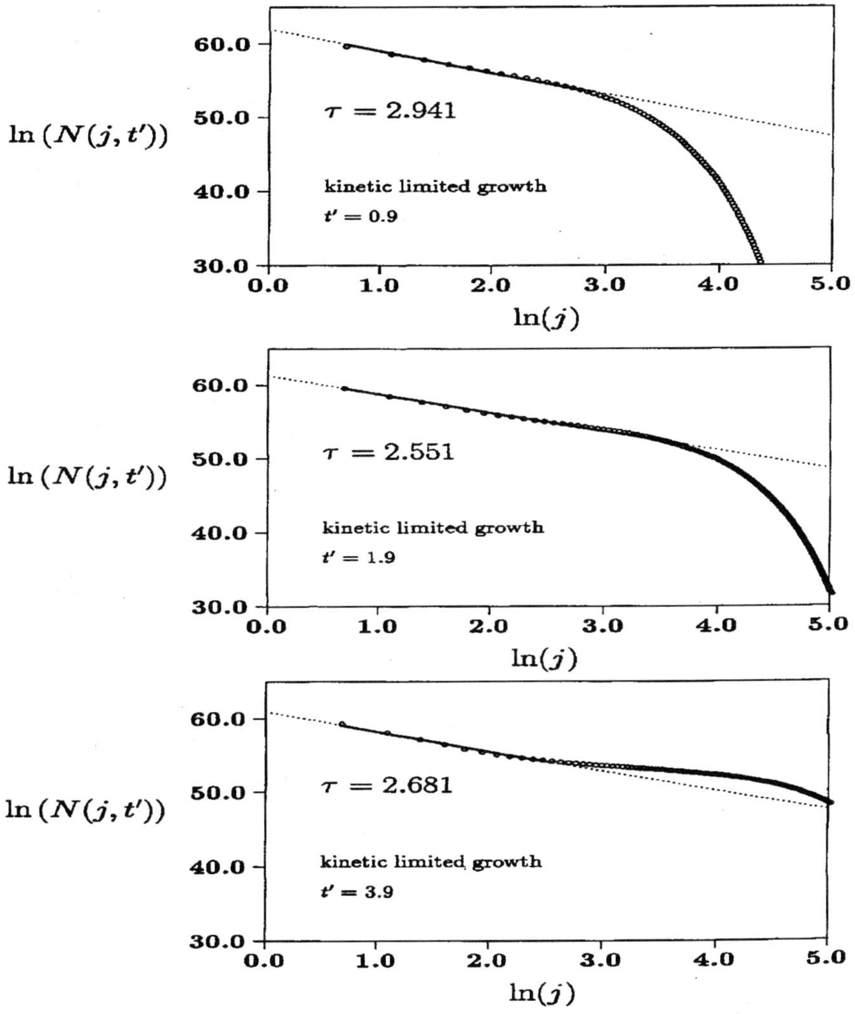

Fig. 10.7 $N(j, t)$ vs j curves for kinetically limited growth shown in coordinates $\ln N(j)$ vs $\ln j$ for different moments of time ($t' = 0.9$, 1.9, 3.9). For cluster sizes less than a value j_{max} the distributions obtained numerically (circles) can be well-fitted by dependencies of the form $N(j) \propto j^{-\tau}$ with different values of τ (full, respectively, dotted lines). In the range of cluster sizes, where the fit is sufficiently accurate the distribution is approximated by a full curve. The respective values of τ are given in the figures. The initial supersaturation was chosen equal to $[(\Delta\mu/k_BT) = \ln 10]$

To give an answer to this question, on Figs. 10.7 (a – c) the $N(j,t)$ vs j curves are shown for an initial supersaturation ($\Delta\mu/k_BT = \ln 10$) for different moments of time ($t' = 0.9$, 1.9, 3.9) in coordinates $\ln N(j)$ vs $\ln j$. It can be seen that for cluster sizes $j \leq j_{max}$ the distributions obtained numerically (circles) can be well–fitted by such a dependence (full, respectively, dotted curves). Hereby the highest value of the size of the clusters, j_{max}, for which such a fit is possible, as well as the value of the exponent τ change with time. This result – the explicite time dependence of τ – gives an additional confirmation that Fisher's or similar static models are inappropriate for the interpretation of cluster size distributions observed experimentally.

In addition to the mentioned time dependencies, the values of τ and j_{max} may vary also in dependence on the initial supersaturation. In Figs. 10.8 a and b, these values of τ and j_{max} are shown as functions of time for kinetic limited growth and different values of the initial supersaturation in the system. It is easily verified that the values of τ are shifted to higher values with increasing supersaturation. Moreover, for the whole range of supersaturations considered j_{max} increases first followed by a sudden decrease.

The origin of this sharp decrease of j_{max} is to be found in the formation of the second maximum in the $N(j,t)$–curves (cf. Fig. 9.1). Once the second maximum appears, curves, of the form as given by Eq.(10.58), cannot describe the shape of the distribution in the whole range of j–values but may be further applied, nevertheless, to a description of the peak at small cluster sizes (compare, again, Fig. 9.1).

It can be shown [482] that the type of behavior does not depend neither on the mode of growth (at least, for the discussed here two cases) nor on the choice of the expression for the work of cluster formation [483].

10.4.5 Application to Nuclear Multi–Fragmentation Processes

The results of the solution of the set of kinetic equations outlined in the preceding section show that the nucleation–growth model allows in a kinetic approach a straightforward explanation of cluster (or droplet) size distributions of the type $N(j) \propto j^{-\tau}$ with different values of τ in the range ($1 \leq \tau \leq 6$) as well as of the evolution of bimodal cluster size distributions.

However, the calculations have been carried out so far for isothermal conditions and a fixed volume of the system. Both conditions are obviously

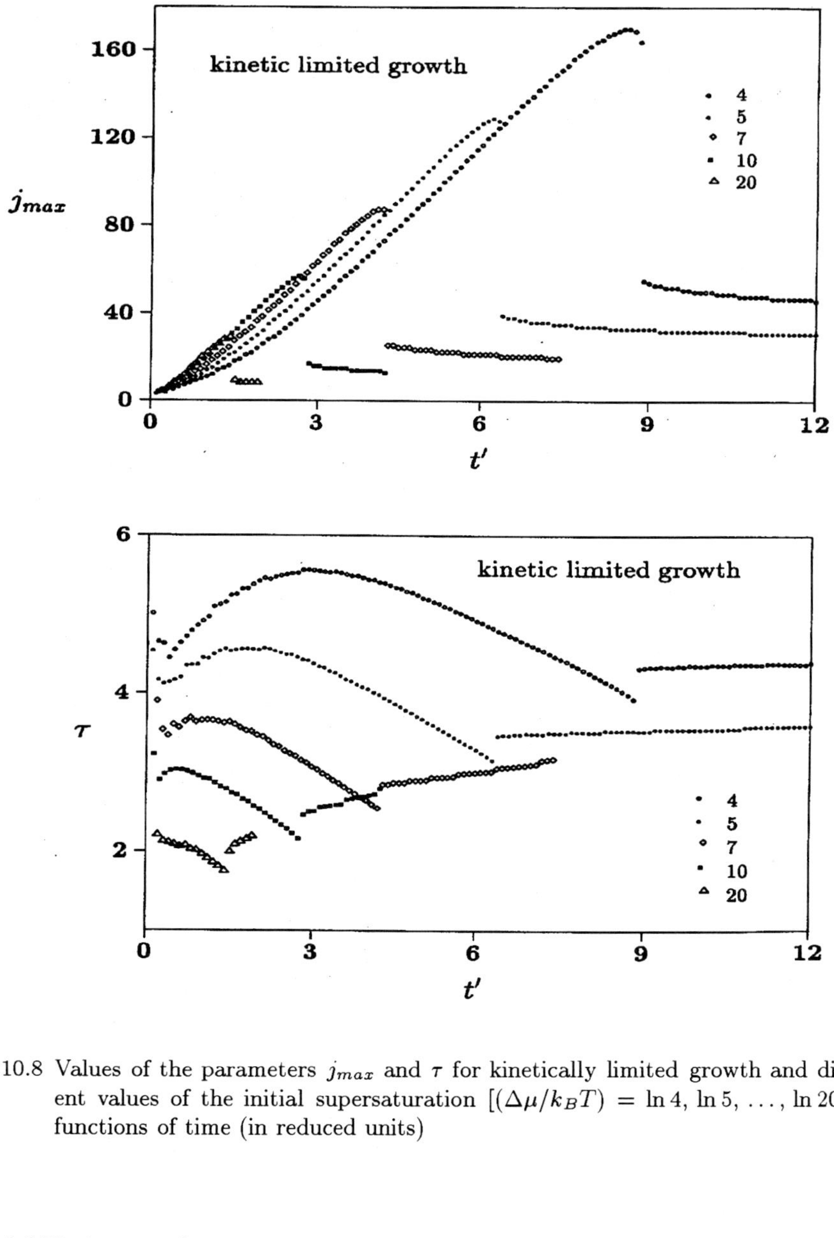

Fig. 10.8 Values of the parameters j_{max} and τ for kinetically limited growth and different values of the initial supersaturation $[(\Delta\mu/k_BT) = \ln 4, \ln 5, \ldots, \ln 20]$ as functions of time (in reduced units)

not fulfilled in real experiments on nuclear collisions. Therefore, it has to be analyzed how the results obtained can be applied to an interpretation of fragment size distributions measured as the result of nuclear collision processes.

In line with the widely expressed opinion, we may assume that in nuclear collisions the system of nucleons is compressed initially to a considerable degree. The potential energy stored in the compression is released in the subsequent expansion process of the initially hot ensemble of nucleons. This expansion of the nucleonic system is accompanied by a decrease in temperature leading to the possibility of clustering processes of the nucleons.

On the other hand, the expansion results also in a decrease of the average density of the system. It follows that aggregation processes become less probable with time and that certain states of the evolution of the system are frozen–in, which are observed later, may be modified to some extent, experimentally. In lattice–gas, percolation or statistical fragmentation models a similar choice of the appropriate fragment size distribution is carried out by the selection of the so–called freeze–out volume or freeze–out density (see, e.g., [65, 179, 411, 412]).

In this way, the different distributions obtained for varying initial supersaturations and fixed values of temperature can be considered to represent a spectrum of possible configurations capable to be observed experimentally. Which of the possible distributions will be realized in a given experimental situation depends on the value of the incident energy, thermodynamic and kinetic properties of nuclear matter, determining, in particular, the ratio of the characteristic time scales for aggregation and expansion.

These qualitative arguments are reconfirmed by the detailed analysis of clustering in expanding gases outlined in detail in chapter 11.

Moreover, as seen from the results of the computer calculations, with an increase of the initial supersaturation (respectively, incident energy in nuclear collisions) the coefficient τ is shifted to lower values. On the other hand, the rate of expansion is increased by increasing the incident energy and, therefore, the time interval, where intensive aggregation processes may occur, decreases. This effect may result in an increase of τ as seen from the dependence of τ on time as obtained in the numerical solutions of the set of equations describing nucleation and growth (see Figs. 10.8 and [482]). Both opposite tendencies in the dependence of τ on the incident energy may give, thus, the key for a kinetic interpretation of the minimum of τ in dependence on the incident energy observed experimentally [322, 378, 411, 412].

The kinetic interpretation of nuclear fragmentation in terms of a nucleation–growth model, developed here, is also in agreement with the experimental observation that bimodal fragment distributions are observed for relatively low incident energies only (in the range, of course, where multifragmentation occurs) accompanied by relatively low expansion rates of the ensemble of nucleons [306, 376]. With an increase of the incident energy the second maximum disappears. This effect can be interpreted by assuming that the time interval the cluster ensemble is allowed to evolve is too short to lead to the second extremum, due to fast expansion processes.

Similarly, also the decrease with increasing incident energy of the maximal value of nucleons in the intermediate mass fragments observed experimentally and the increase of their multiplicity (see, e.g., [395]) can be interpreted in such a way (see also the results of the numerical studies concerning the energy dependence of fragment size distributions in multifragmentation by Latora et al. [301]).

As mentioned, a more detailed analysis of these problems by considering nucleation–growth processes in expanding matter is given in chapter 11 (see also [484, 485]).

10.4.6 Discussion

In the present section, a detailed analysis of possible shapes of cluster size distributions is given evolving by nucleation and growth processes from thermodynamically unstable or metastable homogeneous initial states. The results of the analysis are of interest for the understanding of the course of phase transformation processes, in general (like nucleation in supersaturated vapors or solutions (e.g. [180, 187, 584, 585]) or, for example, clustering in atomic and molecular expanding gases [518]. Here in addition to generally valid results, in particular, the possible applications for a theoretical interpretation of the shape of fragment size distributions are discussed which are observed experimentally as the result of heavy–ion collision processes. In this way, nuclear multifragmentation is interpreted as a selforganization process in a higly excited nucleonic non–equilibrium system.

As already mentioned, in the present study we are interested mainly in the verification of the principal applicability of nucleation–growth models to the interpretation of nuclear fragmentation. A detailed consideration of particular features of nucleonic systems like binding energies, collision cross sections, finite size effects, Coulomb interactions etc. in dense nucleonic systems is not attempted. In principle, such properties may be calculated in the framework of quantum statistical many–body approaches. However, a final solution of all the problems involved in such a task is, at present, out of reach.

One reason for the widespread interest in nuclear fragmentation processes consists in the possibility to derive conclusions concerning the thermodynamic properties of nuclear matter from the results of fragmentation experiments and their dependence on the energy of the collisions and other constraints. This possibility is retained also in the kinetic interpretation, analyzed here, taking into account, for example, the dependence of the values of τ on the initial supersaturation. However, as shown, in the derivation of such conclusions one has to take into account explicitely the dynamics of evolution of the ensemble of nucleons.

This dynamically based way of derivation of the properties of nuclear matter is less straightforward as compared with the traditionally applied but not generally applicable methods based on Fisher's droplet model. On the other hand, the amount of information concerning the properties of nuclear matter, one can obtain in this way, seems to be much larger and more precise than in the traditional approach.

11 Nucleation and Growth in Freely Expanding Gases

J. Schmelzer, G. Röpke

11.1 Introduction

In recent times, condensation processes in expanding systems have become of particular interest for a variety of applications (cf., e.g., Refs. [32, 105, 189, 230, 273, 378, 482, 518]). The examples range from condensation processes in the atmosphere to phase transitions in the early universe.

In the theoretical analysis of such processes a number of additional problems has to be solved beyond the proper description of the nucleation process itself. In expanding systems the boundary conditions for the nucleation–growth process are changing with time. The description of the process in terms of the steady–state nucleation rate of classical nucleation theory is, therefore, inappropriate. The situation is here similar to the description of the kinetics of nucleation with varying boundary conditions (e.g. nucleation and growth at oscillating supersaturation [2], crystallization of glasses in heating processes [187] etc.). Moreover, a proper determination of the changes of the state of the system and its characteristics in the course of the expansion is required.

In the present paper we develop a general model allowing to describe the basic features of clustering in expanding matter, in general, and gases, in particular. The model assumptions do not involve any serious restrictions concerning the equation of state of the gas (ambient phase, in general) and the evolving liquid (newly formed) phase. It is supposed only that with respect to co–moving reference frames at any place of the expanding system a local thermodynamic equilibrium is established at any time. In particular, we do not assume from the very beginning that the expansion proceeds isentropically. This way the possibility is opened to study in addition to other thermodynamic characteristics also the dependence of the entropy and other thermodynamic state functions on time. Hereby special emphasis will be given to the effect of clustering on the entropy.

The assumption of a local equilibrium is, on the other side, the basic precondition for an application of the standard kinetic theory of nucleation and growth. This way, both the model of expansion as well as the theoretical approach to nucleation, applied here, are formulated at the same level of approximation of the real process.

As mentioned, the development of the basic equations, governing clustering in expanding gases, is shown to be possible widely independent on any particular assumptions concerning the equation of state of the system under consideration. Additional information is required, however, when the set of basic equations is to be solved analytically or numerically.

In the derivation of consequences from the model, here the assumptions concerning the thermodynamic properties of the substance considered are chosen for an illustration as simple as possible. Nevertheless, qualitatively the results are expected to hold also for real expansion processes. Moreover, prescriptions are given, how specific properties of different systems may be incorporated adequately into the theory.

The chapter is organized as follows. In section 11.2 the model is developed based on the analysis of thermodynamic aspects of the expansion process. The thermodynamic properties of a gas in free expansion are analyzed in section 11.3 under the additional assumption that clustering processes do not occur. In section 11.4 the set of kinetic equations is formulated applied for the description of clustering. Results of the solution of the set of kinetic equations and consequences are presented in section 11.5. Possible applications and future developments are discussed in section 11.6.

11.2 Free Expansion of Gases. The Model

11.2.1 General Thermodynamic Aspects

Utilizing classical thermodynamics, the process of free expansion may be analyzed if its initial and final states are thermodynamic equilibrium states [269, 295, 437]. Since in the process of free adiabatic expansion from one of these equilibrium states to another one no work is performed and no heat is released by the system, the total internal energy U remains constant, i.e. the relation

$$U_1(T_1, V_1) = U_2(T_2, V_2) \tag{11.1}$$

holds. In Eq.(11.1), T is the absolute temperature and V is the volume of the system under consideration. The numbers refer to the initial (1) and the final (2) equilibrium states of the process. In particular, for a perfect gas the internal energy is a function only of temperature [269, 295, 437]. In the process of free expansion of a perfect gas from one equilibrium state to another one, the temperature in the final state is, consequently, the same as in the initial one.

The details of the process of evolution of the gas between both considered states may be highly complex. Its detailed description is, however, not required if only a comparison between the initial and final equilibrium states is attempted. The situation becomes, however, quite different if clustering in a freely expanding gas is considered. Here the process of nucleation and growth is determined by the way the system evolves in time. To obtain an adequate description of clustering, at least, the main features of the evolution of the system have to be incorporated into the description.

Here we want to develop a model which reflects such basic features of free adiabatic expansion of gases. The model has the advantage that it allows the application of thermodynamic methods as well as of the basic equations of the theory of nucleation and growth. This property of the model determines, on the other side, the range of applicability. It cannot describe features which go beyond the developed mean field type approximation.

The basic characteristics of clustering in free expansion of gases, which have to be incorporated in any realistic model of this process, are the following:

• The free expansion of gases is accompanied by a transformation of a part of its internal energy into mechanical energy of directed collective motion of the gas (cf., e. g., Refs. [32, 105]). We may write thus

$$U = U^{(therm)} + U^{(flow)} . \tag{11.2}$$

The thermodynamic properties of the gas are determined by the first contribution, $U^{(therm)}$. This term describes the energy of chaotic thermal motion including the average potential energy of interaction of the gas particles.

The second term, $U^{(flow)}$, accounts for the part of the initial value of the internal energy transformed into energy of macroscopic flow, i.e., energy of directed macroscopic motion.

- The decrease of the thermal part of the energy, $U^{(therm)}$, in the expansion results in a decrease of the temperature of the system.

- For real gases, there exists an additional factor affecting the value of temperature in free expansion. It consists in the dependence of the internal energy on the volume of the gas. This effect occurs already if different equilibrium states of the gas are compared. It is connected with a redistribution of the energy between average kinetic and potential energy terms of the thermal motion.

- By the change in temperature, the gas may be transferred into thermodynamic non–equilibrium states. In such non–equilibrium states, clustering processes may result in the evolution of the new, liquid phase.

- Clustering processes are accompanied by a release of latent heat. This release of the latent heat has to be taken into account, in addition, in the determination of the actual values of temperature (of course, if such a quantity may be defined at all for the system as a whole).

In order to get an accurate description of these different features we proceed in the following way. Instead of the real expanding gas cloud we consider first an idealized model system – a gas in thermal equilibrium enclosed by a membrane. The basic equations describing the expansion process are established first for this idealized model system. In the analysis of this model system, a model parameter occurs. This model parameter will be determined in a next step in such a way as to get an appropriate description of the expansion of the real gas cloud.

11.2.2 The Idealized Model

The basic features of free expansion of gases, undergoing simultaneously a first–order phase transformation, are reflected in the following simplified model.

Let us assume, we have some amount of a gas enclosed by a membrane impenetrable for the gas. The system is considered to be of spherical shape. Moreover, it is supposed that the membrane can be deformed arbitrarily without doing any work on it. The size of the gas cloud (or the radius of the membrane in the model approach) we denote by R. At time t, each part of the membrane is moving with a velocity $u(t)$ in radial direction with respect to the center of the sphere.

The gas inside the sphere in the simplified model approach is assumed to be at any moment of time in a state near thermodynamic equilibrium. This property allows us to assign definite values to the thermodynamic parameters of the gas like chemical potential μ, temperature T, pressure p, particle density c, mass density ρ etc. Moreover, as we will see later, this assumption is also the basis for the determination of the coefficients of aggregation in the kinetic model of clustering (cf. section 11.4).

To describe the change of temperature due to the conversion of chaotic thermal motion into directed macroscopic flow, we assign a mass $M^{(*)}$ to the membrane. This mass has to be considered so far as a model parameter which has to be determined appropriately in order to relate the results of the model analysis to the behavior of real systems. As shown somewhat later it is of the order of the total mass of the gas.

Quantitatively, the following relations hold:

- The work performed by the gas in a time interval dt on the membrane in the course of the expansion is given by

$$dA = 4\pi R^2(t)p(t)dR \, . \tag{11.3}$$

Here $R(t)$ is the current radius of the gas cloud (the radius of the membrane in the model), and $p(t)$ is the actual value of the pressure of the gas phase. With

$$u(t) = \frac{dR}{dt} \tag{11.4}$$

we may write equivalently

$$dA = 4\pi R^2(t)p(t)u(t)dt \, . \tag{11.5}$$

Here $u(t)$ is the velocity of radial expansion of the membrane (or the external boundary of the gas cloud).

The work performed by the gas on the membrane results in a change of the kinetic energy $U^{(mem)}$ of the membrane (in the model) or, equivalently, in a variation of the energy of collective flow, $U^{(flow)}$. We get

$$dU^{(mem)} = dU^{(flow)} = 4\pi R^2(t)p(t)u(t)dt \; . \tag{11.6}$$

• The work performed on the membrane leads, on the other side, to a variation of its velocity. Utilizing Eq.(11.6) and

$$U^{(flow)} = U^{(mem)} = \frac{M^{(*)}}{2}u^2(t) \; , \tag{11.7}$$

we obtain

$$du(t) = \frac{4\pi R^2(t)p(t)}{M^{(*)}}dt \; . \tag{11.8}$$

This way, an equation for the determination of the velocity of expansion $u(t)$ of the gas cloud is found.

• By assumption, the freely expanding gas does no work on other bodies and is adiabatically isolated from the environment. We have, consequently,

$$dU^{(therm)} = -dU^{(flow)} \; . \tag{11.9}$$

• In general, the system will consist of two phases, the liquid and the gas. Utilizing the relation

$$U^{(therm)} = U^{(therm)}_{gas} + U^{(therm)}_{liquid} \; , \tag{11.10}$$

we obtain with Eqs.(11.6), (11.10) and the relation $dU^{(therm)} = C_v dT$ the following result

$$\begin{aligned} dT^{(flow)} &= \frac{1}{\left[C_v^{(gas)} + C_v^{(liquid)}\right]} dU^{(therm)} \\ &= -\frac{4\pi R^2(t)p(t)u(t)}{\left[C_v^{(gas)} + C_v^{(liquid)}\right]} dt \; . \end{aligned} \tag{11.11}$$

$dT^{(flow)}$ is the change of temperature due to a conversion of thermal energy into energy of collective flow. By C_v the specific heats of the two phases are denoted. We have to take here the value of the specific heat at constant volume. Changes of temperature due to variations of the volume of the system are incorporated in the next step.

• For a real gas, in addition, the change of temperature $dT^{(vol)}$ due to the change of the volume of the gas has to be accounted for. Changes of the temperature of the system due to a conversion of internal energy into energy of collective flow have been taken into consideration already. What is left is to calculate the change of temperature dT due to a change of the volume dV for constant values of the internal energy $U^{(therm)}$. Hereby the number of single particles, N, in the gas is assumed to be constant. With

$$\begin{aligned} U_{liquid}^{(therm)}(T,V) + U_{gas}^{(therm)}(T,V) &= U_{liquid}^{(therm)}(T+dT, V+dV) \\ &+ U_{gas}^{(therm)}(T+dT, V+dV) \end{aligned} \tag{11.12}$$

we have after a Taylor expansion

$$\begin{aligned} \left(\frac{\partial U_{liquid}^{(therm)}}{\partial T}\right)_V dT + \left(\frac{\partial U_{liquid}^{(therm)}}{\partial V}\right)_T dV &= \\ = - \left[\left(\frac{\partial U_{gas}^{(therm)}}{\partial T}\right)_V dT + \left(\frac{\partial U_{gas}^{(therm)}}{\partial V}\right)_T dV\right] . \end{aligned} \tag{11.13}$$

Neglecting variations of the internal energy of the liquid phase due to changes of the total volume of the gas phase, we get

$$dT^{(vol)} = -\frac{1}{\left[C_v^{(gas)} + C_v^{(liquid)}\right]} \left(\frac{\partial U_{gas}^{(therm)}}{\partial V}\right)_{T,N} dV . \tag{11.14}$$

Applying well–known thermodynamic relationships (cf. Refs. [437, 269, 295]) to replace $(\partial U/\partial V)$ by the thermal equation of state of the gas, we obtain, finally,

$$dT^{(vol)} = -\frac{1}{\left[C_v^{(gas)} + C_v^{(liquid)}\right]} \left[T\left(\frac{\partial p}{\partial T}\right)_{V,N} - p\right] dV . \tag{11.15}$$

As it has to be the case, for a perfect gas $dT^{(vol)}$ is equal to zero.

• The net change in temperature in the system due to both effects may be written as

$$dT = dT^{(flow)} + dT^{(vol)} \tag{11.16}$$

or

$$dT = -\frac{4\pi R^2(t)u(t)}{\left[C_v^{(gas)} + C_v^{(liquid)}\right]}\left[T\left(\frac{\partial p}{\partial T}\right)_{V,N}\right] dt\ . \tag{11.17}$$

• Eq.(11.17) describes the change in temperature in an expanding gas in the absence of clustering processes. Clustering and the resulting release of latent heat is reflected by an additional term in the expression for dT of the form

$$dT^{(clust)} = -\frac{\hat{q}}{\left[C_v^{(gas)} + C_v^{(liquid)}\right]} dN\ . \tag{11.18}$$

Here $\hat{q}$ is the heat of evaporation per particle ($\hat{q} > 0$), N the number of gas particles in the system.

Eqs.(11.17) and (11.18) yield, finally,

$$dT = -\frac{1}{\left[C_v^{(gas)} + C_v^{(liquid)}\right]}\left\{\hat{q}\frac{dN}{dt} + \right. \tag{11.19}$$

$$\left. + 4\pi R^2(t)u(t)\left[T\left(\frac{\partial p}{\partial T}\right)_{V,N}\right]\right\} dt\ .$$

This way, all except one (for N) of the basic relationships describing variations of the state of the system in free expansion are established. However, the expressions contain the parameter $M^{(*)}$ unknown so far. It will be determined in a next step.

11.2.3 Determination of the Model Parameter

The derivation of the basic equations governing the free expansion of gases was carried out by considering a simplified model system. In order to apply the results to real expansion processes, we have to determine appropriately the parameter $M^{(*)}$. This procedure can be carried out in the following way.

The change of the kinetic energy of collective motion of a spherical layer of the gas cloud during a time interval Δt can be written as

$$d\Delta U^{(flow)} = \Delta M(r)\,[u(r)\Delta u(r)]\ . \tag{11.20}$$

Here $\Delta M(r)$ is the mass of the considered layer, $u(r)$ its velocity and $\Delta u(r)$ the change of the velocity in the time interval Δt. The variable r specifies the location of the considered layer, i.e., its distance from the center of the system. The outer boundary of the gas corresponds to $r = R$.

The condition of homogeneity of the gas in the course of the expansion is given by

$$u(r) = \left(\frac{r}{R}\right) u(R)\ , \qquad \Delta u(r) = \left(\frac{r}{R}\right) \Delta u(R)\ . \tag{11.21}$$

Thus, we may write

$$d\Delta U^{(flow)} = \Delta M(r) \left(\frac{r}{R}\right)^2 \Delta \left(\frac{u^2(R)}{2}\right)\ . \tag{11.22}$$

$\Delta M(r)$ is the mass of a layer. It can be expressed as

$$\Delta M(r) = 4\pi r^2 \rho dr\ , \tag{11.23}$$

where ρ is the mass density of the gas.

A substitution of this relation into Eq.(11.22) results in

$$d\Delta U^{(flow)} = 4\pi\rho\Delta \left(\frac{u^2(R)}{2}\right) \left(\frac{r^4}{R^2}\right) dr\ . \tag{11.24}$$

Finally, an integration over the whole volume of the gas (going over afterwards to the limit $\Delta t \to 0$) yields

$$dU^{(flow)} = \frac{3}{5} d\left[\frac{M}{2} u^2(R)\right] . \tag{11.25}$$

A comparison with Eq.(11.7) shows that a coincidence is reached if the parameter $M^{(*)}$ is set equal to

$$M^{(*)} = \frac{3}{5} M . \tag{11.26}$$

We will use this value for self–consistency in the model assumptions, i.e. the kinetic energy of the membrane (in the model) has to be equal to the kinetic energy of the collective flow of the gas in the real expansion process.

Note that so far no specific assumptions concerning the equation of state of the gas and the condensing phase have been made except the assumption allowing the transition from Eq.(11.13) to Eq.(11.14). It follows that the equations derived (or slightly modified versions) are applicable for a wide class of expansion processes and not only for the expansion of gas clouds. The liquid–gas transition (condensation of gases) is consequently only one of the possible phase transformation processes which may be studied based on the set of equations derived above.

11.3 Some Preliminary Consequences

According to the model analysis, provided the state of the system (radius R, velocity of expansion u, temperature T, pressure p, number of single particles N) is known at some moment of time, its further evolution is determined by the following set of equations (Eqs.(11.4), (11.8), (11.26) and (11.19))

$$dR(t) = u(t) dt , \tag{11.27}$$

$$du(t) = \frac{20\pi}{3} \frac{R^2(t) p(t)}{M} dt , \tag{11.28}$$

$$dT(t) = -\frac{1}{\left\{\hat{c}_v^{(gas)} N(t) + \hat{c}_v^{(liquid)} \left[N_0 - N(t)\right]\right\}} \left\{\hat{q}\frac{dN}{dt} + \right. \tag{11.29}$$

$$\left. + 4\pi R^2(t) u(t) \left[T(t) \left(\frac{\partial p}{\partial T}\right)_V\right]\right\} dt$$

and via the thermal equation of state of the system.

As a first application of the theoretical approach, we consider the process of free expansion of a gas for the case that clustering processes are negligible. To proceed with the analysis, the thermodynamic properties of the system under consideration have to be specified. For an illustration of the basic qualitative features of the model, we choose them as simple as possible. These assumptions can be easily made more realistic if particular applications are analyzed. However, we expect that the basic features of the process are modelled adequately even by applying such simplified model assumptions concerning the equation of state of both considered phases.

For illustration purposes, it is assumed here and below that the gas is a perfect one. This condition implies that the internal energy of the gas may be expressed as

$$U_{gas}^{(therm)} = \frac{f}{2} N k_B T \; . \tag{11.30}$$

In Eq.(11.30), f is the number of degrees of freedom of the molecules of the gas, k_B is the Boltzmann constant. Lateron, we will set f equal to $f = 3$ retaining only translational degrees of freedom.

Moreover, the specific heats per particle of the gas at constant volume c_v and constant pressure c_p have then the form [269, 295, 437]

$$\hat{c}_v^{(gas)} = \frac{f}{2} k_B \; , \quad \hat{c}_p^{(gas)} = \frac{(f+2)}{2} k_B \; , \quad \hat{c}_p^{(gas)} - \hat{c}_v^{(gas)} = k_B \; . \tag{11.31}$$

Here k_B is Boltzmann's constant, again.

With such notation, $C_v^{(gas)}$ may be expressed as

$$C_v^{(gas)} = N \hat{c}_v^{(gas)} \; , \qquad \hat{c}_v^{(gas)} = \frac{f}{2} k_B \; . \tag{11.32}$$

Considering the ensemble of clusters in the vapor as a mixture of perfect gases, we have, moreover,

$$p(t) = \frac{N_{tot}^{(cl)}(t) k_B T(t)}{V(t)} , \qquad V(t) = \frac{4\pi}{3} R^3(t) . \tag{11.33}$$

By $N_{tot}^{(cl)}$ the total number of clusters (including single particles) in the system is denoted.

The internal energy of the evolving liquid was taken here as independent of cluster size and the specific heat as independent of temperature. We have in this case

$$U_{liquid}^{(therm)} = C_v^{(liquid)} T + \text{constant} , \tag{11.34}$$

$$C_v^{(liquid)} = [N_0 - N(t)] \, \hat{c}_v^{(liquid)} . \tag{11.35}$$

Here N_0 is the total number of single particles (atoms or molecules forming the new phase) present in the system. This number includes the gas particles as well as the particles bound in clusters of different sizes. It is a parameter describing the properties of the gas cloud.

The missing equations for the determination of $N(t)$, $N_{tot}^{(cl)}$ or, more generally, of the cluster size distribution function $f(j,t)$ will be derived later based on the kinetic description of nucleation and growth processes (cf. section 11.4).

In the considered here limiting case (no aggregation, i.e., $N(t) = N_0 =$ const.), Eq.(11.29) is simplified to

$$dT(t) = -\frac{1}{\left[\hat{c}_v^{(gas)} N_0\right]} \left\{ 4\pi R^2(t) u(t) \left[T(t) \left(\frac{\partial p}{\partial T} \right)_{V,N} \right] \right\} dt . \tag{11.36}$$

The equation of state of the gas reads then

$$p(t) = \frac{N_0 k_B T(t)}{V(t)} . \tag{11.37}$$

In the absence of clustering processes ($N = N_0 =$ constant), the differential equation for the change of temperature in the system Eq.(11.29) may be rewritten in the form

$$\frac{dT}{T} = -\frac{dV}{V}\left(\frac{k_B}{\hat{c}_v^{(gas)}}\right) \tag{11.38}$$

or

$$d[\ln T] = -[d\ln V]\left(\frac{\hat{c}_p^{(gas)} - \hat{c}_v^{(gas)}}{\hat{c}_v^{(gas)}}\right) . \tag{11.39}$$

By solving this equation, we obtain (utilizing in addition Eq.(11.31))

$$TV^{\delta-1} = \text{constant} , \qquad pV^{\delta} = \text{constant} \tag{11.40}$$

with

$$\delta = \frac{c_p^{(gas)}}{c_v^{(gas)}} = 1 + \frac{2}{f} , \tag{11.41}$$

i.e., the well–known equations of state for an adiabatic reversible expansion of a perfect gas (cf. Refs. [269, 295, 437]). If clustering processes are absent in the system, our model leads, consequently, to the conclusion that the process of free expansion proceeds isentropically. This way, the model obeys a property usually assigned to hydrodynamic models of expansion at the later stages of this process [32, 105, 295].

The mentioned result is, of course, not accidental. If the mechanical motion of the membrane (in the simplified model) is reversed at some moment of time, the system will return to the initial state without any additional work required from external bodies. But this property is equivalent to constancy of the entropy being the measure of irreversibility of a process.

Partially similar results as outlined above could be obtained also by assuming from the very beginning an isentropic homogeneous expansion of the gas cloud. However, the method applied here has the advantage that it is formulated independently of this assumption. It is applicable as well for the case that clustering processes occur in the system. In latter case, the entropy is expected to change with time. This expectation is confirmed by the results of the analysis given somewhat later.

Consequently, if clustering processes take place in the system the assumption of constancy of entropy cannot be taken any more as the basis for the description of expansion of the gas cloud. Moreover, in our approach we have obtained simultaneously an explicit expression for the rate of expansion of the gas. Such an expression is absolutely essential since the outcome of the nucleation–growth process in an expanding gas is determined by the ratio of the characteristic times of expansion and aggregation, respectively.

Proceeding with the analysis, a substitution of $p \propto V^{-\delta}$ into Eq.(11.28) gives further

$$\frac{du}{dt} \propto R^{-[1+(6/f)]} . \tag{11.42}$$

It follows that the rate of change of the velocity of expansion decreases with increasing size of the gas cloud. This property allows us to obtain some estimates concerning the characteristics of a freely expanding gas (in the absence of clustering).

Assuming, according to Eq.(11.42), a nearly constant value of the velocity of expansion at the later stages of this process

$$u \cong u_0 \cong \text{constant} , \tag{11.43}$$

the radius of the gas cloud may be expressed as

$$R(t) \cong u_0(t + t_0) . \tag{11.44}$$

With these relations, Eqs.(11.31) and (11.40) we obtain

$$T(t) \cong \frac{\text{constant}}{(t+t_0)^{3(\delta-1)}} \cong \frac{\text{constant}}{(t+t_0)^{6/f}} , \tag{11.45}$$

$$p(t) \cong \frac{\text{constant}}{(t+t_0)^{3\delta}} \cong \frac{\text{constant}}{(t+t_0)^{3(1+2/f)}} . \tag{11.46}$$

The initial state, specified by a subscript "0", is assumed to consist of a highly compressed gas at rest. The energy of the gas at this state, $U_0^{(gas)}$, and the temperature, T_0, are connected by (cf. Eq.(11.30))

$$U_0^{(gas)} = \frac{f}{2} N_0 k_B T_0 \, . \tag{11.47}$$

If Eq.(11.45) is extended beyond its strict validity up to the initial state, we obtain from Eqs.(11.45) and (11.47)

$$T(t) \cong \left(\frac{2^{(6/f)+1}}{f} \right) \left(\frac{U_0^{(gas)}}{N_0 k_B} \right) \frac{1}{\left(1 + \frac{t}{t_0}\right)^{6/f}} \, . \tag{11.48}$$

If the single particles have translational degrees of freedom only ($f = 3$), this relation is reduced to

$$T(t) \propto \frac{1}{\left(1 + \frac{t}{t_0}\right)^2} \qquad \text{for} \qquad f = 3 \, . \tag{11.49}$$

Asymptotically, for large times, this result is in aggreement with consequences from a hydrodynamic model of expansion of a perfect gas cloud developed by Bondorf et al. [63] (cf. also [105]).

At the same level of approximation, we obtain for the time dependence of the particle concentration $c(t)$

$$c(t) = \frac{c(t_0)}{\left[\frac{1}{2} \left(1 + \frac{t}{t_0}\right) \right]^3} \, . \tag{11.50}$$

The dependencies of the energies of thermal and collective motion are given by

$$U^{(flow)} \cong U_0^{(gas)} \left(1 - \frac{T}{T_0}\right) , \qquad U^{(therm)} \cong U_0^{(gas)} \left(\frac{T}{T_0}\right) , \tag{11.51}$$

where the ratio T/T_0 is determined by

$$\frac{T(t)}{T_0} \cong \frac{1}{\left[\frac{1}{2}\left(1+\frac{t}{t_0}\right)\right]^{6/f}} . \tag{11.52}$$

This way, approximative analytic expressions for the dependence of a number of characteristic parameters of the gas cloud in dependence on time have been derived. More correct dependencies as compared with these estimates can be obtained from the numerical solution of the set of equations describing the free expansion of gases. Some results will be presented later in section 11.5.

However, in reality the expansion is accompanied, in general, by clustering processes. As a prerequisite for a proper description of the expansion of the gas cloud, the missing equations for $N(t)$ and $N_{tot}^{(cl)}(t)$ have to be determined. This task will be carried out in the subsequent section.

11.4 Kinetic Equations for the Description of Nucleation and Growth

Going over to a kinetic description of clustering, the state of the system is described in addition to thermodynamic parameters by a distribution function with respect to cluster sizes $f(j,t)$. The function $f(j,t)$ represents the number (or number density in a continuous description) of clusters (dimers, trimers, etc.) of the newly evolving phase per unit volume containing j single particles. The single particles or basic building units of the new phase may consist of atoms, molecules or even complex aggregates with a definite stoichiometric composition [510].

The basic kinetic equations describing nucleation and growth may be expressed in the form (cf. e.g. Refs. [483, 510, 511, 516, 517])

$$\begin{aligned}\frac{\partial f(j,t)}{\partial t} &= w_{j-1,j}^{(+)} \{f(j-1,t)- \\ &- f(j,t)\exp\left[\frac{\Delta G(j)-\Delta G(j-1)}{k_B T}\right]\} +\end{aligned} \tag{11.53}$$

$$+ w_{j,j+1}^{(+)} \{-f(j,t)+$$
$$+ f(j+1,t)\exp\left[\frac{\Delta G(j+1)-\Delta G(j)}{k_B T}\right]\Bigg\} .$$

Here by $w_{j-1,j}^{(+)}$ and $w_{j,j+1}^{(+)}$ the coefficients of aggregation are denoted. These coefficients have the meaning of the average number of single particles aggregated to a cluster consisting of $j-1$ (or j) such particles per unit time. For condensation processes in the gas phase, these coefficients may be written as (cf. Refs. [187, 225])

$$w_{j,j+1}^{(+)} = c(t)\left(\frac{3v_\alpha}{4\pi}\right)^{2/3}\left(\frac{8\pi k_B T}{m}\right)^{1/2} j^{2/3} g(j) , \qquad (11.54)$$

$$c(t) = \frac{N(t)}{V(t)} , \qquad g(j) = \left(1+\frac{1}{j^{1/3}}\right)^2\left(1+\frac{1}{j}\right)^{1/2} . \qquad (11.55)$$

Here $v_\alpha = 1/c_\alpha$ is the volume occupied by one particle and c_α is the volume density of condensing particles in the liquid phase, m is the mass of a gas particle. The term $g(j)$ accounts for the motion of clusters with sizes $j \geq 2$. For sufficiently large cluster sizes this function tends to unity.

The expressions for $w^{(+)}$ applied here are derived under the assumption of an internal thermodynamic equilibrium of the gas phase. In this way, the choice of these expressions is, again, self–consistent with the model assumptions discussed in section 11.2.

For the so–called work of cluster formation $\Delta G(j)$ we choose the following relation (cf. Refs. [150, 273, 511])

$$\Delta G(j) = -j\Delta\mu + \alpha_2 j^{2/3} + k_B T\tau \ln j , \qquad (11.56)$$

$$\alpha_2 = 4\pi\sigma\left(\frac{3v_\alpha}{4\pi}\right)^{2/3} . \qquad (11.57)$$

In Eq.(11.56), $\Delta\mu$ is the difference in the chemical potential per particle in the gas and the liquid phases, respectively, σ is the surface tension. The actual value of the parameter τ introduced with Eq.(11.56) depends on specific properties of the substance considered. It can vary in the range $2.0 < \tau < 2.5$ [150, 255, 529].

The difference in the chemical potential per particle in the gas (μ_β) and the liquid (μ_α) can be written as

$$\Delta\mu = \mu_\beta - \mu_\alpha = \mu_\beta(T, c) - \mu_\beta(T, c_{eq}^{(\infty)}) \; . \tag{11.58}$$

Hereby the equilibrium condition for a stable coexistence of both phases at a planar interface $\mu_\alpha(p, T) = \mu_\beta(c_{eq}^{(\infty)}, T)$ was employed. This relation Eq.(11.58) allows us to express $\Delta\mu$ exclusively via characteristics of the gas.

By $c_{eq}^{(\infty)}(T)$ the equilibrium concentration of the gas particles in stable coexistence with the liquid at a planar interface is denoted. Clustering processes may occur for $\Delta\mu/(k_B T) > 0$, i.e. when the concentration of the gas particles exceeds this equilibrium value.

A substitution of the expression for the aggregation coefficients and the work of cluster formation into Eq.(11.53) yields

$$\begin{aligned} \frac{\partial f(j,t)}{\partial t} &= \alpha_1(T) j^{2/3} g(j) \left\{ \frac{c}{c_{eq}^{(\infty)}(T)} \left(\frac{j-1}{j} \right)^{2/3} \frac{g(j-1)}{g(j)} f(j-1,t) \right. \\ &+ \left(\frac{j+1}{j} \right)^{\tau} \exp\left(\frac{2\alpha_2}{3k_B T (j+1)^{1/3}} \right) f(j+1,t) - \\ &- \left[\frac{c}{c_{eq}^{(\infty)}(T)} + \left(\frac{j-1}{j} \right)^{2/3} \frac{g(j-1)}{g(j)} \left(\frac{j}{j-1} \right)^{\tau} \cdot \right. \\ &\left. \left. \cdot \exp\left(\frac{2\alpha_2}{3k_B T j^{1/3}} \right) \right] f(j,t) \right\} \; . \end{aligned} \tag{11.59}$$

Hereby the approximations

$$\begin{aligned} \Delta G(j+1) - \Delta G(j) &\cong -\Delta\mu + \frac{2\alpha_2}{3(j+1)^{1/3}} + \\ &+ k_B T \ln\left(\frac{j+1}{j} \right) , \end{aligned} \tag{11.60}$$

$$\Delta G(j) - \Delta G(j-1) \cong -\Delta\mu + \frac{2\alpha_2}{3j^{1/3}} + k_B T \ln\left(\frac{j}{j-1}\right) \qquad (11.61)$$

were used. The application of this approximation may be omitted in the numerical computations.

Moreover, for simplicity of the notations a parameter $\alpha_1(T)$ was introduced. It is determined as

$$\alpha_1(T) = \left(\frac{3v_\alpha}{4\pi}\right)^{2/3} \left(\frac{8\pi k_B T}{m}\right)^{1/2} c_{eq}^{(\infty)}(T) \,. \qquad (11.62)$$

The change of the chemical potential $\Delta\mu$ is expressed as

$$\Delta\mu(T) = k_B T \ln\left[\frac{c}{c_{eq}^{(\infty)}(T)}\right] . \qquad (11.63)$$

Here, again, the perfect gas law is employed.

The temperature dependence of the equilibrium concentration can be obtained either directly from experiment or from the Clausius–Clapeyron equation. According to this relation, the equilibrium pressure $p_{eq}^{(\infty)}(T)$ varies with temperature as [269, 295, 437]

$$\frac{d[p_{eq}^{(\infty)}(T)]}{dT} = \frac{\hat{q}}{T\left(\hat{v}_{gas} - \hat{v}_{liquid}\right)} = \frac{\hat{q}}{T\hat{v}_{gas}} \frac{1}{\left[1 - \dfrac{\hat{v}_{liquid}}{\hat{v}_{gas}}\right]} . \qquad (11.64)$$

Here $\hat{q} > 0$ is the heat of vaporization per particle, by $\hat{v}_{liquid}$ and $\hat{v}_{gas}$ the volumes per particle in both phases are denoted.

By applying the perfect gas law to express $\hat{v}_{gas}$ in the right hand side of Eq.(11.64) by the pressure and the temperature of the gas, we obtain further

$$\frac{d \ln\left[p_{eq}^{(\infty)}\right]}{dT} = \frac{\hat{q}}{k_B T^2} \frac{1}{\left[1 - \dfrac{\hat{v}_{liquid}}{\hat{v}_{gas}}\right]} . \qquad (11.65)$$

Again, utilizing the perfect gas law, written in the form $p = ck_BT$ (cf. Eqs.(11.37) and (11.54)), we have

$$\frac{d\ln\left[c_{eq}^{(\infty)}(T)\right]}{dT} \cong \frac{1}{T}\left\{\frac{\hat{q}}{k_BT\left[1-\dfrac{\hat{v}_{liquid}}{\hat{v}_{gas}}\right]}-1\right\} . \tag{11.66}$$

This equation is the most general thermodynamic relationship (assuming perfect gas behavior) determining the change of the equilibrium vapor pressure with temperature. It can be solved analytically when additional approximations are applied.

In general, the relation $v_{gas} \gg v_{liquid}$ is fulfilled. Moreover, when the latent heat of the transition $\hat{q}$ is approximately independent of temperature, the solution of this equation reads

$$c_{eq}^{(\infty)}(T) = c_{eq}^{(\infty)}(\widetilde{T})\left(\frac{\widetilde{T}}{T}\right)\exp\left\{-\frac{\hat{q}}{k_B}\left[\frac{1}{T}-\frac{1}{\widetilde{T}}\right]\right\} . \tag{11.67}$$

Here $\widetilde{T}$ is the temperature in an arbitrarily chosen reference state.

This expression Eq.(11.67) we will apply here for a determination of the temperature dependence of $c_{eq}^{(\infty)}(T)$ in solving the kinetic equations Eqs.(11.59).

Eqs.(11.50) and (11.67) allow us to estimate how the supersaturation (cf. Eq.(11.63)) varies in the course of the expansion. We obtain by utilizing in addition also Eq.(11.52)

$$\left[\frac{c(t)}{c_{eq}^{(\infty)}(t)}\right] = \left[\frac{c(t_0)}{c_{eq}^{(\infty)}(t_0)}\right]\left\{\frac{\exp\left\{\dfrac{\hat{q}}{k_BT_0}\left[\left[\dfrac{1}{2}\left(1+\dfrac{t}{t_0}\right)\right]^{6/f}-1\right]\right\}}{\left[\dfrac{1}{2}\left(1+\dfrac{t}{t_0}\right)\right]^{3[1+(2/f)]}}\right\}$$

As mentioned f is the number of degrees of freedom of the single particles. The ratio $(c/c_{eq}^{(\infty)})$ is an increasing function of time, if the inequality

$$\frac{t}{t_0} > 2\left[\left(\frac{k_BT_0}{\hat{q}}\right)\left(1+\frac{f}{2}\right)\right]^{f/6} - 1 \tag{11.68}$$

is fulfilled. It follows from these estimates that widely independent of the properties of the initial state of the gas, in the course of the expansion always states are passed where clustering processes may occur (cf. [105]).

Eq.(11.59) holds for cluster sizes $j \geq 2$. The number of single particles has to be determined from the mass balance equation

$$N(t) + \sum_{j=2}^{\infty} j f(j,t) = N_0 = \text{constant}\,. \tag{11.69}$$

The equation for the total number of clusters in the system is determined similarly as

$$N_{tot}^{(cl)}(t) = N(t) + \sum_{j=2}^{\infty} f(j,t)\,. \tag{11.70}$$

These relations give the missing so far equations needed for a description of the state of the gas (in addition to Eqs.(11.27) – (11.33)). This way, the whole set of kinetic equations required for the description of clustering in expanding gases is established.

11.5 Method and Results of the Solution of the Kinetic Equations

To solve the set of differential equations describing clustering in expanding gases, in addition to the thermodynamic properties of the liquid and the gas phases the initial conditions have to be specified. We proceed in the following way.

We select a reference state in the unstable region of the phase diagram. This state is characterized by a value of temperature $\widetilde{T}$ and a volume $\widetilde{V}$ (or radius $\widetilde{R}$). Once these values are chosen, other thermodynamic parameters of the reference state may be determined, like equilibrium particle density $c_{eq}^{(\infty)}(\widetilde{T})$, particle density $c_\alpha(\widetilde{T})$ and specific heat $\hat{c}_v^{(liquid)}(\widetilde{T})$ of the liquid phase, the heat of evaporation $\hat{q}(\widetilde{T})$, the surface tension $\sigma(\widetilde{T})$. The latter four quantities (c_α, $\hat{c}_v^{(liquid)}$, $\hat{q}$, σ) we consider here approximately as independent of temperature. The values of these parameters are chosen thus to be equal to the respective values in the reference state. Such a choice is motivated by the argument that clustering processes will proceed mainly in the vicinity of the reference state. If required, this approximation may be improved easily by chosing the appropriate temperature dependencies for the considered applications.

The value of the concentration of the gas particles in the reference state we take here equal to $\left[\widetilde{c}_0/c_{eq}^{(\infty)}\right] = 10$. Knowing the volume $\widetilde{V}$ and the mass of one gas particle, m, the total number of gas particles in the system, N_0, and their total mass, M, may be determined via

$$N_0 = \widetilde{c}_0 \widetilde{V} \; , \qquad M = N_0 m \; . \tag{11.71}$$

Based on this reference state, we may reach a number of initial states inside and beyond the region of thermodynamic instability by assigning appropriate negative values to the initial velocity $\widetilde{u}$ ($\widetilde{u} < 0$). The absolute value of the velocity is chosen by setting the quantity $U^{(flow)}$ equal to some part or multiple χ of the internal energy of the gas in the reference state, i.e.

$$U^{(flow)} = \chi \widetilde{U}_{gas}^{(therm)} \; . \tag{11.72}$$

With Eqs.(11.7), (11.26) and (11.30), we have

$$\widetilde{u} = -\sqrt{\chi \left(\frac{5f}{3}\right) \frac{N_0 k_B \widetilde{T}}{M}} = -\sqrt{\chi \left(\frac{5f}{3}\right) \frac{k_B \widetilde{T}}{m}} \; . \tag{11.73}$$

Starting with such a reference state, the gas will be further compressed initially. In this compression stage, the calculations are carried out inhibiting clustering processes (i.e. by setting $(\partial f(j,t)/\partial t)$ equal to zero). The initial state is reached once $u(t)$ becomes equal to zero. It is characterized by an initial value of the internal energy $U_0^{(gas)}$ of the gas cloud given by

$$U_0^{(gas)} = (1 + \chi) \, \widetilde{U}_{gas}^{(therm)} \; . \tag{11.74}$$

By varying χ for otherwise identical initial conditions in the reference state, we may study thus the dependence of the clustering process on the excitation energy of the initial state. Such an analysis is of interest for a number of applications in atomic or nuclear clustering processes (cf. Refs. [189, 378, 482, 518]).

In the calculations we chose the following values of the parameters for the reference state (the parameter values are taken as to refer to some extent to the condensation of water vapor [156, 288]):

$$k_B = 1.38 \cdot 10^{-23}\mathrm{JK}^{-1}\ , \quad \widetilde{T} = 293\mathrm{K}\ , \quad \widetilde{V} = 1\mathrm{m}^3\ ,$$

$$c_{eq}^{(\infty)}(\widetilde{T}) = 5.78 \cdot 10^{23}\mathrm{m}^{-3}\ , \quad f = 3\ , \quad m = 2.99 \cdot 10^{-26}\mathrm{kg}\ ,$$

$$\hat{c}_v^{(gas)} = \frac{3}{2}k_B\ , \quad \sigma(\widetilde{T}) = 7.3 \cdot 10^{-2}\mathrm{Jm}^{-2}\ , \tag{11.75}$$

$$c_\alpha(\widetilde{T}) = 3.35 \cdot 10^{28}\mathrm{m}^{-3}\ , \quad \hat{c}_v^{(liquid)}(\widetilde{T}) = 1.26 \cdot 10^{-22}\mathrm{JK}^{-1}\ ,$$

$$\hat{q}(\widetilde{T}) = 7.35 \cdot 10^{-20}\mathrm{J}\ .$$

By setting $f = 3$, rotational degrees of freedom of the single particles are neglected here.

The time is given in dimensionless units defined by (cf. Eq.(11.62))

$$t' = \frac{t}{\tau_1(\widetilde{T})}\ , \qquad \tau_1(T) = \frac{1}{\alpha_1(T)}\ . \tag{11.76}$$

The quantity $\tau_1(T) = 1/\alpha_1(T)$ can be considered generally as a characteristic time scale of clustering. For the considered set of parameters its value in the reference state equals $\tau_1(\widetilde{T}) = 2.53 \cdot 10^{-8}$ s.

Various aspects of the evolution of the expanding gas cloud are illustrated in the figures. In Fig. 11.1, the dependence of the ratio $[c(t')/c_{eq}^{(eq)}(t')]$ on time) the change of the size of the expanding gas cloud $[R^3(t')]$ and its velocity of expansion $u(t')$, the variation of the temperature of the gas $[T(t')]$, the change of the entropy of the system $[S(t')]$ and the variation of the coefficient δ are given for a value of the initial excitation energy of the gas corresponding to $\chi = 0.01$. It is evident that the velocity of expansion u approaches, indeed, a constant value for large times. The approach to the asymptotic value proceeds more rapidly for larger values of the initial excitation energy.

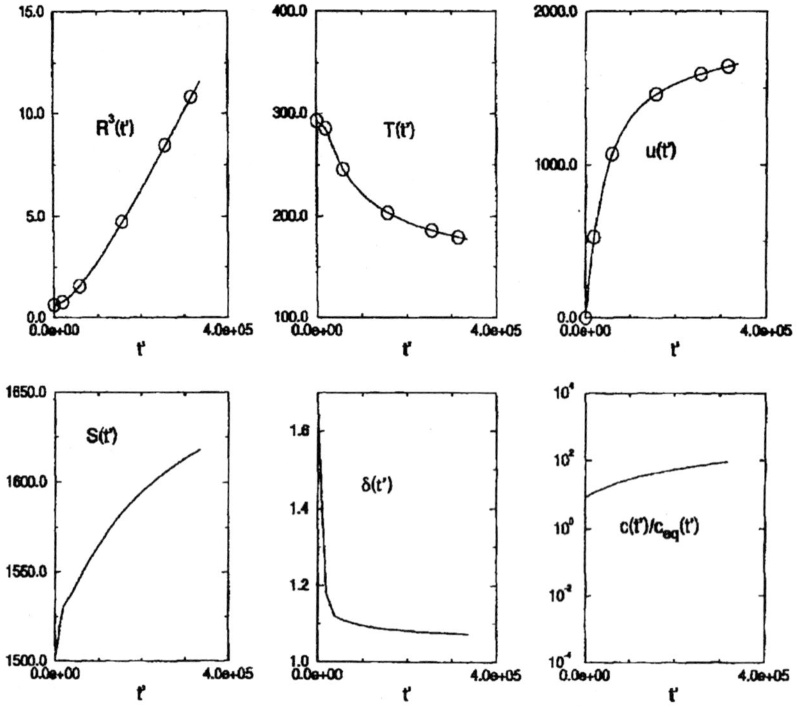

Fig. 11.1 Dependence of the temperature $T(t')$, the volume of the gas cloud $R^3(t')$, the velocity of expansion $u(t')$, the ratio of the actual vs equilibrium concentrations of the nucleonic gas $[c(t')/c_{eq}(t')]$, the entropy of the system $S(t')$ and the Poisson coefficient $\delta(t')$ on time (in reduced units). For further details see [484, 485]

The entropy is calculated here in the following way. We start with the expression for the Helmholtz free energy for an ensemble of clusters in the vapor (cf. Refs. [544, 503, 336] and chapter 3). This relation reads (in our notations)

$$F(T, V, \{\mathbf{f(j, t')}\}) = \sum_{j=1}^{N_0} f(j, t') \Bigg\{ \varphi_j(T) + \\ + k_B T \left[\ln \left(\frac{f(j, t')}{V} \lambda_j^3 \right) - 1 \right] \Bigg\} . \tag{11.77}$$

Here and below the dependence of the thermodynamic functions on the actual cluster size distribution is specified by

$$\{\mathbf{f(j,t')}\} = \{f(1,t'), f(2,t'), \ldots, f(N_0,t')\} \ . \tag{11.78}$$

By λ_j the de Broglie wave length of a cluster of size j is denoted. It is given by

$$\lambda_j(T) = \frac{1}{j^{1/2}} \left(\frac{h^2}{2\pi m k_B T} \right)^{1/2} \ . \tag{11.79}$$

h is Planck's constant.

Moreover, with $\varphi_j(T)$ in Eq.(11.86) the binding energy of a cluster of size j is introduced. For single particles this quantity must be equal to zero. In general, it depends on the cluster size. The ratio $[\varphi_j(T)/j]$ approaches the constant macroscopic value for large j.

The thermal equation of state can be obtained from

$$p = -\left(\frac{\partial F}{\partial V} \right)_{T,\{\mathbf{f(j,t')}\}} \ . \tag{11.80}$$

We get

$$p = \frac{k_B T}{V} \sum_{j=1}^{N_0} f(j,t') = \frac{k_B T}{V} N_{tot}^{(cl)} \ , \tag{11.81}$$

i.e., the equation of state of a mixture of perfect gases.

In Eq.(11.86), by $N_{tot}^{(cl)}$ the total number of particles in the considered volume of the system is denoted, i.e.

$$N_{tot}^{(cl)} = \sum_{j=1}^{N_0} f(j,t') \ . \tag{11.82}$$

Finally, for the chemical potential of the gas we find (in the absence of clustering)

$$\mu_\beta\left(T,\frac{N}{V}\right)=\left(\frac{\partial F}{\partial N}\right)_{V,T}=k_BT\ln\left(\frac{N}{V}\lambda_1^3\right)\ . \tag{11.83}$$

As a special case, we obtain

$$\mu_{eq}^{(\infty)}(T,c_{eq}^{(\infty)})=k_BT\ln\left[\lambda_1^3c_{eq}^{(\infty)}(T)\right]\ . \tag{11.84}$$

The entropy is determined via

$$S(T,V,\{\mathbf{f(j,t')}\})=-\left(\frac{\partial F}{\partial T}\right)_{V,\{\mathbf{f(j,t')}\}} \tag{11.85}$$

resulting initially in

$$S(T,p,\{\mathbf{f(j,t')}\})=k_B\sum_{j=1}^{N_0}f(j,t')\left[\frac{5}{2}-\ln\left(\frac{\lambda_j^3(T)f(j,t')p}{N_{tot}^{(cl)}k_BT}\right)\right]- \\ -\sum_{j=2}^{N_0}f(j,t')\frac{\partial\varphi_j(T)}{\partial T}\ . \tag{11.86}$$

For the internal energy ($U=F+TS$) of the considered state we have, in addition,

$$U=\sum_{j=1}^{N_0}f(j,t')\left(\frac{3}{2}k_BT+\varphi_j(T)-T\frac{\partial\varphi_j}{\partial T}\right)\ . \tag{11.87}$$

In the calculations for the case of the absence of clustering we set $f(1,t')=N_0$ and $f(j\geq 2,t')=0$ and the respective relations may be evaluated immediately. In general, we have to find an expression for $\varphi_j(T)$ and the partial derivative $[\partial\varphi_j(T)/\partial T]$.

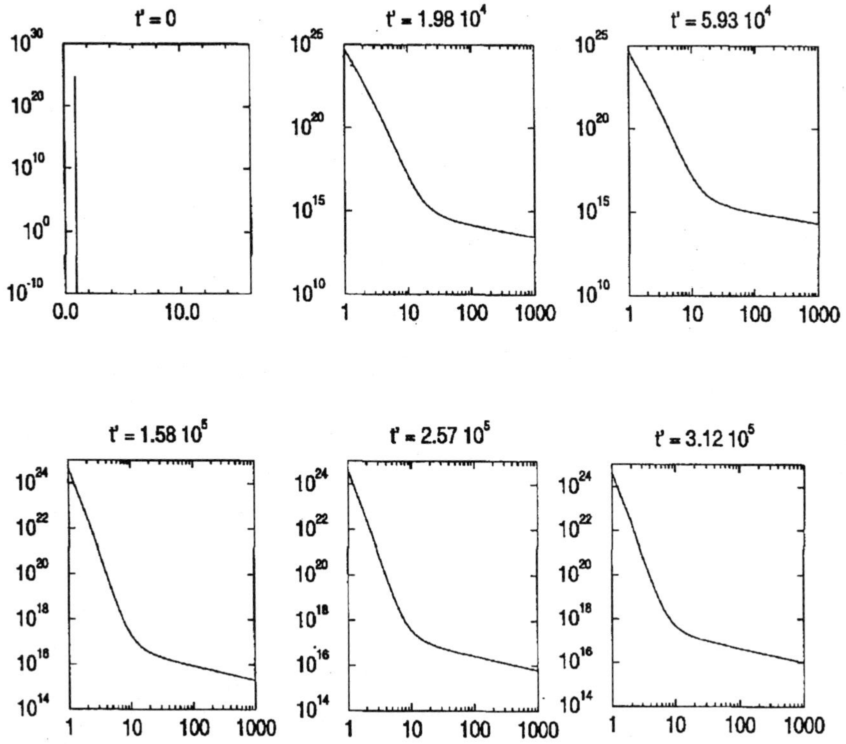

Fig. 11.2 Evolution of the cluster size distribution function $f(j,t')$. The different curves refer to different moments of time (the respective values are given in the figures). For large times, a practically time–independent asymptotic distribution is established.

In accordance with already mentioned simplifying assumptions concerning the thermodynamic properties of the model substance considered, we assume φ_j as independent of cluster size. We may write, thus, approximately (cf. chapter 3 and [502]),

$$\begin{aligned} \varphi_j(T) &\cong 0 && \text{for} \quad j = 1 \\ \varphi_j &\cong \mu_{eq}^{(\infty)}(T) && \text{for} \quad j \geq 2 \end{aligned} \tag{11.88}$$

and

$$\left(\frac{\partial \varphi_j(T)}{\partial T}\right) \cong j k_B \left(\frac{\mu_{eq}^{(\infty)}(T) + \hat{q}}{k_B T} - \frac{5}{2}\right) . \tag{11.89}$$

Thus we obtain, finally,

$$S(T, p, \{\mathbf{f(j, t')}\}) = k_B \left\{ \sum_{j=1}^{N_0} f(j, t') \left[\frac{5}{2} - \ln\left(\frac{\lambda_j^3(T) f(j, t') p}{N_{tot}^{(cl)} k_B T}\right)\right] - \right.$$
$$\left. - \sum_{j=1}^{N_0} (1 - \delta_{1,j}) j f(j, t') \left(\frac{\mu_{eq}^{(\infty)}(T) + \hat{q}}{k_B T} - \frac{5}{2}\right)\right\} , \tag{11.90}$$

$$U(T, p, \{\mathbf{f(j, t')}\}) = \sum_{j=1}^{N_0} f(j, t') \left\{\frac{3}{2} k_B T + (1 - \delta_{1,j}) \left[\frac{5}{2} k_B T - \hat{q} n\right]\right\}$$

Here by $\delta_{1,j}$ the Kronecker symbol is denoted ($\delta_{1,j} = 1$ for $j = 1$, $\delta_{1,j} = 0$ otherwise).

The parameter δ, also presented in the figures, is determined as follows. As outlined in section 11.3, in the absence of clustering the expansion is characterized by (cf. Eq.(11.40))

$$T V^{\delta - 1} = \text{constant} , \qquad p V^{\delta} = \text{constant} , \tag{11.91}$$

$$\delta = 1 + \frac{2}{f} , \qquad S = \text{constant} . \tag{11.92}$$

In other words, the variations of temperature and volume or of pressure and volume are interrelated by

$$\frac{dT}{T} + (\delta - 1)\frac{dV}{V} = 0 , \qquad \frac{dp}{p} + \delta \frac{dV}{V} = 0 . \tag{11.93}$$

In the stage, when clustering processes occur, the thermal equation of state can be approximated by dependencies like Eqs.(11.91) or (11.93), again, as it is the case for a reversible adiabatic expansion of a one–component gas (cf. Eq.(11.40)). However, δ differs from the value $\delta = 1 + 2/f$. It decreases with time and approaches, for the considered times of expansion, the value $\delta \cong 1.08$. In the numerical computations, δ is determined via (cf. Eq.(11.93)) as

$$\delta = -\left(\frac{V}{p}\right)\frac{dp}{dV} . \tag{11.94}$$

It has been demonstrated [484, 485] that the evolution of the cluster size distribution in dependence on time is qualitatively independent of the values of the energy of the initial state or the excitation energy of the gas cloud.

In the first stages, monotonically decreasing distributions develop. These dependencies can be well–approximated (in logarithmic coordinates) by straight lines with different slopes. It follows that in the initial stages of the clustering process the cluster size distribution may be expressed in the form

$$f(j,t) \propto j^{-\tau_{eff}} . \tag{11.95}$$

Hereby the value of τ_{eff} depends, in general, on time and on the initial excitation energy of the gas (cf. also Refs. [482, 483]).

In later stages of the process a second peak develops. The clusters belonging to this part of the distribution are capable to a further growth. However, due to the expansion and cooling of the system, such further growth processes are inhibited considerably and some relatively time–independent asymptotic distribution is established. The formation of the second peak is accompanied by a steepening of the distribution at relatively small cluster sizes. In the range $j < j_{max}^{(\tau)}$ the curves are well–approximated by dependencies of the form as expressed by Eq.(11.95), again.

It has been suggested in the literature [5, 425] that lognormal distributions of the form

$$w(j) = \frac{1}{(j - \bar{j})\bar{\sigma}\sqrt{2\pi}} \exp\left\{-\frac{[\ln(j - \bar{j}) - \bar{\omega}]^2}{2\bar{\sigma}^2}\right\} \tag{11.96}$$

may give an adequate fit of cluster size distributions evolving by nucleation and growth processes. In Eq.(11.96), $\bar{j}$, $\bar{\sigma}$ and $\bar{\omega}$ are parameters of the distribution. In our case such a fit is not a good approximation.

The typical features of the expansion process can be understood as follows: For higher values of the initial excitation energy of the gas, the velocity of expansion is also large. As a consequence, the states where intensive clustering processes may occur are passed quickly. Vice versa, slow expansion rates result in extended stages of the expansion where clustering takes place. Consequently, more advanced stages of the evolution of the cluster size distribution, as found for stationary boundary conditions (cf. also Refs. [482, 483, 471, 322]), may be reached.

Quantitatively, this argumentation may be expressed in the following way. Similarly to Eq.(11.76) we may introduce a characteristic time scale of expansion τ_2. For this purpose, we write down the expression for the relative rate of change of the radius of the gas cloud for the reference state as

$$\left.\frac{dR}{R}\right|_{R=\widetilde{R}} = \left.\frac{u}{R}\right|_{u=\widetilde{u};R=\widetilde{R}} dt = \pm\frac{dt}{\tau_2(\widetilde{T})} , \qquad \tau_2(\widetilde{T}) = \frac{\widetilde{R}}{|\widetilde{u}|} . \tag{11.97}$$

With Eq.(11.73) we have

$$\tau_2(\widetilde{T}) = \sqrt{\frac{3\widetilde{R}^2 m}{5f\chi k_B\widetilde{T}}} . \tag{11.98}$$

Steep distributions have to be expected when the inequality $\tau_1(\widetilde{T}) \gg \tau_2(\widetilde{T})$ holds. In the opposite case, bimodal distributions may develop.

Finally, let us analyze how the characteristic time scales both of expansion and aggregation behave with time. With Eqs.(11.62), (11.67) and (11.76), the characteristic time scale for clustering processes τ_1 may be written as

$$\left[\frac{\tau_1(t)}{\tau_1[\widetilde{T}(\widetilde{t})]}\right] \cong \left(\frac{T}{\widetilde{T}}\right)^{1/2} \exp\left[\frac{q}{R_g\widetilde{T}}\left(\frac{\widetilde{T}}{T} - 1\right)\right] . \tag{11.99}$$

Utilizing Eq.(11.52) this relation yields

$$\left[\frac{\tau_1(t)}{\tau_1[\widetilde{T}(\widetilde{t})]}\right] \cong \left(\frac{1+\widetilde{t}/t_0}{1+t/t_0}\right)^{3/f} \exp\left\{\frac{q}{R_g\widetilde{T}}\left[\left(\frac{1+t/t_0}{1+\widetilde{t}/t_0}\right)^{6/f} - 1\right]\right\} .$$

For the ratio of the characteristic times of expansion, we have, on the other side (cf. Eqs.(11.43), (11.44) and (11.97)),

$$\left[\frac{\tau_2(t)}{\tau_2[\widetilde{T}(\widetilde{t})]}\right] = \frac{t+t_0}{\widetilde{t}+t_0} \, . \tag{11.100}$$

Compared with the exponential increase of the characteristic time of clustering (or the exponential decrease of the aggregation rates), this linear increase of the characteristic time of expansion is relatively insignificant. In this way, clustering processes loose its importance with time and the cluster size distributions become frozen in. This is the reason why for large times the cluster size distributions remain nearly the same as illustrated, for example, in Fig. 11.2.

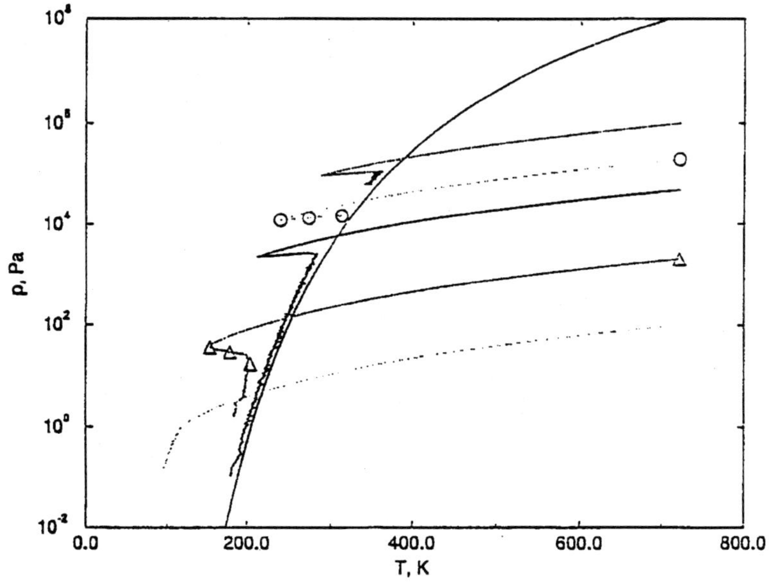

Fig. 11.3 Evolution of an expanding gas as obtained by stochastic simulation (for the details see chapter 3 and [289]. The stochastic approach allows a more detailed analysis of the expansion process, in particular, the initial penetration into the unstable region and the buffering of this process. Qualitatively the same results are obtained also by the solution of the set of kinetic equations as developed in the present chapter [484].

Some further insight into the course of the expansion process can be obtained by an analysis of the evolution of the system in a (p,T)–diagram (cf. Fig. 11.3 and [484]). It can be seen that the path of the system in the region of unstable states is widely independent of the initial excitation energy. It goes parallel to the equilibrium $p = p(T)$ dependence determined by the Clausius–Clapeyron equation Eq.(11.64). In the region, where clustering processes, do not occur, a linear dependence of pressure versus temperature is found in agreement with the analytical approximative solutions given in section 11.3.

A quite similar behaviour was observed in experiments on gas expansion [154]. In mentioned experiments a relation of the form

$$\Delta T \cong \alpha \left(-\frac{dT}{dt} \right)^{\beta} , \qquad \alpha \cong 10.5\,\mathrm{K} , \qquad \beta \cong 0.125 \tag{11.101}$$

was found. Here ΔT is the initial penetration of the system into the unstable region in the course of the expansion. The cooling rate is given in K/μs. This result resembles similar dependencies in glass science, where the vitrification (or glass) temperature depends also on the cooling rate of the glassforming melt (Bartenev–Ritland equation, cf. [187]). However, in contrast to glass vitrification, where the state of the system becomes frozen in in cooling, in expansion of gases clustering processes are continuing to proceed. Larger deviations from equilibrium are buffered by more intensive aggregation processes and the subsequent evolution follows nearly the same path independent of the initial excitation energy or the initial penetration of the system into the unstable state.

11.6 Discussion

The main result of the present paper consists in the formulation of a general model for the description of clustering in free expansion of gases or liquid–like systems. The approach avoids the widely applied assumption of constancy of entropy of the system and, in particular, the application of Poisson's equation for the description of the thermal equation of state of the gas. As it is evident from the results of the theory, developed here, both assumptions can be considered only as poor approximations to the real processes. Only in the limiting case when clustering processes are suppressed both mentioned conditions are shown to be fulfilled.

The basic equations of the theory outlined here are formulated widely independent of specific thermodynamic properties of the both phases involved. As one possible application, the case of clustering in an expanding gas is considered in more detail. Since we are interested here mainly in a qualitative insight into characteristic features of this process, the thermodynamic properties of both phases were taken as simple as possible. However, already at this stage a number of interesting conclusions could be drawn (spectrum of cluster size distributions and its dependence on the excitation energy, path of the system in the (p, T)–phase space, dependence of the degree of penetration into the unstable states on initial excitation energy and buffering of this process by aggregation processes, behavior of entropy and Poisson's exponent δ). In application to experiments on clustering in atomic or molecular expanding gases [518, 189], the model can be applied, for example, to formulate theoretically the boundary and initial conditions required to produce the most appropriate shapes, for a desired application, of cluster size distributions in a given experiment. As mentioned, the outcome of a given expansion process will be determined to a large degree by the value of the ratio of the characteristic time scales for clustering and expansion, respectively.

Dependencies as obtained from the solution of the set of equations as discussed in section 11.5 have been observed, as another possible application, in multifragmentation processes in heavy–ion collisions [113, 378, 411]. Computer simulations of this process by quantum molecular dynamics methods [301, 323] result in similar distributions for the intermediate mass fragments as obtained here. This way, the results give an additional confirmation that nuclear multifragmentation can be interpreted as a result of clustering in a first–order phase transition (see also Ref. [482]). It is verified, moreover, that this mechanism of multifragmentation is effective in a broad range of initial conditions widely independent of the excitation energy of the nucleonic system (cf. Refs. [105, 214]). Moreover, there is no need in the present approach of introducing ad hoc values for a so-called freeze–out volume or for a freeze–out density (cf. e.g. Refs. [113, 411]). Such effects are incorporated directly in the kinetic model by the qualitatively different dependence of the characteristic time scales of clustering and expansion on time (see section 11.5).

Due to its general character, the model developed in this paper can be applied for a variety of further processes involving clustering in expanding systems. By chosing appropriately the thermodynamic properties of the system under consideration, the model may be applied also to the description of bubble formation in expanding liquids (cf. Refs. [56, 238, 316, 317]), liquid–like systems (cf., e.g., Refs. [32, 105]) or even to processes of phase transformations in the evolution of the universe (cf. Ref. [230]).

In order to get a quantitative agreement between experiment and theory, a precise description of the thermodynamic properties of the respective systems has to be given. It requires the determination of the thermodynamic functions of the ensemble of clusters in the ambient phase for the respective system or, at least, the knowledge of the thermal equation of state and some additional thermodynamic properties required for the solution of the set of basic equations of the theory. Such a more precise description may include also the choice of more accurate expressions for the work of cluster formation to incorporate effects like magic numbers [189], of possible transformations of the preferred structure of the clusters in dependence on the size [389], the effect of electric fields on the aggregation rates etc. Such modifications may be incorporated easily into the general scheme as outlined above.

12 Cluster Formation in Expanding Hot and Dense Nuclear Matter

G. Röpke

The formation of clusters in expanding hot and dense matter is an interesting phenomenon, for instance, in order to describe compressed liquids expanding through a nozzle, or expanding nuclear matter produced by heavy–ion collisions. Of particular interest are processes where a phase instability region is reached. Different scenarios such as the freeze out model, the coalescence model, the multifragmentation model are introduced to describe the cluster distribution as a result of this non–equilibrium process. As an example, the formation of clusters in expanding nuclear matter is considered here, generated in heavy–ion reactions or in the early stage of the evolution of the Universe.

12.1 Quasi–Equilibrium in Nuclear Matter

The expansion of hot and dense nuclear matter is a non–equilibrium process. Appropriate concepts of quantum statistics have to be used to describe the formation of bound states which are genuine quantum states. Within a general approach to non–equilibrium it is possible to consider the hydrodynamic evolution where relaxation processes are fast to establish local equilibrium as well as the kinetic evolution described by time–dependent A–particle distribution functions. We first present the quasi–equilibrium to demonstrate the relevance of corrections due to the medium.

12.1.1 Non–Equilibrium Statistical Mechanics

A many–particle system in non–equilibrium is described by the statistical operator $\rho(t)$. It follows from the solution of the Liouville–von Neumann equation

$$\frac{\partial}{\partial t}\rho(t) - \frac{i}{\hbar}[H, \rho(t)] = 0 \tag{12.1}$$

and imposed boundary conditions

$$\langle A_i \rangle^t = \mathrm{Tr}[\rho(t) A_i] \ . \tag{12.2}$$

The boundary conditions can give us only a restricted information about the system. We can introduce a relevant statistical operator which reproduces the imposed boundary condition at maximum information entropy

$$\rho_{\mathrm{rel}}(t) = Z_{\mathrm{rel}}^{-1}(t) \exp\left[-\sum_i F_i(t) A_i\right] \ . \tag{12.3}$$

The thermodynamic parameters $F_i(t)$ are Lagrange parameters to be determined from the self–consistency conditions

$$\mathrm{Tr}[\rho_{\mathrm{rel}}(t) A_i] = \langle A_i \rangle^t \ . \tag{12.4}$$

As shown by Zubarev [592], the boundary conditions can be incorporated into the Liouville–von Neumann equation so that

$$\rho(t) = \epsilon \int_{-\infty}^{t} dt' U(t, t') \rho_{\mathrm{rel}}(t') U(t', t) \tag{12.5}$$

with the time evolution operator $U(t, t') = \exp[iH(t-t')/\hbar]$. The limit $\epsilon \to 0$ has to be taken after the thermodynamic limit.

Cluster Distribution Functions

In particular, we are interested in the distribution function $f_{A\alpha}(\vec{r},\vec{p},t)$ of clusters $(A\alpha)$, which is the Wigner transform of the reduced density matrix in momentum representation,

$$f_{A\alpha}(\vec{r},\vec{p},t) = \int \frac{d^3k}{(2\pi\hbar)^3} e^{i\vec{k}\cdot\vec{r}/\hbar} \langle a^+_{A\alpha,p-k/2} a_{A\alpha,p+k/2}\rangle^t \,, \tag{12.6}$$

with the creation operator for a correlated A–particle states [455], see Eq. (4.49) and following,

$$a^+_{A\alpha p}(t) = \sum_{1\ldots A} \psi^{*\mathrm{corr}}_{A\alpha p}(1\ldots A,t) a^+_1 \ldots a^+_A \,, \tag{12.7}$$

and the corresponding annihilation operator. The operators a^+_i, a_i are creation or annihilation operators, respectively, of single-particle states described by the quantum numbers momentum, spin and isospin, $i = \{\vec{p}_i, \sigma_i, \tau_i\}$.

This approach implies that for a detailed description the set of relevant observables is given by the A–particle distribution $a^+_1 \ldots a^+_A a_{A'} \ldots a_{1'}$. The corresponding correlation functions

$$\langle a^+_1 \ldots a^+_A a_{A'} \ldots a_{1'}\rangle^t \tag{12.8}$$

can be related to two–time Green functions in the A-particle channel. In equilibrium, bound states $(A\alpha\vec{p})$ are obtained from the poles of the A-particle Green function, the corresponding spectral function has a δ–like singularity. The effect of the medium leads to a shift and a broadening of these singularities in the spectral function. Furthermore, correlations are also formed in scattering states. As an example, let us consider resonances. If the interaction is weakened, bound states can be transformed into resonances. This transition should be smooth, without discontinuities in the physical properties. The weakening of the interaction may occur if the interaction with the medium is included.

As detailed in Chapter 4, the wave function $\psi^{\mathrm{corr}}_{A\alpha p}$ shell be taken as optimum basis set of quantum states, i.e. of states with small damping rates. We proposed to use the solution of the cluster-mean field approach. If we consider homogeneous systems, the total momentum $\vec{p}$ is a conserved quantum number as already given above. Then, the dependence on $\vec{r}$ is only as on a parameter characterizing the state of the homogeneous system. The relevant observables are

$$f_{A\alpha}(\vec{r},\vec{p},t) = \langle a^+_{A\alpha p} a_{A\alpha p} \rangle^t_r \ . \tag{12.9}$$

In strongly inhomogeneous systems, the optimum states have to be constructed from the solution of a wave equation with an $\vec{r}$ dependent mean field.

Nucleon Systems

To apply this formalism to nuclear matter, we have to specify the Hamiltonian

$$H = \sum_1 E(1) a_1^+ a_1 + \frac{1}{2} \sum_{12,1'2'} V(12,1'2') a_1^+ a_2^+ a_{2'} a_{1'} \ , \tag{12.10}$$

where a pair interaction potential V has been introduced. In contrast to the Coulomb interaction, the nucleon–nucleon interaction is not a fundamental force but an effective interaction, which should be derived from QCD. Since the nucleons are composite particles consisting of quarks and gluons, the effective interaction can be introduced which reproduces properties like scattering phase shifts in a limited range of energy where no excitations of inner degrees of freedom occur. The same situation is known from the introduction of an effective atom–atom interaction. At present, different effective nucleon–nucleon potentials have been parametrized such as the Paris potential or the Bonn potential. We will use a separable version [423] of such potentials of the form

$$V(12,1'2') = \lambda w_{12}(k_{12}) w_{1'2'}(k'_{12}) \delta_{p_1+p_2, p'_1+p'_2} \ , \tag{12.11}$$

where the parameters for the coupling strength λ and for the form factors $w(k)$ as function of the relative momentum $\vec{k}_{12} = (\vec{p}_1 - \vec{p}_2)/2$ in the different channels can be found in [423].

We describe expanding quantum system as a non–equilibrium process, that is characterized at the beginning by a state in thermodynamic equilibrium, but confined in a finite volume. In the first stage, we assume local thermodynamic equilibrium with density, velocity and temperature depending on space and time. This quasi–equilibrium is characterized by the averages of particle, momentum and energy densities. In a later stage, the distribution functions of the different clusters are taken to characterize the non–equilibrium state. This means that different relevant statistical operators can be used to construct the non–equilibrium statistical operator at different times. Note that an averaged mean field is not able to describe the fragmentation of nuclear matter where the homogineity of the solution is not realized. Instead of introducing large clusters to describe the formation of fragments, one can also consider an inhomogeneous system, working only with lower order nonequilibrium distribution functions such as the space–dependent single–particle density.

12.1.2 Quasi–Equilibrium Distribution

We consider the non–equilibrium evolution based on a relevant statistical operator, which is constructed from the local densities of conserved quantities such as particle numbers, momentum and energy. This so–called quasi–equilibrium is also denoted as local thermodynamic equilibrium. We can derive the hydrodynamic description on this basis as shown, e.g., in [592]. In the special case considered here we are interested in the composition of this dense nucleonic system and the formation of clusters.

The relevant statistical operator for the local thermodynamic equilibrium distribution reads

$$\begin{aligned}\rho_{\mathrm{rel}}(t) = \; & Z_{\mathrm{rel}}^{-1}(t)\, \exp\left\{ -\int d\vec{r}\, \beta(\vec{r},t) \left[H(\vec{r}) - \vec{v}(\vec{r},t)\cdot\vec{P}(\vec{r}) \right.\right. \\ & \left.\left. -\mu(\vec{r},t) N(\vec{r}) \right] \right\}, \qquad (12.12)\end{aligned}$$

where the thermodynamic parameters $\beta(\vec{r},t), \vec{v}(\vec{r},t), \mu(\vec{r},t)$ are determined by the mean values of the densities of energy $\langle H(\vec{r})\rangle^t$, momentum $\langle \vec{P}(\vec{r})\rangle^t$, and particle number $\langle N(\vec{r})\rangle^t$, respectively. An index specifying different species will be dropped. Being in general a functional of the total behaviour in coordinate space, we assume that for $\rho_{\mathrm{rel}}(t)$ a gradient expansion can be performed and a local density approximation is possible. Then, the dependence on $\vec{r}$ must not be taken into account. Furthermore, we perform a boost transformation to a reference frame where the average of the momentum is equal to zero.

In the high density region, the ordinary thermal model becomes not applicable because of density corrections due to the interaction of the cluster with the surrounding matter. The in–medium corrections to the energy and wave function of a cluster lead to an interesting effect: at high densities, bound states disappear because of Pauli quenching. Beyond the so-called Mott density $\rho_{A\alpha}^{\rm Mott}(T)$ [448, 449], the abundances of the corresponding clusters decrease, and in the high density limit all bound states are dissolved so that a degenerate Fermi liquid of quasiparticles remains.

Within the cluster–mean field approximation, the abundances of nuclei are obtained from the equation of state, see also Eq. (4.25),

$$\rho(T,\mu_p,\mu_n) = \sum_{A=1}^{\infty} A \sum_{\alpha}{}^{(b)} g_{A\alpha} \int_{p>P_{A\alpha}^{\rm Mott}} f_A(E_{A\alpha p}) \frac{d^3p}{(2\pi\hbar)^3} , \qquad (12.13)$$

which is derived from a systematic Green function approach to the many nucleon system, if only the correlations leading to bound states are taken into account. $\rho = \rho_p + \rho_n$ denotes the baryonic density, μ_p and μ_n are the chemical potentials of protons and neutrons, respectively. The summation over the internal quantum number α runs only over the bound state part $E_{A\alpha p}$ of the energy spectrum of the A nucleon cluster, $g_{A\alpha}$ is the degeneration factor, $\beta = 1/k_B T$ the inverse temperature, and

$$f_A(E) = \left(\exp\left\{\beta\left[E - Z\mu_p - (A-Z)\mu_n\right]\right\} - (-1)^A\right)^{-1} \qquad (12.14)$$

is the cluster distribution function.

The energies $E_{A\alpha p}$ are obtained from a Bethe–Goldstone equation which accounts for in–medium effects. The cluster-mean field approximation which considers the A-particle system under the influence of a mean field generated by the correlated medium was given in Chap. 4. The energy eigenvalue of the cluster $(A\alpha\vec{p})$ is shifted according to

$$E_{A\alpha p}(\rho) = E^0_{A\alpha p}(\rho) + \vec{p}\cdot\vec{v}/2 + \Delta E^{\rm SE}_{A\alpha p} + \Delta E^{\rm Pauli}_{A\alpha p} . \qquad (12.15)$$

The interaction in the many–nucleon system consists of two contribution, the short-range nucleon–nucleon interaction $V(ij,i'j')$ and the long-range

Coulomb interaction. The energy shifts due to the Coulomb interaction will be treated separately considering screening effects what leads to a density dependent cluster energy $E^0_{A\alpha p}(\rho)$. Furthermore, $\vec{v} = \vec{p}/[Zm_p + (A-Z)m_n]$ is the the center–of–mass velocity of the cluster.

Neglecting clustering of the medium, within first order perturbation theory we have from Eq. (4.35) for the cluster self–energy shift

$$\Delta E^{\mathrm{SE}}_{A\alpha p} = \sum_{1\ldots A} \left|\phi^0_{A\alpha p}(1\ldots A)\right|^2 \times \sum_{i=1}^{A}\sum_{j} \left[V(ij,ij) - V(ij,ji)\right] \tilde{f}_1(j)\,, \tag{12.16}$$

and for the Pauli quenching term

$$\Delta E^{\mathrm{Pauli}}_{A\alpha p} = -\sum_{1\ldots A}\sum_{1'\ldots A'} \phi^0_{A\alpha p}(1\ldots A) V(ij,i'j') \times \tilde{f}_1(i)\phi^0_{A\alpha p}(1'\ldots A')\,. \tag{12.17}$$

$E^0_{A\alpha p}$ and $\phi^0_{A\alpha p}$ are the energy eigenvalue and the wave function, respectively, of the unperturbed cluster. The influence of the medium is given by the Fermi distribution function normalized to the local density $\tilde{\rho}_i$ at the cluster position, with $i = p, n$ specifying protons or neutrons, respectively. In the non–degenerate case we have

$$\tilde{f}_1(i) = \frac{1}{2}\tilde{\rho}_i \left(\frac{2\pi\hbar^2}{m_i k_B T}\right)^{3/2} \exp\left(-\beta\hbar^2 p_i^2/2m_i\right)\,. \tag{12.18}$$

The continuum edge of scattering states for a system of A unbound nucleons with total momentum $\vec{p}$ is shifted due to the self-energy shifts. The bound state energies show an additional shift due to the Pauli blocking, which depends on the density. The Mott density $\rho^{\mathrm{Mott}}_{A\alpha p}(T)$ where the bound state $(A\alpha\vec{p})$ merges with the continuum of scattering states and is dissolved, follows from the relation

$$E_{A\alpha p}(\rho^{\mathrm{Mott}}_{A\alpha p}) - \sum_{i}^{A} \Delta^{\mathrm{SE}}_1(i, \vec{p}_i = m_i\vec{v}) - \vec{p}\cdot\vec{v}/2 = 0 \tag{12.19}$$

At fixed temperature and density, a value $P_{A\alpha}^{\rm Mott}(T,\rho) > 0$ may occur at which the relation (12.19) is fulfilled. Then, clusters $(A\alpha\vec{p})$ with total momentum p below this Mott momentum are destroyed. The summation over only bound states (clusters) in Eq. (12.12) is realized if the integration over the region $p < P_{A\alpha}^{\rm Mott}$ in momentum space, where the respective clusters cannot exist, is excluded. Note that for $A > 2$ the stability of the cluster $(A\alpha\vec{p})$ not only with respect to the decay into free nucleons, but also with respect to other decay channels has to be taken into account.

Whithin a more sophisticated approach to the equation of state (12.13) and the composition of hot and dense nuclear matter, correlations in the medium should be included. Also the contributions of correlations from scattering states should be taken into account, as detailed for the two–particle correlations in Chapter 4.

12.1.3 Evaluation of the In–Medium Correction for Nucleonic Clusters

To outline the effects of medium corrections on the abundances of clusters, model calculations will be performed [444]. Starting from the interaction potential $V(12,1'2')$ which contains the long range Coulomb interaction as well as the short range nucleon-nucleon interaction, ab initio calculations of the energy values including the calculation of the energy eigenvalues and wave functions of the isolated clusters $(A\alpha\vec{p})$ are rather involved. We will use empirical data for the binding energies and the radii of the nuclei. Furthermore, the single-particle self-energy contribution $\Delta_1^{\rm SE}(1)$ due to the nucleon-nucleon interaction is estimated using a Skyrme type zero–range effective interaction [551] fitted to nuclear data. Global charge neutrality has to be considered in the case of infinite nuclear matter. For the treatment of the correlations due to the Coulomb interaction we can use the standard many-particle theory of charged particle systems, see [267].

Coulomb correlations

The main effects of the Coulomb correlations is described by screening effects what leads to two contribution, the modification of the cluster energies $E_{A\alpha p}^0(\rho)$ and the nucleon density variations within the screening sphere around a cluster. To give an estimate of the contributions due to the Coulomb interaction, we adopt the well-known Debye-Thomas-Fermi approach. The charge density $\rho_c(\vec{r})$ of a plasma consisting of protons with mean density ρ_p, neutrons with mean density ρ_n, a neutralizing homogeneous background of degenerate electrons with charge density $-e\rho_p$, and a cluster at $\vec{r} = 0$ with mass number A, radius $R_A = r_0 A^{1/3}$ ($r_0 = 1.2$ fm) and proton number Z, is given by

$$\begin{aligned} \rho_c(r) &= e(Z\rho_0/A + \rho_D(r) - \rho_p) \quad && r < R_A \, , \\ \rho_c(r) &= e(\rho_D(r) - \rho_p) && r > R_A \, , \end{aligned} \tag{12.20}$$

ρ_0 is the nuclear matter density. The density $\rho_D(r)$ of the screening cloud, which is formed by free or clustered protons, is determined by the screened potential $V_D(r)$

$$\begin{aligned} \rho_D(r) &= \rho_p(\beta, \mu_p - V_D(r)) = \rho_p - V_D(r) \frac{\partial}{\partial \mu_p} \rho_p(\beta, \mu_p) \\ &= \rho_p - \frac{\kappa^2}{2\pi e^2} V_D(r) \, , \end{aligned} \tag{12.21}$$

from which the value of κ is derived. The screened potential $V_D(r)$ is obtained from the Poisson equation, and the solution of the Poisson-Boltzmann equation yields

$$\begin{aligned} V_D(r) &= -\frac{2\pi}{3} e^2 (Z\rho_0/A + \rho_D(0) - \rho_p) r^2 + c_1, \; r < R_A \, , \\ V_D(r) &= c_2 \frac{1}{r} \, \mathrm{e}^{-\kappa r}, \qquad\qquad r > R_A \, , \end{aligned} \tag{12.22}$$

where $\rho_D(r)$ was replaced by $\rho_D(0)$ within the cluster. The constants c_1 and c_2 should be calculated self-consistently from the condition of continuity at $r = R_A$.

Two effects are obtained from the density distribution (12.21):

(i) In addition to the Coulomb energy of the isolated nucleus (i.e., $\rho = 0$), a self–energy shift due to the finite nucleon density ρ is obtained which amounts to half of the Coulomb interaction energy of the cluster with its surrounding:

$$\begin{aligned} &E^0_{A\alpha p}(\rho) - E^0_{A\alpha p}(0) = \\ &-\frac{1}{2}(Ze)^2 \kappa (1+\gamma) \left[\frac{1}{1+\kappa R_A} + \kappa R_A \left(\frac{2}{5} \frac{1}{1+\kappa R_A} + \frac{3}{35} \right) \right] , \end{aligned} \tag{12.23}$$

where

$$\gamma = \frac{\rho_D(0) - \rho_p}{\rho_0 Z/A}$$
$$= \frac{\rho_p A}{\rho_0 Z}\left[\exp\left\{-\frac{e^2 Z}{k_B T R_A}(1+\gamma)\left(\frac{1}{1+\kappa R_A}+\frac{1}{2}\right)\right\}-1\right] . \quad (12.24)$$

This effect makes the large clusters more stable.

(ii) The density $\rho_D(R_A)$ of the surrounding charged particles at the surface of the cluster is reduced

$$\rho_D(R_A) = \rho_p \exp\left(-\frac{e^2}{4\pi\rho_p R_A} Z \frac{1+\gamma}{1+\kappa R_A}\right) = \tilde{\rho}_p . \quad (12.25)$$

This effect will diminish the in–medium corrections due to the short–range nucleon–nucleon interaction especially for large clusters. In Eqs. (12.16), (12.17), the effective chemical potential in the nucleonic distribution function $\tilde{f}(i)$ is determined by the local nucleonic density.

Note that a simplified, but not very accurate description of the effects of Coulomb correlations is given by the Wigner–Seitz cell method where all protons are removed from a sphere with the radius $R_{\rm WS} = Z R_A \rho_0 / A \rho_p$.

Single–particle self–energy shift

The single–particle self–energy contribution due to the nucleon–nucleon interaction

$$\Delta_1^{\rm SE}(1) = \sum_2 [V(12,12) - V(12,21)] \tilde{f}_1(2) \quad (12.26)$$

can be estimated using a Skyrme–type zero–range interaction $V(\rho_p, \rho_n)\delta(\vec{r}_1 - \vec{r}_2)$ [551] so that $\Delta_1^{\rm SE}(1)$ is independent of momentum and temperature. This interaction contribution to the density of energy is parametrized according to

$$V(\rho_p, \rho_n) = \frac{3}{8} t_0 \rho^2 + \frac{1}{16} t_3 \rho^3 + \frac{3}{80}\left(\frac{3\pi}{2}\right)^{2/3} (3t_1 + 5t_2)\rho^{8/3}$$
$$+ \left(\frac{\rho_p - \rho_n}{\rho}\right)^2 \left[\frac{1}{6} t_2 \left(\frac{3\pi^2}{2}\right)^{2/3} \rho^{8/3} - \frac{1}{4} t_0 \rho^2 \left(\frac{1}{2} + x_0\right) - \frac{1}{16} t_3 \rho^3\right] \quad (12.27)$$

with $t_0 = -1057.3$ MeV fm^3, $t_1 = 235.9$ MeV fm^5, $t_2 = -100$ MeV fm^5, $t_3=$ 14463.5 MeV fm^6, $x_0 = 0.56$. From this expression, the single–particle shifts follow as

$$\Delta_1^{\mathrm{SE}}(i) = \partial V(\rho_p, \rho_n)/\partial \rho_i \,. \tag{12.28}$$

Energy shift of small clusters

For small clusters ($A \leq 16$), the self–energy shift is given by $\Delta E^{\mathrm{SE}}_{A\alpha p} = \sum_i^A \Delta_i(\tilde{\rho}_p, \tilde{\rho}_n)$ which is determined by the effective nuclear matter density at the cluster $(A\alpha\vec{p})$. In the Pauli quenching term (12.17), the interaction potential can be eliminated if the cluster wave function $\phi^0_{A\alpha p}(1...A)$ is approximated by the antisymmetrized product of single-particle wave functions $\varphi_\nu(i)$ [444],

$$\Delta E^{\mathrm{Pauli}}_{A\alpha p} = -\sum_\nu^{\mathrm{occ.}} \sum_1 |\varphi_\nu(1)|^2 (E_\nu - E_1) \tilde{f}(\vec{p}_1 - m_1 \vec{v}) \,. \tag{12.29}$$

The single–particle energies E_ν are related to the total energy $E^0_{A\alpha p}$ and the binary interaction energy V according to $\sum E_\nu = E^0_{A\alpha p} + V$. The expression (12.29) has been evaluated for $4 < A \leq 16$ by using the wave functions and energy levels of a harmonic oscillator potential [60] $V(r) = V_0 + m\omega^2 r^2/2$, with $\hbar\omega = 40 A^{-1/3}$ MeV. The parameter V_0 has been adjusted to the empirical values of the cluster binding energies $E^0_{A,0}(\rho)$

$$V_0 = \frac{2}{A}(E^0_{A,0} + 3\hbar\omega) - \frac{15}{4}\hbar\omega \,. \tag{12.30}$$

For $A \leq 4$, the bound state wave function $\Phi^0_{A,0}$ was also approximated by a product of Gaussian functions $\varphi_\nu(p) \propto \exp(-p^2/2k^2)$ with the range parameter

$$k^2 = 4 m_i |E^0_{A,0}| / A\hbar^2 \,. \tag{12.31}$$

Eliminating the binary interaction energy V we have $E_\nu = 5E^0_{A,0}/A$, and we obtain

$$\Delta E^{\mathrm{Pauli}}_{A0p} = \frac{1}{2A} \sum_i^A \tilde{\rho}_i \Lambda_i^3 |E^0_{A,0}| \delta^{-3/2} \mathrm{e}^{-x_i/2\delta} \times \left[5 + 3/\delta + x_i(\delta - 1)/\delta^2\right] \tag{12.32}$$

with $\delta = 1 + 2|E^0_{A,0}|/AT$; $x_i = m_i v^2/k_B T$, $\Lambda_i^2 = 2\pi\hbar^2/m_i k_B T$. A similar result is obtained for $\Delta E^{\rm Pauli}_{A0p}$ for $4 < A \leq 16$, see [444].

Energy shifts for large clusters

The total energy shift of a cluster $(A\alpha\vec{p})$

$$\begin{aligned}\Delta E_{A0p}(\rho) &= \Delta E^{\rm SE}_{A0p} + \Delta E^{\rm Pauli}_{A0p} \\ &= \sum_{\nu}^{\rm occ.} \sum_{22'} \varphi^*_\nu(2)\varphi_\nu(2') \sum_{11'} \tilde{f}(\vec{p}_1 - m_1\vec{v}) \\ &\times [V(12,1'2') - V(12,2'1')][\delta_{11'}\delta_{22'} - \sum_{\nu'}^{\rm occ.} \varphi^*_{\nu'}(1)\varphi_{\nu'}(1')] \end{aligned} \tag{12.33}$$

has a simple structure if within a homogeneous Fermi gas model the wave functions $\varphi_\nu(i)$ are taken as momentum eigenfunctions. At not too high temperatures, a strong compensation between the self-energy and the Pauli quenching term results, Thus large clusters remain nearly unshifted. Applying the homogeneous Fermi gas model to a cluster with density $\rho_i^A(r)$ for the protons or neutrons, respectively, we obtain

$$\begin{aligned}\Delta E_{A0p}(\rho) &= \sum_{i=p,n} \int d^3r \Delta_1^{\rm SE}(\rho_p^A(r), \rho_n^A(r))\tilde{\rho}_i \\ &\times \int_{\Lambda_i p_F(\rho_i^A(r))}^{\infty} \frac{y dy}{2\pi x_i} \left[{\rm e}^{-(y-x_i)^2/4\pi} - {\rm e}^{-(y+x_i)^2/4\pi} \right] \end{aligned} \tag{12.34}$$

with $p_F(\rho_i) = (3\pi^2\rho_i)^{1/3}$, $x_i = m_i v \Lambda_i/\hbar \approx p\Lambda/A$. An analysis of experimental data was performed [77] to fit the following density distribution function

$$\rho^A(r) = \frac{3A}{4\pi R^3} \frac{1}{1 + (\pi b/R)^2} \left[\frac{1}{1 + {\rm e}^{(r-R)/b}} + \frac{1}{1 + {\rm e}^{(-r-R)/b}} - 1 \right] . \tag{12.35}$$

Appropriate values for $A > 16$ are $R = 1.05 A^{1/3}$ fm, $b = 0.57$ fm. Furthermore we take $\rho_p^A(r) = \rho^A(r) Z/A$.

The Mott momentum

After having determined the shifts of the bound state energies, we can ask whether or not the cluster is stable with respect to decomposition into smaller clusters. As discussed above, because of the Pauli quenching the binding energy of a cluster is lowered with increasing density, and the bound states merging with the continuum of scattering states are dissolved. The Mott momentum for the decay of the cluster into free particles is given for $A \leq 4$ by

$$\Delta E^{\text{Pauli}}_{A0P^{\text{Mott}}} + E^{0}_{A,0} = 0 \tag{12.36}$$

and the corresponding expression for $4 < A \leq 16$. This condition (12.36) may be replaced by further restrictions in the p space if further decay modes are considered. Especially, the disintegration of larger clusters including the strongly bound α particles is of interest, where the energy shift of the α particles has to be determined with the actual density $\tilde{\rho}_i$. This effect is of importance especially in the region of weakly bound nuclei with $6 \leq A \leq 11$ where also the stability of the clusters with respect to the decay into other fragments as deuterons and tritons/ ^{3}He has to be checked in order to determine $P^{\text{Mott}}_{A\alpha}$.

Excited states

The summation over all excited cluster states α in Eq. (12.13) can be replaced by an integral after introducing the density of states [60]

$$D_A(E) = \frac{1}{12}\left(\frac{\pi^2}{a}\right)^{1/4} E^{-5/4}\exp(2\sqrt{aE}) \tag{12.37}$$

with $a = A/15\ \text{MeV}^{-1}$ for the homogeneous Fermi gas model and arbitrary values Z. Then the abundance of the clusters with mass number A is given by [444]

$$X_A = \rho_A/\rho = \frac{A^3}{2\pi^2\rho\Lambda^3}\int_{P^{\text{Mott}}_{a,0}\Lambda/A}^{\infty} x^2 dx \int_{E^A_{\text{min}}}^{E^A_{\text{max}}} dE D_A(E)$$
$$\times \mathrm{e}^{-(E^0_{A,0}(\rho)+E+\Delta E_{A,0}(\rho,x)+Ak_BTx^2/4\pi - Z\mu_p-(A-z)\mu_n)/k_BT}. \tag{12.38}$$

In this formula, it was assumed that Coulomb corrections, self-energy and Pauli blocking corrections to the cluster energy do not strongly depend on the excitation state . A lower bond $E^A_{\rm min} = E_F/A$ with E_F – nuclear matter Fermi energy, was introduced to take into account that below this energy the density of states (12.37) is not applicable, and the discrete structure of the excitation spectrum of the clusters should be considered. The upper bond $E^A_{\rm max}(x)$ is introduced into (12.38) to take into account that excited clusters may become instable with respect to the decay into smaller fragments as given, e. g., by Eq. (12.36).

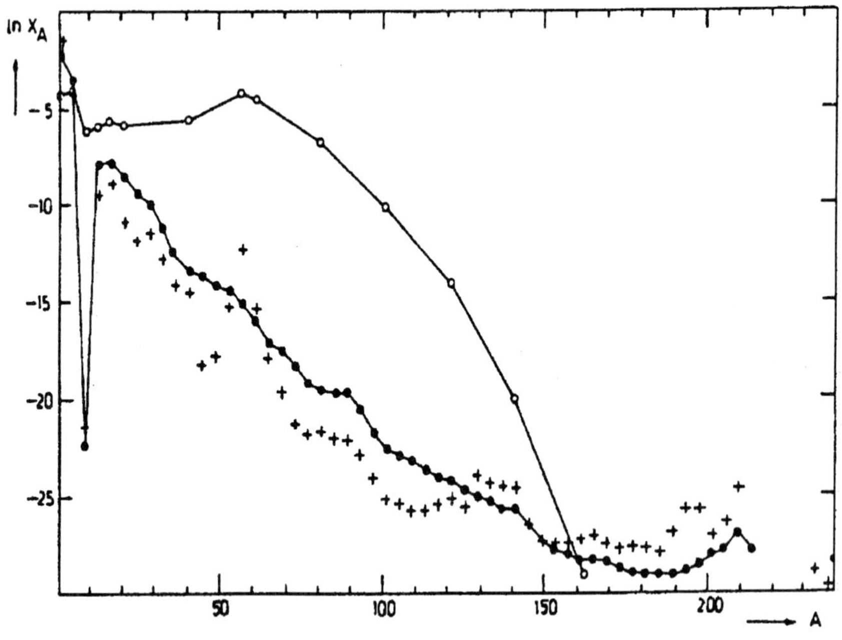

Fig. 12.1 Abundances of elements, $A = 4n$. Crosses: averaged solar elemental abundances $\bar{X}_A = (X_{A-1} + X_A + X_{A+1} + X_{A+2})/4$ according to Ref. [80]. Open circles: ordinary thermal model calculation of X_A neglecting in–medium corrections for $k_BT = 5$ MeV, $\rho = 0.016\,\text{fm}^{-3}$. Dots: quantum statistical calculation of X_A for the same parameters. For $A < 12$, the averaged calculated abundances $\bar{X}_A$ are shown. Connecting lines are guides to the eye.

Model calculation

As an example, a calculation of the abundances $X_A = \rho_A/\rho$ of clusters has been performed for the parameter values $k_BT = 5$ MeV, nucleon density $\rho = 0.016$ fm^{-3} and $\rho_p = \rho_n = \rho/2$ [453]. From the equation of state (12.13) the values $\mu_P = -13.23$ MeV, $\mu_n = -13.4$ MeV were obtained by a self-consistent calculation. For $E^0_{A,0}(0)$, the empirical values of the binding energies of the most abundant isobars have been taken.

At these densities, Coulomb corrections are of importance especially for the heavier clusters. The screening parameter $\kappa = 0.135$ fm^{-1} (12.21) is determined from $\partial\rho_p/\partial\mu_p \approx 0.0005$ MeV^{-1} fm^{-3}. The well-known Debye shift $-\kappa(Ze)^2/2$ is modified according to (12.23) because of the finite size of the nuclei. A typical value for $E^0_{A\alpha p}(\rho) - E^0_{A\alpha p}(0) = -188.0$ MeV is obtained for $A = 100$, $Z^* = 46$.

As a consequence of the Coulomb correction (12.23) to the bound state energy, for given mass number A of a cluster the proton number Z^* for which the binding energy of the cluster takes the minimum value differs from the case $\rho = 0$ and is shifted towards the more symmetric nuclei. Thus, the isotopic composition of the elements in hot dense matter may strongly differ from the usual one.

The proton density is decreasing near a cluster because of the Coulomb correlations. According to Eq. (12.25), e.g., the value $\tilde{\rho}_{100} = 0.00371$ fm^{-3} was obtained for $A = 100$, $Z^* = 46$. The shifts $\Delta E_{A\alpha p}$ were evaluated for the most bound isobar Z^*. For instance, a value $\Delta E_{A\alpha p}(\rho) = -38.2$ MeV was obtained for $A = 100$ according to Eq. (12.34).

For all clusters with $2 \leq A \leq 11$, the Pauli quenching term $\Delta E^{\mathrm{Pauli}}_{A,0}$ exceeds the binding energy at given parameter values of T, ρ, so that a value $P^{\mathrm{Mott}}_{A,0} > 0$ was obtained and the region of p -integration is reduced.

The results for clusters with mass numbers $A = 4n, n = 3, 4, \ldots$, are shown in Fig. 12.1. A steep decrease of the calculated abundances X_A for values A up to $A \approx 100$ and a flattening for $A > 100$ is seen; furthermore, there are oscillations due to shell effects in the energy values $E^0_{A,0}(0)$. For the light metal abundances a strong depletion as a consequence of the Pauli quenching is obtained. Especially the weakly bound clusters with $6 \leq A \leq 11$ are very sensitive to in–medium corrections and are destroyed because of the Pauli blocking effect.

For comparison, results of a calculation neglecting in–medium corrections, i.e., $\Delta E_{A\alpha p} = 0$ in Eq.(12.13), corresponding to the ordinary thermal model are presented for the same values of temperature and density.

It is clearly seen that the account of in–medium corrections strongly modifies the shape of the cluster distribution function. The use of a simple chemical model based on free cluster properties is not justified already at relatively low nuclear matter densities. The effect of the medium on bound state energies leads to essential effects at densities of the order of 0.01 fm^{-3}.

It should be mentioned that the results for local thermodynamic equilibrium have been derived here in a certain approximation. In particular, the contributions of scattering states have not been taken into account. For the special case $A = 2$, a more sophisticated treatment is possible and has been given in Chapter 4.

Comparing the results of the ordinary thermal model with the evaluation of abundances including in–medium effects, a strong depletion of light cluster abundances is obtained as consequence of the Pauli quenching, see Ref. [596]. Coulomb correlations are of importance in evaluating the heavier element abundances. In spite of the approximations in evaluating the in–medium energy shifts $\Delta E_{A\alpha p}$, we conclude that the abundances may become strongly modified in high density matter if in–medium effects are taken into account. This is of interest in connection with the theory of heavy ion collisions as well as with astrophysical and cosmological problems.

12.2 Time Evolution of Distribution Functions

The quasi–equilibrium described in the previous section cannot give the time evolution of the distribution functions which needs the full non–equilibrium statistical operator. We discuss some of the typical approaches to calculate the time–dependent cluster distribution functions.

12.2.1 General Approach

The time–dependent distribution functions are defined according to

$$f_{A\alpha}(\vec{r}\vec{p}t) = \int \frac{d^3k}{(2\pi\hbar)^3} e^{i\vec{k}\cdot\vec{r}/\hbar} \langle a^+_{A\alpha,\vec{p}-\vec{k}/2} a_{A\alpha,\vec{p}+\vec{k}/2} \rangle^t . \qquad (12.39)$$

The average has to be performed with the non–equilibrium statistical operator

$$\rho(t) = \epsilon \int_{-\infty}^{t} dt_0 e^{\epsilon(t_0 - t)} U(t, t_0) \rho_{\rm rel}(t_0) U^+(t, t_0) . \tag{12.40}$$

Having the non–equilibrium statistical operator to our disposal, we are able to evaluate time–dependent mean values such as

$$\frac{\partial}{\partial t} f_{A\alpha}(\vec{r}\vec{p}t) = \int \frac{d^3k}{(2\pi\hbar)^3} e^{i\vec{k}\cdot\vec{r}/\hbar} {\rm Tr} \left\{ \rho(t) \frac{i}{\hbar} \left[H, a^+_{A\alpha,\vec{p}-\vec{k}/2} a_{A\alpha,\vec{p}+\vec{k}/2} \right] \right\} . \tag{12.41}$$

An important problem is the determination of the relevant statistical operator $\rho_{\rm rel}(t_0) = e^{-S(t_0)}$. Different forms are possible. All the different variants give the same result if the limit $\varepsilon \to 0$ is performed in the final expression. However, different results arise in finite order perturbation theory. The finer the description in the relevant statistical operator the better works a Markov approximation where memory effects are neglected.

A particular version for the relevant statistical operator appropriate to the problem of clustering in dense matter has been discussed in Chapter 4. To reproduce the distribution function at time t_0 the cluster decomposition of the entropy

$$S(t_0) = \sum_{A\alpha p} s_{A\alpha p}(t_0) a^+_{A\alpha p} a_{A\alpha p} \tag{12.42}$$

has been considered. The thermodynamic parameters $s_{A\alpha p}(t_0)$ have to be determined from the given mean values $f_{A\alpha}(\vec{r}\vec{p}t)$.

This way we arrive at a highly sophisticated self–consistent problem. The time evolution of the thermodynamic parameters is determined by the time evolution of the distribution functions, which, in turn, is determined by the statistical operator containing the thermodynamic parameters. In principle, one has to eliminate the time variation of the thermodynamic parameters.

Instead of solving this involved problem, we also can try to determine the distribution functions in some approximations, for instance from a local equilibrium. Such an approach may be possible in certain stage of the evolution, where the exact time evolution agrees with the time evolution of a reduced distribution. In particular, the hydrodynamic description will be accepted as long as it is not in conflict with the kinetic description.

12.2.2 Single–Particle Approach

To present the approach to non–equilibrium for expanding nuclear matter, we first consider as a simple example the case where all correlations in the medium are neglected. Then, the relevant characteristics of the non–equilibrium state are the single–particle distributions

$$f_1(11', t_0) = \langle a_1^+ a_{1'} \rangle_0^t \tag{12.43}$$

and the relevant statistical operator is found as

$$\rho_{\mathrm{rel}}(t_0) = \exp\{-\sum_{11'} s_1(11', t_0) a_1^+ a_{1'}\} . \tag{12.44}$$

In this special case, the relation between the thermodynamic parameters and the given mean values (equation of state) can be given explicitly after diagonalizing

$$s_1(11', t_0) = \ln\left[\frac{1}{f_1(11', t_0)} - 1\right] . \tag{12.45}$$

In classical statistics, this relevant statistical operator corresponds to a product of single–particle distributions.

Now we have to construct the non–equilibrium statistical operator according to

$$\rho(t) = \epsilon \int_{-\infty}^{t} dt_0 \mathrm{e}^{\epsilon(t_0 - t)} U(t, t_0) \rho_{\mathrm{rel}}(t_0) U^+(t, t_0) . \tag{12.46}$$

This operator allows us to evaluate the time variation of the averages

$$\frac{\partial}{\partial t} f_1(11', t_0) = \mathrm{Tr}\left\{\rho(t) \frac{i}{\hbar}\left[H, a^+_{\vec{p}-\vec{k}/2} a_{\vec{p}+\vec{k}/2}\right]\right\} \tag{12.47}$$

in the kinetic equations. Then we are able to obtain the time dependence of the thermodynamic parameters.

The solution of this scheme is simple in the case of the ideal quantum gas, where the Hamiltonian is given by

$$H^0 = \sum_1 E(1) a_1^+ a_1 . \tag{12.48}$$

The action of the time evolution operators can be solved with the result

$$U(t-t_0)\exp\left[\sum_{11'} s_1(11',t_0)a_1^+ a_{1'}\right] U^+(t-t_0)$$
$$=\exp\left[\sum_{11'} s_1(11',t)a_1^+ a_{1'}\right] \tag{12.49}$$

with

$$s_1(11',t)=s_1(11',t_0)\exp[i(E(1)-E(1'))(t-t_0)/\hbar] \tag{12.50}$$

and describes the free motion. The thermodynamic parameters can be eliminated, and the solution is directly related to the initial state.

Kinetic Equation for the Single–Particle Distribution Function

The inclusion of interaction has the consequence that no rigorous expression for $\rho(t)$ can explicitly be given. However, we can perform some approximations well known from perturbation theory.

The time evolution is determined not only by the kinetic part, but contains also an interaction part. The kinetic equation reads

$$\frac{\partial}{\partial t} f_1(\vec{r},\vec{p},t) = \mathrm{Tr}\left\{\rho(t)\frac{i}{\hbar}[H, n_1(\vec{r},\vec{p})]\right\}$$
$$= -D_1^{\text{Vlasov}}(\vec{r},\vec{p},t) + I_1^{\text{coll}}(\vec{r},\vec{p},t) \tag{12.51}$$

where the statistical operator after partial integration

$$\rho(t)=\rho_{\text{rel}}(t)-\int_{-\infty}^{t} dt' \mathrm{e}^{\epsilon(t'-t)}\frac{d}{dt'}\{U(t,t')\rho_{\text{rel}}(t')U^\dagger(t,t')\} \tag{12.52}$$

has been used to separate the mean–field (Vlasov) and the collision term, see Eq. (4.57). The Vlasov term

$$D_1^{\text{Vlasov}}(\vec{r},\vec{p},t) = -\frac{i}{\hbar}\langle[H, n_1(\vec{r},\vec{p})]\rangle^t_{\text{rel}} \tag{12.53}$$

describes the reversible time evolution. The collision term

$$I_1^{\rm coll}(\vec{r},\vec{p},t) = -\int_{-\infty}^{t} dt' {\rm e}^{\epsilon(t'-t)} {\rm Tr}\left\{\frac{i}{\hbar}[H, n_1(\vec{r},\vec{p})]\right.$$
$$\left.\times\, {\rm e}^{iH(t'-t)/\hbar}\left(\frac{i}{\hbar}[H,\rho_{\rm rel}(t')] + \frac{\partial}{\partial t'}\rho_{\rm rel}(t')\right){\rm e}^{-iH(t'-t)/\hbar}\right\} \qquad (12.54)$$

gives rise to irreversible behaviour and describes scattering processes between the particles.

12.2.3 Mean–Field Approximation: The Vlasov Equation

Considering the lowest order in the interaction V we drop the collision term. In the remaining kinetic equation in mean–field approximation we have for the Vlasov term

$$D_1^{\rm Vlasov}(\vec{r},\vec{p},t) = -\int \frac{d^3k}{(2\pi\hbar)^3} {\rm e}^{i\vec{k}\cdot\vec{r}/\hbar} \langle \frac{i}{\hbar}[H, n_1(\vec{p},\vec{k})]\rangle^t_{\rm rel}\,. \qquad (12.55)$$

Using Wick's theorem we can express the averages over the relevant statistical operator by the single–particle distribution function

$$\langle \frac{i}{\hbar}[H, n_1(\vec{p},\vec{k})]\rangle^t_{\rm rel} = -\frac{i}{\hbar m}\vec{p}\cdot\vec{k} f_1(\vec{p},\vec{k},t) + \frac{i}{\hbar}\int \frac{d^3p_2 d^3q}{(2\pi\hbar)^6} \qquad (12.56)$$
$$\times\{f_1(\vec{p}+\vec{q}/2,\vec{k}-\vec{q},t) f_1(\vec{p}_2-\vec{q}/2,\vec{q},t)(V(\vec{q}) - V(-\vec{q}-\vec{p}+\vec{p}_2+\vec{k}/2))$$
$$-f_1(\vec{p}-\vec{q}/2,\vec{k}-\vec{q},t) f_1(\vec{p}_2+\vec{q}/2,\vec{q},t)(V(\vec{q}) - V(-\vec{q}+\vec{p}-\vec{p}_2+\vec{k}/2))\}$$

The corresponding kinetic equation in mean–field approximation coincides with the self–consistent TDHF formalism.

A more simple form of the mean–field term is obtained if the exchange terms are dropped and if $V(q)$ is assumed to be a smooth function. We find

$$D_1^{\rm Vlasov}(\vec{r},\vec{p},t) = \frac{\partial E^{\rm qu}(\vec{r},\vec{p},t)}{\partial \vec{p}}\frac{\partial f_1(\vec{r},\vec{p},t)}{\partial \vec{r}} - \frac{\partial E^{\rm qu}(\vec{r},\vec{p},t)}{\partial \vec{r}}\frac{\partial f_1(\vec{r},\vec{p},t)}{\partial \vec{p}} \qquad (12.57)$$

with the quasiparticle energy

$$E^{\mathrm{qu}}(\vec{r},\vec{p},t) = \frac{p^2}{2m} + \int \frac{d^3r_2 d^3p_2}{(2\pi\hbar)^3} V(\vec{r}-\vec{r}_2) f_1(\vec{r}_2,\vec{p}_2,t) \, . \tag{12.58}$$

In this approximation, the system can be treated as a system of free quasiparticles, which follow from a wave equation where the Hamiltonian contains a mean–field contribution. In that approximation, the elimination of the thermodynamic parameters can also be performed explicitly. Of course, the evolution of the system is obtained in an approximation characterized by a averaged potential, neglecting any fluctuations.

For the evaluation of the mean field in nuclear matter one can directly evaluate the intergral over the pair interaction potential and the single–particle distribution function. Sometimes it is more convenient to use a simple parametrization of the mean field in terms of the density as done, e.g., by Skyrme [551], see Eq. (12.27). This parametrization is performed in such a way that in mean field approximation the empirical values for nuclear matter and finite nuclei are reproduced.

The solution of the Vlasov equation can be done using computer simulations. For classical particles effective codes have been worked out. To reproduce also quantum properties, quantum molecular dynamics has been introduced based, e.g., on Gaussian wave packets which get the respective symmetry of the wave function in an approximative way.

12.2.4 Inclusion of Collisions

In higher orders of the interaction the collision term (12.54) has to be taken into account. In the Markov approximation we neglect non–locality in time, i.e. $\rho_{\mathrm{rel}}(t') \approx \rho_{\mathrm{rel}}(t)$, and also disregard non–locality in space, i.e. we do not consider variations with respect to the coordinate $\vec{r}$ of the single–particle distribution function in evaluating the collision process. This implies that the retardation effects $\propto \partial\rho_{\mathrm{rel}}(t')/\partial t$ can be dropped and that the collision term contains only occupation numbers in momentum space

$$n_1(\vec{p}) = a_p^\dagger a_p \, . \tag{12.59}$$

Representing the collision integral as

$$I_1^{\text{coll}}(\vec{r}\vec{p},t) = -\int_{-\infty}^{\infty} dE \int_{-\infty}^{\infty} dE' \int_{-\infty}^{t} dt'$$
$$\times \exp\{(i(E-E')/\hbar+\epsilon)(t'-t)\}$$
$$\mathrm{Tr}\left\{\frac{i}{\hbar}[H,n_1(\vec{p})]\delta(E-H)\frac{i}{\hbar}[H,\rho_{\text{rel}}(t)]\delta(E'-H)\right\}, \qquad (12.60)$$

the integrals over t' and E' can be performed. Replacing $\delta(E-H)$ by $1/\pi\,\mathrm{Im}[1/(E-H+i\epsilon)]$ we obtain a Kubo–Greenwood formula for the collision integral in Markov approximation. Introducing the T–matrix as

$$T(E^+) = V(E-H+i\epsilon)^{-1}(E-H^0+i\epsilon) \qquad (12.61)$$
$$= VG(E^+)G_0^{-1}(E^+)\,,$$

where the index zero denotes no interaction, the optical theorem is obtained as

$$\mathrm{Im}\frac{1}{E-H+i\epsilon} = \frac{1}{V}T(E^+)\mathrm{Im}G_0(E^+)T(E^-)\frac{1}{V}\,. \qquad (12.62)$$

Using the relations $T = V + VG_0T = V + TG_0V$, $[n_1(\vec{p}), G_0] = 0$ and $[\rho_{\text{rel}}(t), G_0] = 0$ we find

$$I_1^{\text{coll}}(\vec{r}\vec{p},t) = \frac{1}{\pi\hbar}\int_{-\infty}^{\infty} dE\ \mathrm{Tr}\left\{[T(E^+),n_1(\vec{p})]\right. \qquad (12.63)$$
$$\left.\mathrm{Im}G_0(E^+)[T(E^-),\rho_{\text{rel}}(t)]\mathrm{Im}G_0(E^+)\right\}\,.$$

The Boltzmann collision integral follows if the on–shell many–particle T–matrix is replaced by the two–particle T–matrix $T(12,1'2')a_1^+a_2^+a_{1'}a_{2'}$ and the Wick theorem is used

$$I_1^{\text{coll}}(\vec{r}\vec{p},t) = \frac{2\pi}{\hbar}\int\frac{d^3p_2d^3p_1'd^3p_2'}{(2\pi\hbar)^6}|T(12,1'2')-T(12,2'1')|^2$$
$$\times\delta(E_1+E_2-E_{1'}-E_{2'})\delta(p_1+p_2-p_{1'}-p_{2'}) \qquad (12.64)$$
$$\times\{f(1)f(2)(1-f(1'))(1-f(2'))-f(1')f(2')(1-f(1))(1-f(2))\}.$$

This so–called Boltzmann–Uehlig–Uhlenbeck collision integral can be introduced into the kinetic equation to improve the mean–field equation and to describe dissipation. For simulations the nucleon–nucleon cross section of about 40 mb can be used to represent the collision term.

In a dense medium the collision cross section will be modified. Taking the influence of an uncorrelated medium into account, the modifications of the cross sections have been evaluated [490].

To improve the single–particle picture by including correlations in the system, one can alternatively either relax the Markov approximation by taking into account memory terms, or one can introduce the correlations as additional degrees of freedom so that a Markov approximation may be justified.

12.2.5 Hydrodynamic Model of Expanding Hot and Dense Matter

The behaviour of expanding hot nuclear matter should be described by a system of generalized kinetic equations of the Vlasov–Uehling–Uhlenbeck type which include not only the in–medium effects such as Pauli blocking but also the formation of clusters. At present such an intention is far from being realized and simpler models are in use. Among them, the so–called hadro–chemical model reflects some of the important features of the kinetic approach by cluster distribution functions. The expansion of hot nuclear matter is described in a relativistic gas dynamical approach while the cluster abundances are calculated in terms of ordinary chemical rate equations. One has to follow the time evolution of expanding hot nuclear matter up to the freeze–out at which the density has become so small that the particles rather seldom collide and cannot maintain the thermal equilibrium.

Nuclear hydrodynamics is based on the conservation of particle number, momentum and energy. For a nonrelativistic system with mass density ρ we have the balance equations

$$\begin{aligned}
&\frac{\partial}{\partial t}\rho + \frac{\partial}{\partial \vec{r}}(\rho\vec{v}) = 0 \,, \\
&\frac{\partial}{\partial t}(\rho\vec{v}) + \frac{\partial}{\partial \vec{r}}(\rho\vec{v}\times\vec{v}) = -\frac{\partial}{\partial \vec{r}}p \,, \\
&\frac{\partial}{\partial t}(\rho e) + \frac{\partial}{\partial \vec{r}}(\rho e\vec{v}) = \frac{\partial}{\partial \vec{r}}(p\vec{v}) \,.
\end{aligned} \tag{12.65}$$

In addition we need the equation of state for the pressure $p = p(\rho, T)$ and for the energy density $e = mv^2/2 + W(\rho, T)$.

Freeze–out Model

This scenario assumes that the system in the early stage is in thermal and chemical equilibrium. The system is dense, and the reaction rates are large, so that at variation of the parameters the system can relax to equilibrium. With decreasing density, the reaction rates are also decreasing, so that at a certain stage it is not more possible to maintain chemical equilibrium, and the equilibrium distribution freezes out at a certain temperature T and density ρ.

Typical values for freeze out parameters for nuclear matter are $\rho = \rho_0/3 \dots \rho_0/5$. At these densities, however, an ideal mixture of clusters (bound states) is not appropriate, and medium effects have to be included.

Statistical Multifragmentation Model

The Statistical Multifragmentation Model [179] describes the disassembly of an expanding system into fragments according to thermodynamic equilibrium (microcanonic ensemble). Afterwards de–excitation of the hot primary fragments occur. The motion of a fragment has a thermal as well as a collective component. The average kinetic energy of a fragment is

$$E_0 = E_{\text{thermal}} + E_{\text{flow}} = \text{const} + c \cdot A \, . \tag{12.66}$$

The thermal component is given by $v_{th}^2/2m = 3k_bT/2$ and hence independent of the mass of the fragment. Due to the additional collective flow, which is proportional to A, the average kinetic energy of a fragment increases with the mass. Under the constraints on the total energy, mass number and charge contained in the break–up volume, statistical weights of various break–up channels are calculated. When the partition has been determined, fragment excitation energies and momenta are generated by a microcanonical treatment. The post–break–up evolution of the system contains the motion of the fragment under the influence of the mutual Coulomb field and the de–excitation of hot fragments.

12.2.6 Inclusion of Correlations

Instead of a model which combines a single–particle description with a composition derived from chemical equilibrium for an ideal mixture, one can try to evaluate the correlations during the dynamical process of expanding nuclear matter.

In–Medium Reaction Rates

An alternative approach to cluster formation is based on the kinetic equations for the single–particle distribution function, the two–particle distribution function, etc. In contrast to the freeze–out scenario, the dynamical aspects are adequately described, see [108]. In the low–density limit, the time evolution of the cluster distribution functions is described by the reaction kinetics as described in Chapter 4. However, for a dense medium, the reaction rates will be modified. The most simple example, the break-up and formation rate of deuterons, is considered in Chapter 4, see also [33, 34].

The Coalescence Model

The coalescence model, see [595] for a review, was introduced to simulate the formation of clusters in a system of nucleons in an approximation which is similar to the sudden approximation in perturbation theory. In its simplest version the coalescence model forms a deuteron from a proton and a neutron (plane wave representation) if their relative momenta are within a certain capture radius p_0, comparable to the deuteron's momentum content. The deuteron cross section can then be computed in terms of the proton and neutron cross sections as

$$\frac{d\sigma_d}{dp^3}(\vec{p}) = \frac{3}{4}\frac{4\pi p_0^3}{3}\frac{d\sigma_p}{dp^3}\left(\frac{\vec{p}}{2}\right)\frac{d\sigma_n}{dp^3}\left(\frac{\vec{p}}{2}\right) \tag{12.67}$$

where 3/4 is the combinatoric factor for angular momentum.

A more advanced coalescence model can be formulated which assumes that neutron and proton are described by wave packets of width σ localized in space around $\bar{\vec{x}}_i$ and in momentum space around $\bar{\vec{p}}_i$

$$\psi_i(\vec{x}) = \frac{1}{(\pi\sigma^2)^{3/4}}\exp\left(-\frac{(\vec{x}-\bar{\vec{x}}_i)^2}{2\sigma_i^2}\right)\exp\left(i\bar{\vec{p}}_i\vec{x}\right) \tag{12.68}$$

and write the deuteron wave function as a product of its center–of–mass motion and its internal motion,

$$\psi_d(\vec{x}_1, \vec{x}_2) = \Phi_{\vec{P},\vec{\bar{R}}}(\vec{R})\phi_d(\vec{r}) \;, \tag{12.69}$$

where

$$\Phi_{\vec{P},\vec{\bar{R}}}(\vec{R}) = \frac{1}{(\pi\Sigma^2)^{3/4}} \exp\left(-\frac{(\vec{R}-\vec{\bar{R}})^2}{2\Sigma_i^2}\right) \exp\left(i\vec{\bar{P}}\vec{R}\right) \tag{12.70}$$

and

$$\phi_d(\vec{r}) = (\pi\alpha^2)^{-3/4} \exp(-r^2/2\alpha^2) \;. \tag{12.71}$$

The coalescence probability, or the deuteron content of the two–particle wave function, can now be computed from the squared overlap

$$C(\vec{x}_n, \vec{k}_n; \vec{x}_p, \vec{k}_p) = |\langle \psi_n \psi_p | \Phi_{\vec{P},\vec{\bar{R}}} \phi_d \rangle|^2. \tag{12.72}$$

Including a factor 3/4 for spin, and neglecting pairs lost to higher clustering, the deuteron number is then

$$\begin{aligned} n_d &= \frac{3}{4} \int d\vec{x}_n d\vec{k}_n f_n(\vec{x}_n, \vec{k}_n) \int d\vec{x}_p d\vec{k}_p f_p(\vec{x}_p, \vec{k}_p) C(\vec{x}_n, \vec{k}_n; \vec{x}_p, \vec{k}_p) \\ &= \frac{3}{4} \sum_{ij} C(\vec{x}_{n_i}, \vec{k}_{n_i}; \vec{x}_{p_j}, \vec{k}_{p_j}) \end{aligned} \tag{12.73}$$

with the sum extending over all appropriate pairs ij in the system. The overlap coalescence is equivalent with the Wigner coalescence under a factorization hypothesis.

The projection on the deuteron wave function has to be performed in a dense medium. The use of a medium modified wave function for the overlap coalescence has been discussed in [499].

Clustering in Quantum Molecular Dynamics Simulations

The basic assumption of the quantum molecular dynamics (QMD) model [3, 4] is that a test wave function of the form

$$\phi_i(\vec{r},t) = (2/L\pi)^{3/4} e^{-(\vec{r}-\vec{r}_i(t))^2/L} e^{i(\vec{r}-\vec{r}_i(t))\vec{p}_i(t)} e^{ip_i^2(t)t/2m} \tag{12.74}$$

is a good approximation to the nuclear wave function. The total n–body wave function is assumed to be the direct product of coherent states, neglecting antisymmetrization,

$$\Phi = \prod_i \phi_i(\vec{r},\vec{r}_i,\vec{p}_i,t) \tag{12.75}$$

The evolution due to wave-packet dynamics is considered. At the final stage of evolution where the density is low so that nucleon-nucleon interaction processes die out, fragments (clusters) are formed if the distance between nucleons is smaller than 3 fm. However, such a prescription can only give an approximate description of the cluster formation process where more detailed properties of the bound states such as binding energy and wave function should be taken into account.

12.2.7 Clustering in a Time–Dependent Mean Field

Correlations in a dense medium are formed during the expansion as a nonequilibrium process occurring in a dense medium. The full quantum statistical treatment of this process is rather involved. It should describe the formation of bound states when the density falls below the Mott density. A redistribution of the occupation numbers of the cluster state is due to collisions, but also as a consequence of the time variation of the mean field. We consider here the latter process.

We will use the instantaneous state approximation. The clusters $(A\alpha\vec{p})$ are considered under the influence of a time-dependent cluster-mean field and are described by an A-particle wave equation with a hamiltonian $H_A^{\mathrm{MF}} = H(t)$. The instantaneous state approximation considers the time derivative of $H(t)$ as a small parameter suitable to perform an expansion. Also step-like (sudden) variations of $H(t)$ at discrete time instants t_i can be considered.

For brevity we use the index ν instead of $(A\alpha\vec{p})$. Considering t as a parameter, the instantaneous eigenfunctions of $H(t)$ follow from

$$H(t)|\phi_\nu(t)\rangle = E_\nu(t)|\phi_\nu(t)\rangle \,. \tag{12.76}$$

The phase of the eigenstates $|\phi_\nu(t)\rangle$ is fixed, e.g., choosing a real eigenfunction in the case of a bound state. If $H(t)$ is a slowly varying function of t, we may expect that

$$|\psi_\nu(t)\rangle = |\phi_\nu(t)\rangle e^{-\frac{i}{\hbar}\int_{t_0}^{t} E_\nu(t')dt'} \,. \tag{12.77}$$

is a good approximate solution of the Schrödinger equation. t_0 is an arbitrary time instant where the phases of the states $|\psi_\nu(t)\rangle$ coincide with that of the basis $|\phi_\nu(t)\rangle$. We will use the wave function $\psi_\nu(1\ldots A,t) = \langle 1\ldots A|\psi_\nu(t)\rangle$ to construct the time dependent operators

$$a_\nu^\dagger(t) = \psi_\nu^*(1\ldots A,t)a_1^\dagger\ldots a_A^\dagger \,. \tag{12.78}$$

The relevant statistical operator is taken as

$$\rho_{\mathrm{rel}}(t) = Z_{\mathrm{rel}}^{-1}(t)e^{-\sum_\nu s_\nu(t)a_\nu^\dagger(t)a_\nu(t)} \tag{12.79}$$

and commutes with $H(t)$ because of (12.76). It does not solve the Liouville–von Neumann equation (12.1) because of the explicite time dependence. However, it can be used to fix the initial condition for the solution of the Liouville–von Neumann equation as $\lim_{t\to-\infty}[\rho(t)-\rho_{\mathrm{rel}}(t)] = 0$. Then the solution reads after partial integration of (12.5)

$$\begin{aligned}
\rho(t) &= \rho_{\mathrm{rel}}(t) - \int_{-\infty}^{t} dt' e^{-\epsilon(t-t')} \\
&\times U(t,t')\left\{\frac{i}{\hbar}\left[H(t'),\rho_{\mathrm{rel}}(t')\right] + \frac{\partial}{\partial t'}\rho_{\mathrm{rel}}(t')\right\}U(t',t) \\
&\approx \rho_{\mathrm{rel}}(t)\left\{1 - \int_{-\infty}^{t} dt' e^{-\epsilon(t-t')}\left[-\sum_\nu \frac{\partial s_\nu(t')}{\partial t'}a_\nu^\dagger(t)a_\nu(t)\right.\right. \\
&\left.\left. + \sum_{\nu\mu} s_\nu(t)a_\mu^\dagger(t)\langle\psi_\mu(t')|\frac{\partial}{\partial t'}\psi_\nu(t')\rangle a_\nu(t) + cc.\right]\right\} \,,
\end{aligned} \tag{12.80}$$

where we used that the commutator of $H(t')$ with $\rho_{\rm rel}(t')$ vanishes. Furthermore we made the approximation that the time evolution operator only shifts the time arguments in the operators. Differentiation of the instantaneous Schrödinger equation gives

$$\langle\psi_\mu(t')|\frac{\partial}{\partial t'}\psi_\nu(t')\rangle = \mathrm{e}^{-\frac{i}{\hbar}\int_{t_0}^{t'}[E_\mu(t'')-E_\nu(t'')]dt''}$$
$$\times\frac{1}{E_\nu(t')-E_\mu(t')}\langle\phi_\mu(t')|\frac{\partial H(t')}{\partial t'}|\phi_\nu(t')\rangle\ . \tag{12.81}$$

In the adiabatic limit, the term containing the time derivative of $H(t)$ can be neglected, and $s_\nu(t)$ does not change with time. Transitions between different states ν are obtained if the time derivative of $H(t)$ is taken into account.

In the sudden approximation where the hamiltonian changes abruptly at the time instant t_i so that $\partial H(t')/\partial t' \propto \delta(t'-t_i)$, the integral over dt' can be carried out, and with $[E_\nu(t_i^-)-E_\mu(t_i^+)]^{-1}\langle\phi_\mu(t_i^+)|[H(t_i^+)-H(t_i^-)]|\phi_\nu(t_i^-)\rangle = \langle\phi_\mu(t_i^+)|\phi_\nu(t_i^-)\rangle$ the projection on the new states is obtained.

In conclusion, the cluster formation in expanding nuclear matter can be described on the mean-field level approximately as an adiabatic evolution without changing the occupation numbers when the mean field changes slowly, but in sudden approximation when the mean field changes rapidly. The basis of models such as the freeze out model or the coalescence model can be given within this approximation. However, the dynamical approach given here can be used to find a more systematic dynamical approach.

A simple example would be the two-particle correlations in a dense medium which are described by the T matrix. As shown in the first section, for a separable interaction the T matrix can be solved including the medium effects in mean-field approximation. The Pauli blocking factors as well as the Hartree-Fock shifts are depending on density and temperature, if local equilibrium is assumed. Within a hydrodynamical approach, the local temperature and density will depend on time, so that also the T matrix is changing with time. Now we can use the instantaneous state approximation to evaluate the occupation numbers of the two-particle states. Of particular interest is the occupation number of the resonance which goes over to a bound state if the density becomes low.

It should be mentioned that also collisions have to be included improving the mean-field approximation. To determine the relevant distribution in the high-density case as the initial condition to solve the Liouville–von Neumann equation, a local thermal equilibrium can be assumed as long as collisions dominate. If the redistribution of occupation numbers due to collisions becomes small, the further evolution is described by the time-dependent mean field. Note that collisions can consistently be included in the nonequilibrium statistical operator approach as shown above.

12.3 Production of Nuclei from Hot Dense Matter

Different situations are of interest where the distribution of elements is formed during the expansion of hot and dense nuclear matter. The early Universe and supernovae collapses are important places where elements are produced. In laboratory experiments, heavy ion collisions are situations where such processes can be studied. We cannot give an exhaustive presentation of these processes here. We only will focus to some topics which are of relevance for the problem of correlations in dense nonequilibrium systems. We first discuss the application of the cluster formation in non-equilibrium to heavy ion collisions and afterwards astrophysical applications.

12.3.1 Element Distributions from Excited Nuclei

The production of nuclei from excited nuclei is a wide field and cannot be discussed here. We only review some items related to quantum statistics of hot and dense matter.

Well known already for a long time is the emmission of nuclei from a compound nucleus. Thermodynamic models have been proven to be applicable so that the spectra of the emmitted particles can be described by a temperature.

The life–time of a compound nucleus is determined by the coupling to continuum states, in particular in the presence of a Coulomb barrier. In order to describe the emission of clusters, preformed correlations in the compound nucleus are of relevance, which should be evaluated taking into account medium effects.

Higher excited nuclei, as produced in heavy–ion reactions, disintegrate on a very short time scale, forming an expanding hot nuclear medium. In a simple approximation, this highly excited nuclear matter is considered as a so–called fireball, which is cooling down during expansion. A more realistic description can be given on the basis of the time evolution of the single–particle distribution function, for what different numerical codes have been worked out to simulate the results of heavy–ion collisions. The hydrodynamical evolution is found by solving the Boltzmann–Uehlig–Uhlenbeck (BUU) equation or treating quantum molecular dynamics (QMD). However, in contrast to the zero–density limit, the medium effects should be considered which lead to a self–energy as well as the medium modified nucleon–nucleon cross section in nuclear matter at finite density, see [7, 490].

In general, an approach based on a single–particle distribution function is not able to describe the formation of clusters consistently, since quantum correlations are averaged out in that approximation. Therefore, the formation of clusters is introduced in that approaches by additional consideration such as a freeze–out of a time–dependent thermodynamic equilibrium or a prescription that defines cluster formation by the distance of nucleons in the phase space. However, such simplified descriptions where, e.g., during the quantum molecular dynamics simulations nucleons at distances smaller than 3 fm are considered to form a cluster, are not very satisfactionary. Some attempts have been performed to elaborate a transport code which includes the formation of correlations on quantum statistical level [108, 109]. Further efforts are necessary to include higher clusters as well as a more detailed description of the effects of the expanding medium.

Besides the formation of clusters, correlations in final states of heavy ion reactions can be measured, e.g., within the Hanbury–Brown Twiss experiments. The inclusion of initial correlations in nuclear matter is a topic which demands further considerations, and should be included consistently on the basis of the treatment of few–body quantum correlations in expanding nuclear matter.

A highly exciting effect observed in heavy ion collisions is multifragmentation of clusters. The maximum energy which can be stored in a droplet is found to be about 5 Mev/A [24]. At higher energies, the droplets fragmentate into smaller clusters and a typical distribution of the final cluster abundances is observed. Recently the dependence of the observed temperature from the excitation energy has been investigated to reconstruct the equation of state in form of the caloric curve [424]. However, different concepts are used to derive a temperature such as the slope parameter of the particle spectra, the isotopic ratio and the occupation of correlated states, and contradictionary results are obtained. It should be stressed out that the use of an equilibrium quantity such as temperature in describing a strongly nonequilibrium process is questionable, and the formation of isotopes and correlations should be considered as a nonequilibrium process occurring in a dense medium. Thus, different distributions will be produced at different times where different values for density and temperature are of relevance. In particular, correlations between scattering states can be established only in the latest stage of the reaction process where the density is sufficiently low so that these loosely correlated states are not destructed.

Further recent problems are, for instance, the ^{3}He to tritium ratio in heavy ion reaction, which cannot be derived from recent calculations of compositions such as statistical multifragmentation and transport codes [426].

12.3.2 Element Distribution in Astrophysics

In the expanding Universe, different processes are going on due to that the distribution of elements change. Starting from the nucleosynthesis in the early stage of the Universe, the chemical evolution continues until now. Burning processes in stars, reactions such as cosmic rays interacting with matter, but also violent processes like supernova explosions etc. lead to further changes in the abundances of the elements as seen at the present time t_0. Recently a lot of information is available on the chemical composition of different stars, galaxies, interstellar and intergalactical matter, cosmic rays at different energies, etc. We will not cover this extended field here, but we will focus only on the relevance of medium modifications for the evolution of the chemical composition. In the standard approach, kinetic equations are considered, with reaction rates taken from the zero density case, what can be justified only if the density of the nuclear medium is small compared with the nuclear matter density.

Expanding Hot Universe

Assuming that in the early times the Universe was homogeneous and isotropic on a sufficiently large scale, the time evolution of a scale factor $R(t)$ for distances in comoving coordinates is given within the Friedman–Robertson–Walker metric according to ($\hbar = c = 1$) [594]

$$ds^2 = dt^2 - R^2(t)\left[\frac{dr^2}{1-\kappa r^2} + r^2(d\theta^2 + \sin^2\theta d\phi^2)\right] . \tag{12.82}$$

With appropriate rescaling of the coordinates, κ can be choosen to be +1, -1, or 0, corresponding to closed, open, or spatially flat geometries. Einstein's equations lead to the Friedman equation

$$H^2 = \left(\frac{\dot{R}}{R}\right)^2 = \frac{8\pi G\rho}{3} - \frac{\kappa}{R^2} + \frac{\Lambda}{3} \tag{12.83}$$

as well as to

$$\frac{\ddot{R}}{R} = \frac{\Lambda}{3} - \frac{4\pi G}{3}(\rho + 3p) , \tag{12.84}$$

where $H(t)$is the Hubble parameter, ρ is the total mass–energy density, p is the isotropic pressure, and Λ is the cosmological constant. The present time t_0 estimate for $H_0 = H(t_0)$ is

$$H_0 = 100h_0 \text{kms}^{-1}\text{Mpc}^{-1} = h_0/(9.78\text{Gyr}) \tag{12.85}$$

Observational bounds give $0.4 < h_0 < 1$.

The cosmological constant, which can be related to the vacuum energy density in quantum field theory, is taken as $\Lambda = 0$. The Friedman equation serves to define the density parameter $\Omega_0 = \Omega(t_0)$

$$\frac{\kappa}{R_0^2} = H_0^2(\Omega_0 - 1) , \tag{12.86}$$

$\Omega_0 = \rho_0/\rho_c$, and the critical density defined as

$$\rho_c = \frac{3H_0^2}{8\pi G} = 1.88 \times 10^{-29} h^2 \text{gcm}^{-3} . \tag{12.87}$$

The three curvature signatures $\kappa = +1, -1$, and 0 correspond to $\Omega_0 > 1, < 1$, and 1. The knowledge of Ω_0 is not accurate. Luminous matter (stars and associated material) contribute $\Omega_{\text{lum}} < 0.01$. The minimum of matter required to explain the flat rotation curves of spiral galaxies amounts to $\Omega_0 \approx 0.1$, while estimations for Ω_0 based upon cluster virial masses suggests $\Omega_0 \approx 0.2 - 0.4$. The highest estimates follow from the peculiar motions of galaxies as $0.1 < \Omega_0 < 2$.

Energy conservation implies that

$$\dot{\rho} = -3(\dot{R}/R)(\rho + p) \tag{12.88}$$

so that for a matter dominated ($p/\rho \ll 1$) Universe $\rho \propto R^{-3}$, while for a radiation–dominated ($p = \rho/3$) Universe $\rho \propto R^{-4}$. So the less singular curvature term κ/R^2 in the Friedman equation can be neglected at early times when R is small. If the Universe expands adiabatically, the entropy per comoving volume ($= R^3 s$) is constant, where the entropy density is given by $s = (\rho + p)/T$ and T is temperature. The energy density of radiation can be expressed (with $\hbar = c = 1$) as

$$\rho_r = \frac{\pi^2}{30} N(T)(kT)^4 , \tag{12.89}$$

where $N(T)$ counts the effectively massless degrees of freedom of bosons and fermions

$$N(T) = \sum_B g_B + \frac{7}{8} \sum_F g_F . \tag{12.90}$$

At temperatures less than about 1 MeV, neutrinos have decoupled from the thermal background, i.e. the weak interaction rates are no longer fast enough compared with the expansion rate to keep neutrinos in equilibrium with the remaining thermal bath consisting of $\gamma, e^{\pm}$. Furthermore, at temperatures $kT < m_e$, by entropy conservation, the ratio of the neutrino temperature to the photon temperature is given by $(T_\nu/T_\gamma)^3 = g_\gamma/(g_\gamma + \frac{7}{8} g_e) = 4/11$.

In the early Universe when $\rho \approx \rho_r$, then $\dot{R} \propto 1/R$, so that $R \propto t^{1/2}$ and $Ht \to 1/2$ as $t \to 0$. The time–temperature realationship at very early times can be found from the above equations

$$t = \frac{2.42}{\sqrt{N(T)}} \left(\frac{1\text{MeV}}{kT} \right)^2 \text{sec} . \tag{12.91}$$

At later times, since the energy density in radiation falls off as R^{-4} and the energy density in non–relativistic matter falls off as R^{-3}, the Universe eventually became matter dominated. The epoch of matter–radiation density equality is determined by equating the matter density equality is determined by equating the matter density at $t_{\text{eq}}, \rho_m = \Omega_0 \rho_c (R_0/R_{\text{eq}})^3$ to the radiation density, $\rho_r = (\pi^2/30)[2 + (21/4)(4/11)^{4/3}](kT_0)^4(R_0/R_{eq})^4$ where T_0 is the present temperature of the microwave background (see below). Solving for (R_0/R_{eq}) gives

$$
\begin{aligned}
&(R_0/R_{eq}) = 2.4 \times 10^4 \Omega_0 h_0^2 \ , \\
&kT_{eq} = 5.6 \Omega_0 h_0^2 \mathrm{eV} \ , \\
&t_{eq} = 3.2 \times 10^{10} (\Omega_0 h_0^2)^{-2} \mathrm{sec} \ .
\end{aligned}
\tag{12.92}
$$

Prior to this epoch the density was dominated by radiation (relativistic particles; see Eq. (12.89)), and at later epochs matter density dominated. Atoms formed at $(R_0/R_{eq}) \approx 1300$, and by $(R_0/R_{eq})_{dec} \approx 1100$ the free electron density was low enough that space became essentially transparent to photons and matter and radiation were decoupled. These are the photons observed in the microwave background today.

The age of the Universe today, t_0, is related to both the Hubble parameter and the value of Ω_0 (still assuming that $\Lambda = 0$). In the Standard Model, $t_0 \gg t_{eq}$ and we can write

$$
t_0 = H_0^{-1} \int_0^1 \left(1 - \Omega_0 + \Omega_0 x^{-1}\right)^{-1/2} dx \ . \tag{12.93}
$$

Constraints on t_0 yield constraints on the combination $\Omega_0 h_0^2$. For example, $t_0 \geq 13 \times 10^9$ yr implies that $\Omega_0 h_0^2 \leq 0.25$ for $h_0 \geq 0.5$, while $t_0 \geq 10^{10}$ yr implies that $\Omega_0 h_0^2 \leq 0.8$ for $h_0 \geq 0.5$ or $\Omega_0 h_0^2 \leq 1.1$ for $h_0 \geq 0.4$.

The present temperature of the microwave background is $T_0 = 2.726 \pm 0.005$ K as measured by COBE [344], and the number density of photons $n_\gamma = (2\zeta(3)/\pi^2)(kT_0)^3 \approx 411 \ \mathrm{cm}^{-3}$. The energy density in photons (for which $g_\gamma = 2$)is $\rho_\gamma = (\pi^2/15)(kT_0)^4$. At the present epoch, $\rho_\gamma = 4.65 \times 10^{-34} \mathrm{gcm}^{-3} = 0.26 \ \mathrm{eVcm}^{-3}$. For nonrelativistic matter (such as baryons) today, the energy density is $\rho_B = m_B n_B$ with $n_B \propto R^{-3}$, so that for most of the history of the Universe n_B/s is constant. Today, the entropy density is related to the photon density by $s = (4/3)(\pi^2/30)[2 + (21/4)(4/11)](kT_0)^3 = 7.0 n_\gamma$.

Homogeneous Big–Bang Nucleosynthesis

The standard model of nucleosynthesis [74], see also Refs. [345, 561] for some review, assumes different processes which are relevant for the recent abundances of elements. Primordial nucleosynthesis is determined by the evolution of the early Universe which is assumed to be uniform and isotropic. The path of the so–called "basic big–bang model" [561] in the $\rho - T$ plane is shown in Fig. 12.2. Below $k_B T \approx 1$ MeV the weak interaction freezes out. At abount 0.1 MeV, nuclei are synthesized, mostly ^{4}He, and a primordial abundance distribution is established where "metals" ($Z > 2$) are almost not abundant. Recent light element abundances are considered as a relict of primordial nucleosynthesis, and especially a primordial ^{4}He mass fraction $Y_{\rm pr} = 0.245 = -0.003$ [574] seems to be consistent with astrophysical observations.

As a result, a relation between the parameter η and the primordial abundances of the light elements can be derived. Observed abundances which can be related to the primordial abundances are consistent with a value $\eta = n_B/n_\gamma$ to $2.8 \times 10^{-10} < \eta < 4.0 \times 10^{-10}$. The parameter η is related to the portion of Ω in baryons

$$\Omega_B = 3.66 \times 10^7 \eta h_0^{-2} (T_0/2.726\ \mathrm{K})^3 \tag{12.94}$$

so that $0.010 < \Omega_B h_0^2 < 0.015$, and hence the Universe cannot be closed by baryons.

Within the standard hot big–bang model of nucleosynthesis [561], the occurrence of heavier elements ($Z > 2$) is due to burning processes in stars after they have been formed in a later stage of the evolution of the Universe. For instance, it is assumed that the neutron–rich heavy nuclei are produced during explosive supernovae collapses (r–process), but the parameters of the astrophysical environment are not exactly specified up to now [345].

Nuclear Matter Nucleosynthesis

Constraints from the standard hot big–bang model of primordial nucleosynthesis [574] point to a low–density Universe. Within this scenario, nucleons alone fail to close the Universe by at least a factor 5.

If the assumption of perfect homogeneity is relaxed [593] and the constraint is investigated which the standard hot big–bang model of primordial nucleosynthesis places upon the amplitude of isothermal density perturbations $\delta = \delta\rho/\bar{\rho}$ ($\bar{\rho}$ being the average nucleon density) a constraint $\delta < 2$ follows [574].

The standard hot big–bang model of primordial nucleosynthesis predicts the existence of metal–deficient matter (population III) which has not been found up to now. The observation that all astrophysical objects are contaminated by heavy nuclei needs therefore a participation of all matter in supernovae collapses during the evolution of the Universe.

These consequences of the standard hot big–bang model of primordial nucleosynthesis are obtained without the account of possible in–medium effects. In order to point out the possibility of primordial nucleosynthesis with nearly empirical values of heavier element ($Z > 2$) abundances, in Fig. 12.2 the parameter values of T and ρ are given where the ^{4}He mass fraction amounts to $Y_{\rm pr} = 0.245$ (the $Y_{\rm pr}$ line). For the sake of simplicity, symmetric nuclear matter $\rho_p = \rho_n$ is considered.

If the abundances of all clusters are considered for values of ρ, T near the $Y_{\rm pr}$ line, a steep decrease of abundances with increasing mass number is obtained as long as the in–medium corrections to the cluster energies can be neglected (the ordinary thermal model). In particular, this refers to the path of the basic big–bang model and corresponds to the metal–deficient matter without heavier elements. This behaviour is drastically changed in that region of the $\rho - T$ plane where in–medium effects become of importance, i.e., near the Mott density as introduced in Refs. [448, 451]. The example presented in Fig. 12.1 shows that in this region cluster abundances can occur which exhibit besides the primordial ^{4}He mass fraction $Y_{\rm pr}$ also nearly empirical values for heavy element abundances.

Of course, as discussed above, the freeze–out concept can only give a first approximation to the dynamical non–equilibrium process of expanding hot dense matter which should be described by a coupled system of kinetic equations. Evaporation of nucleons from the hot clusters, neutron absorption, photodisintegration, fission of heavy nuclei, weak interaction, etc., will modify the abundance distribution as matter cools down. Especially, single particle reactions will determine the finer structures of the final abundance distribution function.

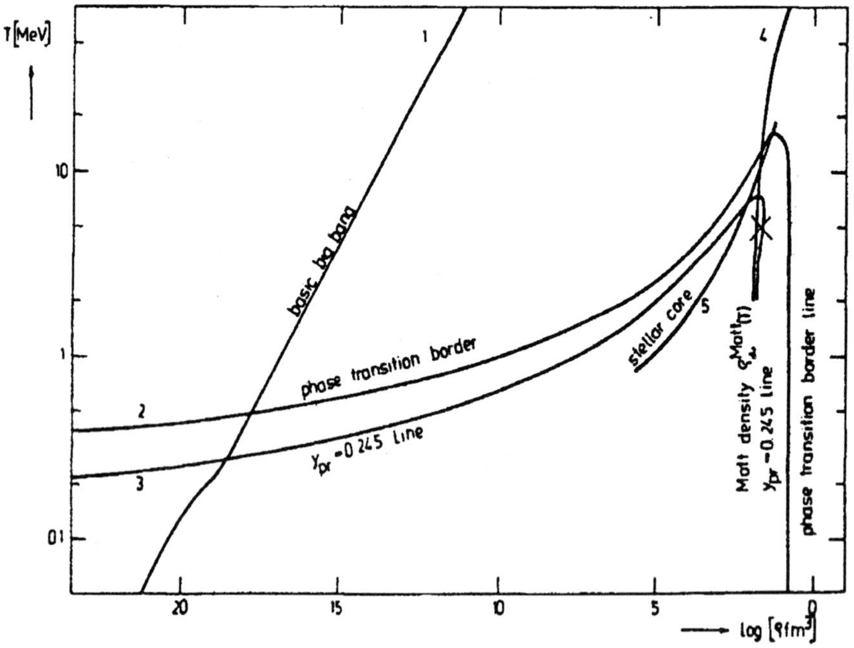

Fig. 12.2 The density–temperature plane for hot dense matter. 1: the path of the basic big–bang model; 2: saturation line of the nuclear matter phase transition; 3: the Y_{pr}–line ($Y_{pr} = 0.245$); 4: the Mott density for ^{4}He; 5: the temperature and density in the center of the stellar core during collapse [10]. The cross denotes the parameter values of the model calculation $k_B T = 5$ MeV, $\rho = 0.016\,\mathrm{fm}^{-3}$.

Of special interest is the question whether the three constraints of the standard model of nucleosynthesis given above (nonclosure, homogeneity, metal deficient matter) can be relaxed if in–medium effects are taken into account.

In Fig. 12.2, the phase transition region [107, 272, 291, 449, 498] as well as the calculated values of the thermodynamic parameters during explosive supernovae collapses according to Refs. [10, 70, 346] are also shown to point out the relevance of high–density effects for astrophysical processes. Recent investigation [345] of the astrophysical site of r–processes in heavy element nucleosynthesis point also into this region of the $\rho - T$ plane.

Different Astrophysical Objects

We have mainly considered the solar elemental abundances. The question arises whether different chemical composition of different astrophysical objects (stars, galaxies, cosmic rays) may reflect different paths in the $\rho - T$ diagram during nucleosythesis.

Measured distributions give metal enhanced or deficient abundances for different stars, a corresponding parametrization by parameters ρ and T has been discussed in [58]. Measured chemical profiles of galaxies also can be parametrized by space dependent parameter values of ρ and T. Another example is the chemical composition of cosmic rays which also can be parametrized by energy dependent parameter values ρ, T, see [450].

12.4 Conclusions

In–medium effects such as Pauli quenching and Coulomb correlations should be taken into account to explain the formation of clusters and correlations not only in heavy ion collisions but also in astrophysics. Especially, equilibrium element abundances can be obtained which exhibit at the same time the primordial mass ratio of ^{4}He as well as a relevant amount of heavier elements (see Fig. 12.1 for a model calculation). Considering the freeze–out concept as a simple approximation, this equilibrium result may be of interest to discuss primordial nucleosynthesis.

A more rigorous treatment of the formation of clusters from hot and dense matter should be based on a systematic nonequilibrium approach to finite excited nuclon systems. The time evolution of the different cluster distribution functions should be considered for interacting asymmetric nuclear matter, where the time evolution of fluctuations at high densities, nucleation and multifragmentation in the instability region as well as reaction kinetics in the low density region should be considered as special cases.

13 Pattern Formation in Segregation Processes in Porous Materials

J. Schmelzer

13.1 Introduction

Despite the variety of research done in the theoretical description of coarsening – the late stage of phase order phase transformations – most of the attempts developed so far deal with a restricted problem only, when the matrix where the phase separation process takes place can be considered as a homogeneous body allowing the formation and the growth of clusters at any place with the same probability and the growth to a practically arbitrary size. Spatial correlations occur in these cases via diffusional or elastic interactions of the precipitates (see, e.g., [225]).

However, in a number of applications precipitation and coarsening takes place in spatially inhomogeneous systems, where the dimension of the pores determine the maximum size of the clusters of the newly evolving phase or result, at least, in a significant inhibition of cluster growth once the sizes of the clusters become comparable with the pore sizes. Particularly interesting applications of such types of processes are segregation in porous materials like vycor glasses and zeolithes, they are of relevance also for the understanding of sintering processes (see, e.g., [168]).

Another application consists in the description of the kinetics of cluster growth and coarsening, when the elastic response of the matrix with respect to cluster growth is accompanied by an increase of the total energy of elastic deformations growing more rapidly than linear with the volume of the clusters. This is generally the case when the diffusion coefficient of the segregating particles exceeds by orders of magnitude the self–diffusion coefficient of the matrix building units (cf. [474, 475, 531]). In these cases also qualitative modifications of the kinetics of coarsening due to cluster matrix interactions occur (see, e.g., [414]), which can be described by the same methods.

In a paper by Slezov, Schmelzer and Möller [509] for the first time the process of Ostwald ripening was analyzed for the case that clusters are formed and grow only inside the pores (in the discussed general meaning) of a porous solid matrix. Hereby it was assumed that the solid can be considered as an Hookean elastic body. This assumption implies that after the cluster has reached the size of the pore the evolving elastic strains become sufficiently high as to prevent immediately any further growth of the clusters. Consequently, also the kinetics of coarsening and the type of asymptotic solutions is different as compared with the results obtained first by Lifshitz and Slezov [310] (cf. also chapter 2).

There exists, however, the alternative possibility that the elastic strains evolving in the course of growth of clusters of a new phase in a solid increase only slowly with increasing cluster size but may reach, nevertheless, so large magnitudes with time as to stop the further growth. In this situation the limiting case considered in [509] is not an adequate description of the process and has to be generalized.

Moreover, in real systems one has to deal not with a monodisperse pore size distribution but with some given non–monodisperse distribution of pores. The particular features of the pore size distribution will affect, of course, the kinetics of coarsening.

Generalizations of the results outlined in [509] for the cases,

(i.) when instead of hard non–deformable pores a system of "weak" pores is considered, i.e., when the elastic strains inhibiting the growth increase only slowly with increasing cluster size;

(ii.) when any particular arbitrary size distributions of "hard" pores exists in the solid and the clusters of the new phase are formed and grow only inside the non–deformable pores,

have been developed by Schmelzer, Möller and Slezov [479]. In the present chapter, the results a summarized briefly, the main attention is directed hereby to the discussion of additional topics which are not reflected to a sufficient degree in the original papers ([509, 479]).

Note, that the mechanism of the influence of elastic strains on coarsening discussed here is different from the type Kawasaki and Enomoto [237] dealt with, where the inhibition, respectively, acceleration of coarsening is due to elastic interactions between the growing or dissolving clusters. In the case considered here, the inhibition is due to the interaction cluster–matrix, which may also qualitatively affect the coarsening kinetics as will be shown in the subsequent analysis (see also [468, 472, 474, 477]).

13.2 Ostwald Ripening in a System of Hard Pores

For a derivation of the equations governing the evolution of an ensemble of clusters in a system of pores with an arbitrary distribution with respect to pore sizes in a solid matrix as a first precondition two basic equations are required, the growth equation for a single cluster with a radius, R, in a pore of size R_o and the distribution function with respect to pore sizes. In this section, we consider the simplest case first that all pores are of the same size R_o.

The velocity of growth of a single cluster in a pore of size R_o was assumed in [509] to be of the form (cf. also Eq.(2.88))

$$\frac{dR}{dt} = \frac{R_{co}^3}{R}\left(\frac{1}{R_c} - \frac{1}{R}\right)\Theta(R_o - R) , \tag{13.1}$$

$$\Theta(R_o - R) = \begin{cases} 1 & \text{for } R_o - R > 0 \\ 0 & \text{for } R_o - R \leq 0 \end{cases} . \tag{13.2}$$

In Eq.(13.1) and furtheron a dimensionless time scale is used (see Eq.(2.87)). By R_c the critical cluster radius is denoted, determined by the following relation (cf. Eqs.(2.7) and (2.8))

$$R_c(t) = \frac{2\sigma}{c_\alpha \Delta\mu} . \tag{13.3}$$

R_{co} is its value at the beginning of the process of competitive growth.

Eq.(13.1) implies that the growth of the cluster is terminated at a radius R_o due to the interactions with the walls of the pore.

To simplify the analysis. we assume that the pores are sufficiently large that initially a Lifshitz–Slezov distribution with respect to cluster sizes (cf.

Eqs.(2.97) – (2.101)) can be established in the system, before the interactions with the walls become dominating in the kinetics of coarsening. Note, however that the method outlined is applicable to any arbitrary initial cluster size distribution.

Since nucleation processes are of no importance for the late stages of the process, the evolution of the cluster size distribution function, $f(R,t)$, can be described by a continuity equation of the form (cf. Eq.(2.21))

$$\frac{\partial f(R,t)}{\partial t} + \frac{\partial}{\partial R}\left[f(R,t)\frac{dR}{dt}\right] = 0 \, . \tag{13.4}$$

In the range $R < R_o$, the solution of Eq.(13.4) can be expressed as (cf. [509])

$$f(R,t) = \frac{N(t_0)}{R_c(t_0)} P_{LS}\left[\frac{R_1(R,t)}{R_c(t_0)}\right]\frac{\partial R_1}{\partial R} \, , \qquad R < R_0 \, . \tag{13.5}$$

t_0 is the moment of time when the interactions with the walls of the pores become of importance for the evolution of the cluster size distribution function. $R_1(R,t)$ is the characteristic solution of the growth equation

$$\frac{dR}{dt} = \frac{R_{co}^3}{R}\left(\frac{1}{R_c} - \frac{1}{R}\right) \, . \tag{13.6}$$

A substitution of Eq.(13.1) into the continuity equation (13.4) yields further

$$f(R,t) = \left(\int_{t_0}^{t} dt\, f(R_o - \varepsilon)\left.\frac{dR}{dt}\right|_{R_o-\varepsilon} \delta(R_o - R)\right)_{\varepsilon\to 0} \, , \tag{13.7}$$

$$R \geq R_o \, .$$

The set of equations has to be solved, now, in a self–consistent way. In general, the determination of $f(R,t)$ is possible by numerical methods, only (see, e.g., [225, 509]). However, an inspection, in particular, of Eq.(13.7) shows that in the course of time a monodisperse distribution of clusters develops with an average cluster size $\langle R\rangle = R_o$.

13.3 Ostwald Ripening in a System of Weak Pores

We consider, now, the opposite situation that starting with some initial value of the cluster size, R_o, the elastic strains begin to inhibit the further growth of the clusters, but assume that the inhibiting effect increases only slowly with an increasing size of the cluster.

In general, the elastic strains inhibiting the growth are functions of the actual cluster radius, R, and the initial pore size, R_o, when elastic strains start to accumulate in the vicinity of the growing cluster. We will assume that their inhibiting effect on cluster growth may be described by a term $\Phi(R, R_o)$ as

$$\frac{dR}{dt} = \frac{R_{co}^3}{R}\left\{\frac{1}{R_c} - \frac{1}{R}\right\} - \Phi(R, R_o) \,. \tag{13.8}$$

By introducing the reduced variables (see [310], chapter 2)

$$u = \frac{R}{R_c(t)} \,, \qquad x = \frac{R_c(t)}{R_{co}} \,, \qquad \tau = \ln\left[\frac{R_c(t)}{R_{co}}\right] \tag{13.9}$$

Eq.(13.8) is transformed to

$$\frac{du^3}{d\tau} = \gamma(\tau)\left\{u - 1 - \frac{u^2x^2\Phi(R, R_o)}{R_{co}}\right\} - u^3 \tag{13.10}$$

with

$$\gamma(\tau) = \left\{\frac{x^2dx}{dt}\right\}^{-1} \,. \tag{13.11}$$

The asymptotic stage of Ostwald ripening is reached if the conditions

$$F(u) = \frac{\partial}{\partial u}F(u) = 0 \,, \tag{13.12}$$

$$F(u) = \gamma(\tau)\left\{u - 1 - \frac{u^2x^2\Phi(R, R_o)}{R_{co}}\right\} - u^3 \tag{13.13}$$

are fulfilled. It means that the function $F(u)$ is tangent to the u–axis at some point $u = u_2$ and has a point of intersection with this axis at $u = -u_1$ (see [310], Fig. 2.2).

Eqs.(13.12) and (13.13) yield

$$u = \frac{3}{2} - \frac{x^2 u^2 \Phi}{2R_{co}} \left\{ \frac{u\Phi_u}{\Phi} - 1 \right\} \tag{13.14}$$

$$\gamma = \frac{\left\{ \frac{3}{2} - \frac{x^2 u^2 \Phi}{2R_{co}} \left[\frac{u\Phi_u}{\Phi} - 1 \right] \right\}^3}{\left\{ \frac{1}{2} - \frac{x^2 u^2 \Phi}{2R_{co}} \left[\frac{u\Phi_u}{\Phi} + 1 \right] \right\}} \tag{13.15}$$

In these equations, the notation $\Phi_u = (\partial\Phi/\partial u)$ is used. By solving both equations the quantities γ and the particular value of u ($u = u_2$) may be obtained. In the limiting case $\Phi = 0$ the well–known Lifshitz–Slezov asymptotic values

$$u_2 = \frac{3}{2}, \qquad \gamma = \frac{27}{4} \tag{13.16}$$

result, again.

These values are, of course, modified by the term Φ and its derivative, characterizing the influence of the pores on the evolution of the cluster ensemble. Provided, as assumed, that these terms are smoothly increasing functions of the cluster size we get in the next approximation instead of Eqs.(13.14) and (13.15) the following expressions

$$u = \frac{3}{2} - \frac{9x^2\Phi}{8R_{co}} \left[\frac{3\Phi_u}{2\Phi} - 1 \right], \tag{13.17}$$

$$\gamma = \frac{\left\{ \frac{3}{2} - \frac{9x^2\Phi}{8R_{co}} \left[\frac{3\Phi_u}{2\Phi} \right] \right\}^3}{\left\{ \frac{1}{2} - \frac{9x^2\Phi}{8R_{co}} \left[\frac{3\Phi_u}{2\Phi} + 1 \right] \right\}}. \tag{13.18}$$

Eqs.(13.17) and (13.18) are obtained by substituting on the right hand side of Eqs.(13.14) and (13.15) for u its value in first-order approximation ($u = 3/2$). Similarly, the terms Φ and Φ_u have to be understood as

$$\Phi(R, R_o) = \Phi(uR_c, R_o) = \Phi(\frac{3R_c}{2}, R_o) \ , \tag{13.19}$$

$$\Phi_u(R, R_o) = R_c \left\{ \frac{\partial \Phi(uR_c, R_o}{\partial (uR_c)} \right\}_{u=\frac{3}{2}} \ . \tag{13.20}$$

Since Φ is an increasing function of u (the inhibiting effect of the pores increases with an increasing size of the clusters of the evolving phase), Eq.(13.18) may be rewritten in the form (the second term in the numerator of Eq. (13.18) may be neglected)

$$x^2 \frac{dx}{dt} = \frac{4}{27} \left\{ 1 - \frac{9x^2 \Phi}{4R_{co}} \left[\frac{3\Phi_u}{2\Phi} + 1 \right] \right\} \ . \tag{13.21}$$

Hereby in addition the definition of $\gamma(\tau)$, as given by Eq.(13.11), was used.

Eq.(13.21) represents a differential equation for the determination of the time evolution of the critical cluster radius R_c.

Once, having solved Eq.(13.21), the time evolution of the critical cluster radius and, therefore, $\gamma(\tau)$, are known then the solutions u_1 and u_2 of the equation $F(u) = 0$ can be determined also.

As it was shown already in [477] in the analysis of a special case these solutions u_1 and u_2 determine the shape of the cluster size distribution function $\varphi(u, \tau)$ in reduced variables (compare [310], Eqs.(13.9)). We have (see for the details [477])

$$\varphi(u, \tau) = N(\tau) P(u) \ , \tag{13.22}$$

$$P(u) = \frac{3u^2 \exp\left[-\frac{3C_1 u}{(u_2 - u)u_2} \right] u_2^{3C_2} u_1^{3C_1}}{(u_2 - u)^{3C_2+2} (u + u_1)^{3C_2+1}} \tag{13.23}$$

with

$$C_1 = \frac{u_2^2}{u_1 + u_2} , \quad C_2 = \frac{u_2(2u_1 + u_2)}{(u_1 + u_2)^2} , \quad C_3 = \frac{u_1^2}{(u_1 + u_2)^2} . \qquad (13.24)$$

In the limiting case $\Phi = 0$ the well-known Lifshitz–Slezov distribution $P_{LS}(u)$

$$P_{LS}(u) = \frac{3^4 \mathrm{e} \exp\left\{ - \dfrac{3}{[2(3/2 - u)]} \right\}}{2^{5/3}(u + 3)^{7/3}(3/2 - u)^{11/3}} \qquad (13.25)$$

is obtained as a special case. However, in general, the parameters u_1 and u_2 are slowly varying functions of time. Consequently, also the distribution function in reduced variables $P(u)$ slowly changes with time.

The outlined method can be applied for any model of cluster growth provided the elastic strains are sufficiently moderately increasing functions of cluster size. For a particular model of cluster growth, developed in [473], the resulting curves for the evolution of the cluster size distribution function and related quantities are given in [477].

13.4 Coarsening in a System of Non–Deformable Pores with a given Pore–Size Distribution

13.4.1 A First Approximation

Whether an arbitrarily chosen cluster with an actual radius R will grow or not depends on the size of the pore it is contained in or, more generally, on the pore size distribution, which we denote by $W(R_o, \langle R_o \rangle)$. The parameter $\langle R_o \rangle$ has the meaning of the average pore size.

In order to apply the results obtained for ensembles of pores of equal size to the description of, at least, some average characteristics of Ostwald ripening in systems of pores of a given distribution an effective average growth rate may be introduced. This effective growth velocity of a cluster of size R

may be obtained by averaging the growth rate with the normalized to one distribution function $W(R_o, \langle R_o \rangle)$, i.e.,

$$\left\langle \frac{dR}{dt} \right\rangle = \int_0^\infty dR_o \left[W(R_o, \langle R_o \rangle) \frac{dR}{dt} \right] \tag{13.26}$$

resulting with Eq.(13.1) in

$$\left\langle \frac{dR}{dt} \right\rangle = \frac{R_{co}^3}{R} \left(\frac{1}{R_c} - \frac{1}{R} \right) \cdot \\ \cdot \int_0^\infty dR_o \left[W(R_o, \langle R_o \rangle) \Theta(R_o - R) \right] . \tag{13.27}$$

As mentioned, the normalization condition

$$\int_0^\infty W(R_o, \langle R_o \rangle) dR_o = 1 \tag{13.28}$$

has to be fulfilled.

It follows from Eq.(13.27) that the effective growth rate $\left\langle \frac{dR}{dt} \right\rangle$ is a function of the quantities R, R_c and $\langle R_o \rangle$, only.

Examples for pore size distributions which may be of relevance for different applications are the Gaussian distribution

$$W(R_o, \langle R_o \rangle) = \frac{\exp\left[-\alpha(\langle R_o \rangle - R_o)^2 \right]}{\int_0^\infty \exp\left[-\alpha(\langle R_o \rangle - R_o)^2 \right]} \tag{13.29}$$

and a distribution given by

$$W(R_o, \langle R_o \rangle) = \frac{(n-1)}{\langle R_o \rangle \left(1 + \frac{R_o}{\langle R_o \rangle} \right)^n} . \tag{13.30}$$

Here, n is some positive number.

The average growth rates are obtained then in the form

$$\left\langle \left(\frac{dR}{dt} \right) \right\rangle = \left(\frac{dR}{dt} \right) \left\{ \frac{1 + \operatorname{erf}\left[\sqrt[2]{\alpha} (\langle R_o \rangle - R) \right]}{1 + \operatorname{erf}(\sqrt[2]{\alpha} \langle R_o \rangle)} \right\} \tag{13.31}$$

for the case given by Eq.(13.29) and

$$\left\langle \left(\frac{dR}{dt} \right) \right\rangle = \frac{\left(\frac{dR}{dt} \right)}{\left(1 + \frac{R}{\langle R_o \rangle} \right)^{n-1}} \tag{13.32}$$

for the distribution described by Eq.(13.30).

The evolution of the cluster size distribution function $f(R,t)$ and related quantities are governed in this approximation by a continuity equation in cluster size space of the form (c.f. [509])

$$\frac{\partial f(R,t)}{\partial t} + \frac{\partial}{\partial R} \left(f(R,t) \left\langle \frac{dR}{dt} \right\rangle \right) = 0 \, . \tag{13.33}$$

The solution of the problem can be formulated, consequently, in the same way as done in section 13.2 for the case of monodispers pore size distributions. R_o has to be replaced, now, by $\langle R_o \rangle$.

On the other hand, provided, initially a Lifshitz–Slezov distribution is established and the effective growth rate may be expressed in the form as given by Eq.(13.8) with a moderately increasing inhibiting term Φ then also the method developed in section 13.3 may be used for a first estimation of the time evolution of some characteristic quantities like the number of clusters, the average and the critical cluster sizes.

However, the transition to an average growth rate performed with Eq.(13.26) implies that the details of the evolution of the cluster size distribution in the system of pores cannot be described adequately by either of both mentioned analytical approaches. Consequently, for a detailed determination of the evolution of the cluster size distribution in a system of pores of a given distribution another method has to be developed which will be discussed in the next section.

13.4.2 General Approach: Description of the Method

We assume that a system of clusters grows in an ensemble of pores with a size distribution characterized by the normalized to one distribution function $W(R_o)$. Each pore contains by assumption not more than one cluster, since processes of coarsening inside one pore may be expected to proceed with a much higher rate as compared with the process in the system as a whole (compare [379]).

We demand further that the pores are sufficiently large to allow initially the establishment of a time–independent distribution with respect to cluster sizes in reduced coordinates, as given by the original Lifshitz–Slezov distribution $P_{LS}(u)$ Eq.(13.25). This assumption supplies us with an universal initial distribution independent of the pecularities of nucleation and growth in the initial stages of the phase transformation. It means, moreover, that cluster and pore size distributions at the beginning of the coarsening process are statistically independent.

Note, that our method is applicable, in principle, also for any other initial distribution.

Since we assume that initially an evolution of the cluster size distribution not influenced by the interactions with the walls is possible the distribution of clusters and the pore size distributions are initially well–separated in cluster size space, i.e., the largest cluster in the distribution is less in size than the smallest pore.

Let N be the number of clusters in a unit volume of the porous solid and dN their ratio with cluster sizes in the interval $R, R + dR$ and in pores of radii $R_o, R_o + dR_o$, then a distribution function $w(R, R_o, t)$ may be introduced, defined by

$$dN = w(R, R_o, t)\, dR\, dR_o\,. \tag{13.34}$$

In order to find the cluster size distribution $f(R, t)$ this equation has to be integrated over all possible pore sizes, i. e.,

$$f(R, t) = \int_0^\infty w(R, R_o, t) dR_o\,. \tag{13.35}$$

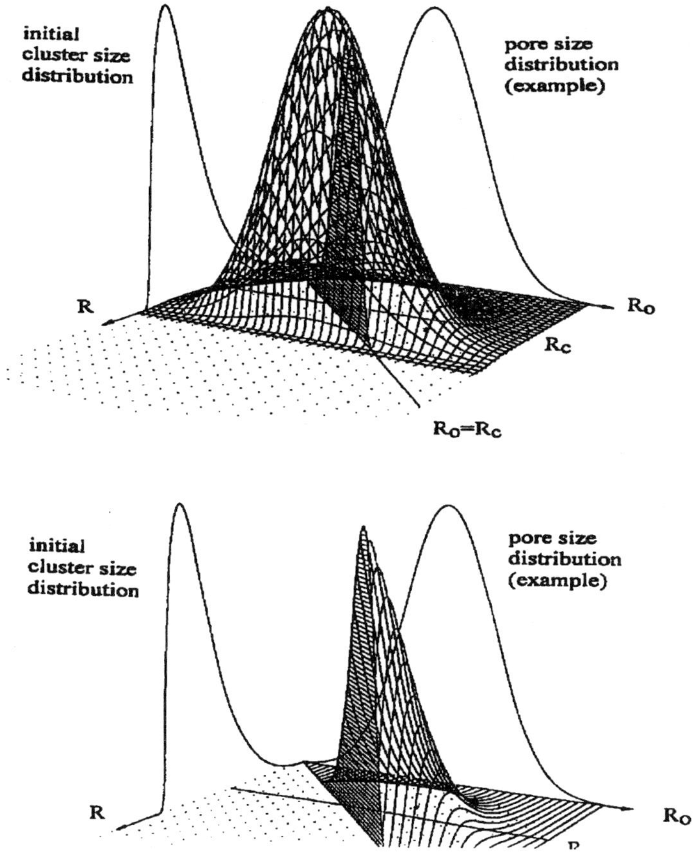

Fig. 13.1 Illustration of the procedure of determination of the evolution of the cluster size distribution in a porous material with a given pore size distribution. (upper part): Both the cluster and the pore distributions do not depend on each other in the initial state. Hence, the size distribution $w(R, R_0)$ depending both on R and R_0 may be written as a product of the pore and cluster size distributions, which now represent marginal distributions. (lower part): No cluster can be larger than the pore it is contained in. Hence, while growing along the R–direction, those clusters reaching the pore size R_o stop to grow immediately, being piled up at $R = R_o$, indicated by the striped plane above this straight line. Crossing this line when carrying out an integration means adding a finite non–differential amount to this integral, depending in its magnitude on the point where the line is crossed. Hence, along this line we have a one–dimensional δ – distribution over the $R - R_o$ – plane.

Because of the assumed statistical independence of pore and cluster radii the distribution function $w(R, R_o, t)$ in the initial stage may be written also in the form

$$w(R, R_o, t) = f(R, t)W(R_o) . \qquad (13.36)$$

An illustration of these statements is given in Fig. 13.1.

If the cluster radius R in a pore of size R_o approaches the limiting value R_o, then the growth is stopped immediately. As illustrated in Fig. 13.1, in the $(R - R_o)$ – plane the straight line $R = R_o$ cannot be trespassed by growing clusters. All clusters contained in pores of size R_o with a radius greater than the critical cluster size will grow only until the pore size is reached.

In the numerical simulation, this effect is accounted for in the following way. Starting with the initial distribution at time t and the growth equation the size distribution at the moment $t + \Delta t$ is calculated. If in some pores the calculated cluster radius R exceeds R_o then its size is redefined to $R = R_o$. So long as R is greater than the actual value of the critical cluster radius these clusters located at $R = R_o$ become immobile, i.e., do not grow or shrink. Consequently, along the line $R = R_o$ a delta–function like distribution is formed with time. It follows that in calculating the total number of clusters the contributions of such delta–function like distributions have to be taken into account in addition to contributions of the other regions of the $R - R_o$ – plane, where the function $w(R, R_o, t)$ behaves regularly.

In the course of time the critical cluster radius increases, in general, and the immobilized temporarily clusters in the pores with a radius $R_o \leq R_c$ start to change their sizes with time (decay), again.

For the numerical determination of the time evolution of the cluster size distribution function, the space of cluster radii is divided into sufficiently small ΔR–intervals. To all clusters belonging to one such interval the same R–value is assigned to and its motion for one time step is determined according to the employed growth equation. After this step, it is checked whether the clusters exceed the size of the pore they are contained in. If this is the case, then the correction procedure is employed as described above. Otherwise, respectively, after the correction procedure an averaging method is employed to reestablish the distribution from the discrete number of intervals, again. Afterwards the same procedure is repeated from the very beginning.

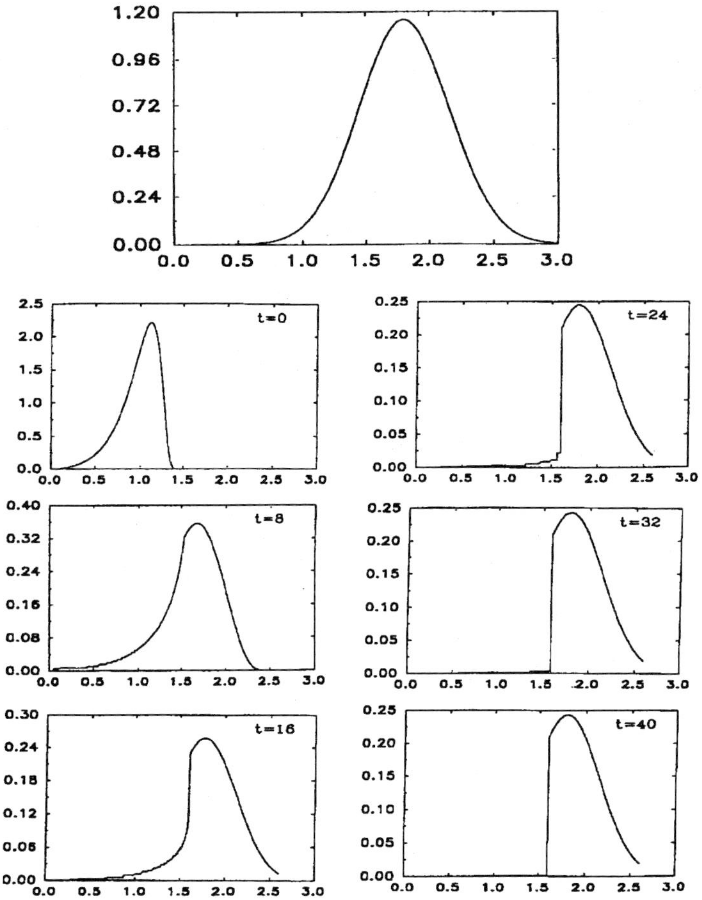

Fig. 13.2 Different stages in the time evolution of the cluster size distribution. The pore-size distribution is shown at the top of the figure. The evolution of the cluster size distribution starts by assumption with the LS-distribution.

Further details of the numerical procedure allowing the determination of the time evolution of the cluster size distribution are given elsewhere [374]. Here we want only to note that in the numerical calculations no use was made of the continuity equation, which was crucial for guarantying the stability of the numerical procedure.

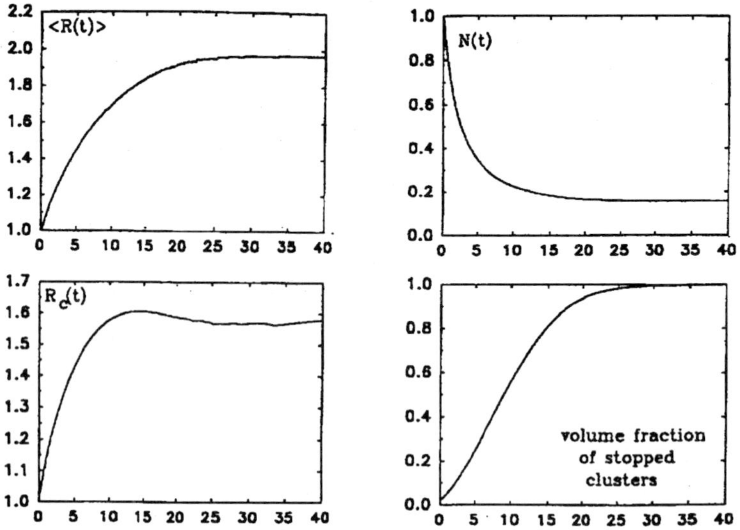

Fig. 13.3 Time evolution of the average cluster radius $\langle R \rangle$, the critical cluster radius R_c, the number of clusters $N(t)$ in the system and the amount of the new phase immobilized in the pores as functions of time for the process shown in Fig. 13.2 as obtained from the numerical calculations.

13.4.3 Results

The results of the numerical determination of the time evolution of the cluster size distribution function in a given distribution with respect to pore sizes are illustrated in Figs. 13.2. Starting always with an initial distribution (corresponding here to the LS–distribution) as depicted in Fig. 13.2 (for $t = 0$) and assuming a pore size distribution as shown at the top of the figure, a number of different stages in the time evolution of the size distribution is shown.

Moreover, also the time dependence of some related quantities is demonstrated like for the average cluster radius, the critical cluster radius, the number of clusters in the system and the ratio of the evolving phase immobilized in the pores.

As it is to be expected, the cluster size distribution in the final state reproduces partly the pore size distribution. The degree of filling depends hereby mainly on the initial density of segregating particles.

13.5 Influence of Stochastic Effects on Coarsening in Porous Materials

In the preceding discussion the process of Ostwald ripening was analyzed on the basic of deterministic growth equations like Eq.(13.1) or Eq.(13.8). By the application of such equations the influence of stochastic effects (like thermal noise) on coarsening cannot be described, in principle.

In ref. [320] it was shown, however, that the incorporation of stochastic effects into the description of Ostwald ripening in homogeneous media results in a broadening of the cluster size distribution function for intermediate time scales, and can be considered, thus, as one of the factors which may account for the gap between the theoretical and experimental results concerning the shape of the cluster size distribution function in Ostwald ripening [353]. Consequently, the same effect is also expected to occur for Ostwald ripening in porous media.

As outlined in detail in refs. [320] and [510] stochastic effects can be accounted for in the description of coarsening by the numerical solution of the set of basic equations underlying classical nucleation theory, i. e.,

$$\frac{\partial N(j,t)}{\partial t} = J(j-1,t) - J(j,t) \,, \tag{13.37}$$

$$J(j,t) = w^{(+)}(j,t)N(j,t) - w^{(-)}(j+1,t)N(j+1,t) \,. \tag{13.38}$$

Here $N(j,t)$ is the number of clusters in the considered volume consisting at time t of j monomers (cf. chapter 2). The coefficients of attachement $w^{(+)}$ and of detachement $w^{(-)}$ reflect the particular mechanism of growth or decay of the clusters.

Considering the case of coarsening in a systems of hard pores of equal size, the stationary distribution for $t \to \infty$ can be obtained from Eqs.(13.37) and (13.38) as

$$N(j-1) = \frac{w^{(-)}(j)}{w^{(+)}(j-1)} N(j) \,, \qquad j = j_{max} \,, \tag{13.39}$$

$$N(j-k) = \frac{w^{(-)}(j)w^{(-)}(j-1)\ldots w^{(-)}(j-(k-1))}{w^{(+)}(j-1)w^{(+)}(j-2)\ldots w^{(+)}(j-k)} N(j) \,. \tag{13.40}$$

Not a delta–function like distribution is established finally, but a stationary distribution with a sharp maximum for clusters having the size of the pores and decreasing rapidly with a decreasing number of particles in the clusters.

13.6 Discussion

With [479, 509] and the present chapter, both analytical and numerical methods have been developed allowing to investigate the kinetics of Ostwald ripening in an ensemble of pores for different pore size distributions and types of response of the walls of the pores with respect to cluster growth. As mentioned already in [509] the notation pore is used with a generalized meaning either as a real object in materials like vycor glasses or zeolites or as a low density region of a homogeneous matrix where clusters may be formed and grow preferentially.

In the examples considered so far it was assumed that the diffusional flow of the segregating particles is not influenced by the structure of the porous material. This simplifying assumption can be replaced in a more accurate description by an investigation of the influence of the particular structure of the matrix on the diffusional flow of the segregating particles. Qualitatively no changes in the kinetics of coarsening as compared with the results outlined here and previously are expected.

Moreover, in all considered cases, the growth is finally stopped by the matrix. In terms of the phenomenological theory of rheology latter assumption corresponds to the assumption of a Kelvin's like body in modelling the rheological properties of the matrix [161].

Of interest would be, of course, also the consideration of the alternative model, where the properties of the matrix are described by Maxwell's or qualitatively similar models of viscous flow. In such cases, for large times the evolution of the cluster ensemble is expected to be determined by the rate of stress relaxation (cf. [531]). First steps in modelling the growth of single clusters under such conditions have been undertaken [375], the generalization to the case of evolution of ensembles of clusters is missing till now.

14 Structure Formation by Spinodal Decomposition in Adiabatically Isolated Systems: A New Class of Dynamic Universality

J. Schmelzer

14.1 Introduction

The process of spinodal decomposition, as analyzed already briefly in chapter 2, is considered conventionally by assuming that the temperature in the system remains constant in the course of the decomposition process. However, both from a principal point of view and with respect to a number of possible technological applications an important question is, how is the picture changed, if isothermal conditions are replaced by the constraint of an adiabatic closure of the system.

Under the condition of an adiabatic closure, the latent heat of the transformation results in an increase of the temperature in the system and, consequently, modifications of the decomposition kinetics are expected to occur. This qualitative difference is due to the fact, that the critical wave number becomes temperature and thus time–dependent and decreases with increasing temperature, respectively, increasing degree of advancement of the decomposition process. In this way, non–linearities enter the description already in the framework of the linear Cahn–Hilliard–Cook theory.

The aim of the present chapter consists in the analysis of these modifications and the characterization of the whole course of spinodal decomposition in adiabatically isolated systems. Since we are interested mainly in qualitative results, we restrict ourselves here to a description of spinodal decomposition at the level of the Cahn–Hilliard–Cook theory to reveal the specific role of the non–linearity arising from the chosen boundary conditions. The results obtained in the framework of the Cahn–Hilliard–Cook theory for isothermal systems serve as a reference for the variations caused by the temperature changes in the system.

Moreover, the system, where the transformation occurs is assumed to be a binary regular solution as discussed by Becker [27], Cahn and Hilliard [78]. Heat conduction processes inside the system are considered to proceed fast as compared with diffusional fluxes. In this way, at any moment of time an internal thermal equilibrium is established.

14.2 Thermodynamic Aspects

The free enthalpy g^* and the potential energy u^* (both quantities referred to one particle) of a regular solution can be expressed as (cf. [27, 78])

$$g^* = \omega^* x(1-x) + k_B T\left[x \ln x + (1-x)\ln(1-x)\right] + \kappa^*(\nabla x)^2 , \tag{14.1}$$

$$u^* = \omega^* x(1-x) + \kappa^*(\nabla x)^2 . \tag{14.2}$$

The molar fraction x in Eqs.(14.1) and (14.2) can be replaced by the volume concentration c via

$$x = \frac{c}{c_t} , \qquad c_t = \frac{N}{V} . \tag{14.3}$$

N is the total number of particles and V the volume of the solution.

The change in the temperature of the adiabatically isolated system due to the evolution of the concentration field $c(\vec{r}, t)$ can be expressed as (see Schmelzer, Milchev [476])

$$\Delta T = \frac{1}{(NC_p^*)} \int \left\{\omega(c-c_0)^2 - \kappa\left[\nabla(c-c_0)\right]^2\right\} dV . \tag{14.4}$$

C_p^* is the specific heat per particle of the solution, moreover, the notations

$$\omega = \frac{\omega^*}{c_t} , \qquad \kappa = \frac{\kappa^*}{c_t} \tag{14.5}$$

are used.

In terms of the spectral function S (cf. section 2.4), the temperature variations may be written as (cf. [321])

$$\Delta T = \left(\frac{V\omega}{C_p^* N}\right) \sum_{\vec{k}} |S(\vec{k},t)|^2 \left(1 - \frac{1}{2}a^2\vec{k}^2\right) . \tag{14.6}$$

Hereby the relation

$$\kappa = \frac{1}{2}\omega a^2 \tag{14.7}$$

was used in addition, valid for the considered regular solutions [78]. The parameter a in Eq.(14.7) is a measure of the intermolecular distance in the solution.

It turns out in this way that the structure function S determines not only the concentration field (cf. Eq.(2.109)) but also the actual temperature of the system. To obtain results from above equations, in this way a self–consistent solution both for the concentration fields and the temperature changes has to be developed.

In application of above derived equations, one should bear in mind that in the Cahn–Hilliard–Cook approach one deals with a coarse–grained free energy, averaged over volume elements considerably exceeding in size the intermolecular distances (lattice constants) (cf. [54, 300]). Similarly, the applied macroscopic concept of a wave length λ looses any meaning if $\lambda = 2\pi/k$ is less than the lattice constant. This condition gives a natural estimate of the upper limit of the number of modes which have to be taken into account in the numerical calculations carried out to solve above given equations.

14.3 Results of Numerical Calculations

Going over to the dimensionless variables (cf. Eqs.(2.120)) $\tilde{t} = t(4\kappa M)/a^4$ and $\tilde{k} = ak$, the basic equation (2.121) describing the time–dependence of the structure factor gets the form (see also [321])

$$\frac{\partial \left\langle S^2(\tilde{k},\tilde{t})\right\rangle}{\partial \tilde{t}} = \tilde{k}^2 \left[\left(\tilde{k}_c^2 - \tilde{k}^2\right) \left\langle S^2(\tilde{k},\tilde{t})\right\rangle + \left(\frac{k_B T}{\omega V}\right)\right] , \tag{14.8}$$

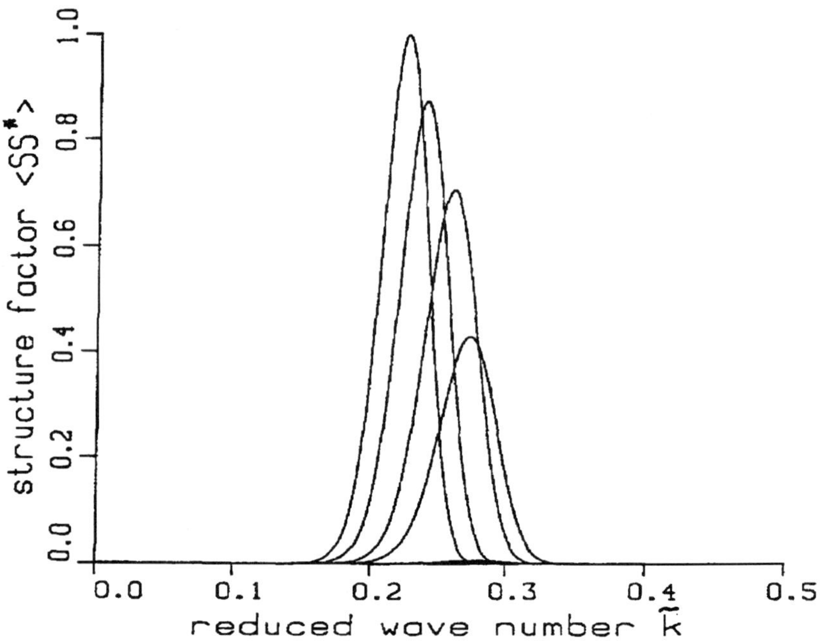

Fig. 14.1 Structure factor $S^2(\tilde{k}, \tilde{t})$ for different moments of time ($\tilde{t}$=3000, 4000, 5000, 7000 and 9000 (in reduced units)). With increase of time a shift of the maximum to lower values of the wave number and an increase of the value of the maximum is observed.

$$\tilde{k}_c^2 = 2 - \Omega_1 T \ , \qquad T = T_0 + \Delta T \ , \tag{14.9}$$

$$\Delta T = \Omega_2 \sum_{\vec{k}} \left\langle S^2(\tilde{k}, \tilde{t}) \right\rangle \left(1 - \frac{\tilde{k}}{2} \right) . \tag{14.10}$$

Ω_1 and Ω_2 are two parameters characterizing specific properties of the solution considered.

In the homogeneous thermodynamically unstable initial state, the temperature T equals T_0 corresponding to an initial value $\tilde{k}_{co}$ of the critical wave number $\tilde{k}_c$. In the course of the decomposition process the temperature increases and $\tilde{k}_c$ decreases.

Results of the numerical evaluation of Eqs.(14.8) – (14.10) are shown in Figs. 14.1 – 14.3. In Fig. 14.1 the structure factor is shown for different moments

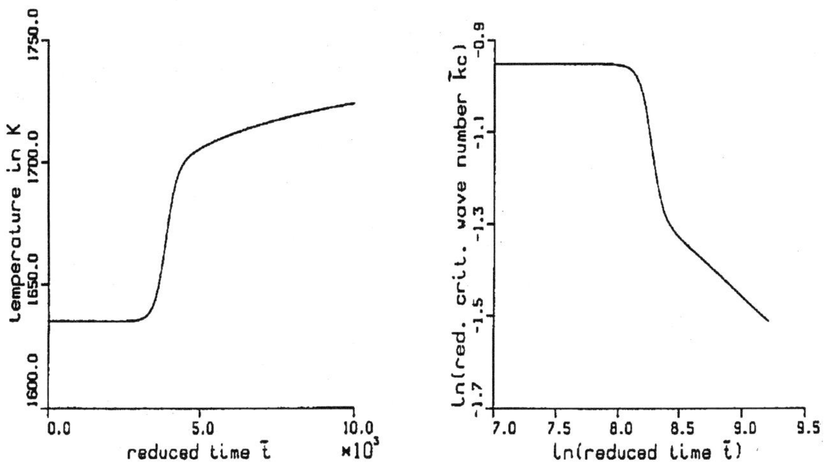

Fig. 14.2 (left side): Temperature vs time curve for spinodal decomposition in adiabatically closed systems. The curve results from the solution of Eqs.(14.8) – (14.10) (for the details of the computation and the values of the parameters see Ludwig, Schmelzer, Milchev [321]); (right side):The dependence of the critical wave number $\tilde{k}_c$ on time in a double logarithmic plot. In the second and third stages of the decomposition process a linear behavior is found indicating the existence of power laws of the form $\tilde{k}_c \sim t^{-\alpha}$. As the result of a linear regression a value $\alpha \sim 1/20$ was found for the second stage of the process while for the third stage $\alpha = 0.245 \sim 1/4$ was obtained. Similar depenencies were observed recently in experimental investigations of spinodal decomposition [547, 486] in sodium silicate glasses.

of time. In contrast to the picture observed for the isothermal case (cf. Fig. 2.4) the maxima of the S^2–curves are shifted to lower values of $\tilde{k}$ with time.

In Fig. 14.2 (left curve) the temperature T is shown as a function of time. Clearly three different stages of the transformation may be distinguished characterised by different types of functional dependencies $T = T(\tilde{t})$.

The first stage of spinodal decomposition is characterized by the stochastic generation of the initial concentration field. Remember that we start the process with a homogenoues initial state here. In experimental investigations, such a stage may not occur since already in any equilibrium state density fluctuations with well–defined properties are present. Such fluctuations will be retained in the process of transformation of the system into the respective non–equilibrium state (cf. also [486]).

This first stage of spinodal decomposition is followed by a stage of independent growth of the different modes. Hereby such modes are amplified in particular corresponding to wave numbers $\tilde{k}_{max} = \tilde{k}(0)/\sqrt{2}$ (cf. Fig. 2.4). This type of evolution goes over then into a stage of moderate competitive growth of the modes characterized by self-similarity and scaling laws.

The behavior of temperature in dependence on time is reflected also in the $\tilde{k}_c = \tilde{k}_c(\tilde{t})$ – curve, shown in logarithmic co-ordinates in Fig. 14.2 (right curve). An analysis of the curve verifies that in the second and third (late) stages of the transformation a scaling behavior for the critical wave number of the form

$$\tilde{k}_c \propto \tilde{t}^{-\alpha} , \qquad \alpha \sim \begin{cases} \frac{3}{2} & \text{second stage} \\ \\ \frac{1}{4} & \text{third stage} \end{cases} \tag{14.11}$$

is established. This type of functional dependence of the critical wave number on time is confirmed both by experimental results [547, 486] as well as by Monte-Carlo simulations of spinodal decomposition in adiabatically closed systems (Milchev, Gerroff, Schmelzer [371]).

The scaling law as expressed by Eq.(14.11) is different from the well-known power law one has to expect according to the Lifshitz–Slezov theory of coarsening (see chapters 2 and 13, [310]). Such a Lifshitz–Slezov behavior is found indeed for isothermal MC-simulations of the same process. It follows that the temperature variations in the system result in a different scaling behavior as compared with the isothermal case, at least, for the considered interval of time.

The existence of a power law for k_c in the late stage of the transformation, where the temperature changes only slightly, suggests that the structure factor also obeys some kind of self-similarity. Thus, we may propose an equation of the form (compare section 2.3, [215, 320])

$$\langle S^2 \rangle = f(t) g(u) , \qquad u = \frac{\tilde{k}}{\tilde{k}_c(\tilde{t})} \tag{14.12}$$

to be valid. Moreover, we demand, that at any moment of time the normalisation condition

$$\int_0^\infty g(u)\, du = 1 \tag{14.13}$$

is fulfilled.

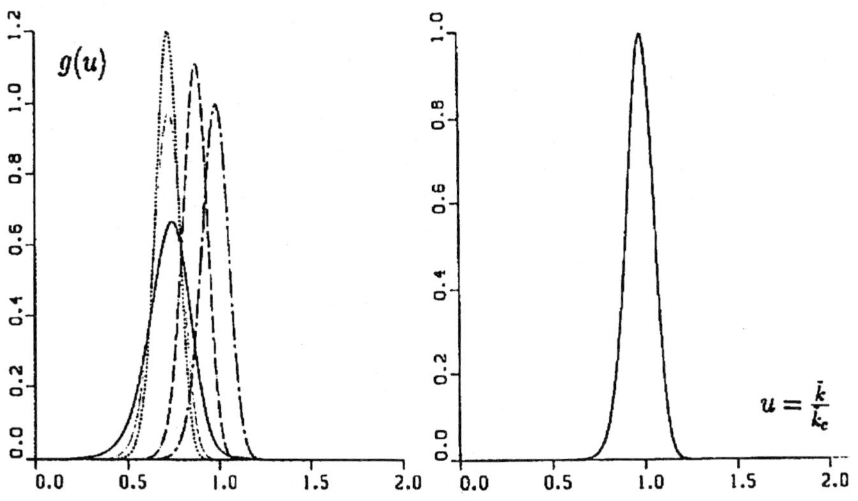

Fig. 14.3 (a): Normalized structure factor $g(u)$ in dependence on the reduced wave number $\tilde{k}/\tilde{k}_c$. The different curves on the left hand side of the figure correspond to $\tilde{t} =$ 1000 (full curve), $\tilde{t} = 2000$ (dotted curve), $\tilde{t} = 3000$ (dashed curve), $\tilde{t} = 4000$ (dashed–dotted curve), $\tilde{t} = 5000$ (double–dashed curve). In the course of the evolution, a time–independent distribution is approached as shown on the right hand side. Here the curves obtained for $\tilde{t} = 5000$ (full curve) and for $\tilde{t} = 9000$ (dotted curve) are shown. As easily verified both curves coincide practically.

In Fig. 14.3 the function $g(u)$, as calculated from the numerical solution of the set of Eqs.(14.8) – (14.10), is shown for different moments of time in the third stage of the transformation. Is can be noticed easily that, indeed, an universal distribution function $g(u)$ is established, which gives a verification of the proposal expressed by Eqs.(14.12) – (14.13).

The results of the numerical solution of the basic equations (14.8) – (14.10) describing decomposition in adiabatically closed systems lead to the conclusion that in analogy to nucleation and growth processes in finite systems ([471], chapter 2) the whole course of the transformation can be divided roughly into three different stages, a first stage of formation and moderate growth of the initial density inhomogeneities, a second stage of rapid increase of these density differences, connected in adiabatic systems with a sharp increase of temperature, and a third stage of reorganisation of the concentration field, characterised by self–similarity and scaling laws. The following section is devoted to a theoretical interpretation of the scaling behavior (see also Schmelzer, Slezov, Milchev [480]).

14.4 Theoretical Interpretation

For the analytical verification of the scaling laws we omit stochastic terms in the Cahn–Hilliard–Cook equation being of significant importance only with respect to the formation of the initial concentration profile. A substitution of Eq.(14.12) into the differential equation Eq.(14.8) yields then

$$\frac{d}{dt}\ln f(t) - \frac{\tilde{k}}{\tilde{k}_c^2}\frac{d\tilde{k}_c}{d\tilde{t}}\frac{d}{du}\ln g(u) = \tilde{k}_c^4 u^2(1-u^2)\ . \tag{14.14}$$

Assuming (as it is verified by the numerical calculations) that

$$\frac{d}{dt}\ln f(t) \cong 0 \tag{14.15}$$

holds we arrive at

$$\frac{d}{dt}\left(\frac{\tilde{k}_c^{-4}}{4}\right)\frac{d}{du}\ln g(u) = u(1-u^2)\ . \tag{14.16}$$

Since the right hand side of Eq.(14.16) is a function only of the reduced variable u, the left hand side of this equation must also be. Consequently, the relation

$$\frac{d}{dt}\left(\frac{\tilde{k}_c^{-4}}{4}\right) = C_1^{-1} \tag{14.17}$$

has to be fulfilled. The general solution of this equation may be written as

$$\tilde{k}_c^4 = \frac{C_1}{4(\tilde{t}+C_1^*)} \tag{14.18}$$

which for large times results in

$$\tilde{k}_c^4 \propto \tilde{t}^{-1} . \tag{14.19}$$

This result immediately confirms the scaling law (14.11) obtained numerically. Moreover, for the determination of the function g(u) we get the equation

$$\frac{d}{du} \ln g(u) = C_1 u(1 - u^2) \tag{14.20}$$

with the general solution

$$g(u) = C_2 \exp \left[C_1 \left(\frac{u^2}{2} - \frac{u^4}{4} \right) \right] \tag{14.21}$$

or, equivalently,

$$g(u) = A \exp \left[-B \left(u^2 - 1 \right)^2 \right] , \tag{14.22}$$

$$A = C_2 \exp \left(\frac{C_1}{4} \right) , \qquad B = \frac{C_1}{4} . \tag{14.23}$$

This result shows in agreement with the numerical solution that, indeed, the maximum of the distribution function $g(u)$ is located at $u = 1$ or $\tilde{k} = \tilde{k}_c$. Moreover, also the shape of the distribution function is reproduced in a correct way.

One of the constants A and B in Eq.(14.22) may be determined from the normalisation condition (14.13). However, it is easily seen from Eqs.(14.8) – (14.10) or

$$2\tilde{k}_c \frac{d\tilde{k}_c}{dt} = -\Omega_1 \frac{dT}{dt} \tag{14.24}$$

that the coefficient C_1 depends on specific properties of the system under consideration (Ω_1 and Ω_2) and is not an universal parameter. In this way, the width of the distribution is well–defined but not universal, it depends on specific properties of the system under consideration, in particular, on the rate of change of the critical wave number.

14.5 Discussion

The theoretical description of spinodal decomposition in the framework of the Cahn–Hilliard–Cook theory comes across with two serious difficulties. The first one consists in the correct determination of the free energy density for the initial thermodynamically unstable state. The second problem is connected with a correct introduction of non–linear terms into the original Cahn–Hilliard–Cook equation, to avoid an unlimited amplification of the density fluctuations. Different attempts have been developed to overcome these difficulties (see, e.g., [54, 300]), however, so far a final solution of these problems is missing.

An analysis of phase separation processes, starting from metastable initial states and proceeding via nucleation and growth, leads to the conclusion that the change of the state of the system (change of supersaturation or thermodynamic driving force of the transformation) resulting from the transition determines the whole scenario of the process [187, 471]. In spinodal decomposition the thermodynamic driving force of segregation is given by the second derivative of the free energy with respect to molar fraction or concentration of one of the components. By analogy and in agreement with the thermodynamic stability conditions, one may expect that variations of this quantity should also determine basically the course of the phase transformation in spinodal decomposition.

In the special case, considered here, the variations of the thermodynamic driving force of the transformation can be calculated relatively easily. They are due in the applied approach exclusively to temperature changes in the system. This is the only factor taken into account and as already mentioned, in analogy to nucleation and growth processes it determines qualitatively the whole scenario of the transformation. Since, as discussed, the change of the driving force of the transformation is expected to be the major factor, governing the phase transformation, a similar scenario should be always expected to occur in experiments and also in theoretical approaches, when the change of the state of the system (or non–linearities in spinodal decomposition) are described in a proper way.

Indeed, Binder [40], Mazenko, Valls and Zannetti [358] found theoretically a $1/4$ power law behavior based on a field theoretical approach to the description of spinodal decomposition. Moreover, it has been shown by Velasco and Toxvaerd [553] in recent molecular dynamics and Monte–Carlo simulations of spinodal decomposition that in binary systems under isothermal conditions in the late stages such (1/4)–power laws are also found, giving a additional support of our hypothesis. While in one–component closed systems under adiabatic constraints, as considered here, the change of the thermodynamic driving force of the transformation is due to the increase in temperature, for isothermal conditions in binary systems it is connected with a change in the composition. The existence of such stages in spinodal decomposition has been predicted in an alternative way also theoretically (cf. [397]) and observed experimentally [547, 486].

The exponent in the power law for the critical wave number k_c may be, of course, different in dependence on the specific mechanism governing the decomposition process at a certain stage. Note that the equations considered do not include hydrodynamic effects (compare [40, 553]) as well as restrictions on the amplitude of the density variations. In this sense, the considered process describes not the evolution of an already formed two–phase system – as in the Lifshitz–Slezov theory of coarsening – but the process of formation of two distinct phases. It can – and will be, in general, – followed by laws of the type $\langle R \rangle \propto t^{1/2}$ and $\langle R \rangle \propto t^{1/3}$ describing independent and competitive growth of ensembles of clusters, respectively (cf. chapters 2, 9, 13 and [547, 486]).

Changes of the kinetics of spinodal decomposition are expected to occur, in particular, also in cases, when not an internal thermal equilibrium is assumed to be established at any moment of time but heat conduction processes are also have to be taken into account explicitely. This topic will be addressed in a future communication.

15 Stochastic Master Equation Approach to Aggregation in Freeway Traffic

R. Mahnke

15.1 Introduction

Fluid–dynamical approaches to traffic flow have been developed long ago starting in the 1950's by Lighthill and Whitham. Since the pioneering work of Prigogine and Herman [432] on the kinetic theory of vehicular traffic cars have been considered as interacting particles. Depending on the number of cars on the road (the density of cars) bound states (so–called car clusters) may become possible. If the density is small the free flow of nearly non–interacting particles is dominant. The flux rate increases linearly with the density. If the density of cars exceeds some critical value a jamming transition takes place. This phase transition separates the low–density situation in which all cars move indepentently at maximal speed from the high–density region in which the formation of car aggregates as bound states reduces the average velocity of cars. The flux rate is decreasing with increasing car density. In a one–dimensional situation (single lane traffic) all cars are stopped if the road is crowded with cars (density equals one). The curves in the flux–density–space are known as fundamental diagram.

Based on several approaches [102, 201, 304, 432, 568, 570] like a cellular automaton model for freeway traffic by Nagel and Schreckenberg [385] (cf. Sect. 6.6) and others [134, 382, 441, 495] the fundamental diagram shows clearly the phase transition from the free jet situation (no aggregation effect) to the car cluster regime with start–stop–waves. It has been shown by Kerner and Konhäuser [250, 251] that for high densities car cluster can spontaneously appear in which the average velocity of cars is considerably lower than in the initial flow and outside the cluster.

The aggregation of particles (e. g. the emergence of car clusters in traffic flow) out of an initially homogeneous situation is well known in physics. Depending on the system under consideration and its control parameters the cluster formation in a supersaturated (unstable) situation has been observed in nuclear physics as well as in other branches. We mention the well–known example of condensation (formation of liquid droplets) in an undercooled water vapour. The formation of bound states as an aggregation process is related to self–organization phenomena [334, 337].

In analogy to common aggregation phenomena such as the formation of liquid droplets in a supersaturated vapour or segregation in solid and liquid solutions, here processes of nucleation, growth and shrinkage of car clusters are considered. As an example we analyze a circular one–lane freeway traffic flow model. The clustering behaviour (known as congestion) in a initially homogeneous traffic stream is described by a newly derived Master equation. The construction of the stochastic equation is given as well as its relationship to other dynamical models. Numerical experiments in heavy traffic with well–explained transition probabilities show the transition from the initial free particle situation (free jet of vehicles) to the final congested cluster state, where one big aggregate of cars is formed. In particular, the dependence of the stable car cluster size as a function of car concentration is investigated. The results are presented both in analytic form as well as numerically as stochastic trajectories.

The stationary solution of the Master equation is derived. It depends on the concentration of cars on the road. This stationary solution corresponds to the long–time limit of the probability distribution function. It shows the stable cluster size as mean number of congested cars. The obtained fundamental diagram in terms of a flow–density relation indicates clearly the different regimes of traffic flow (free jet of cars, coexisting phase of jams and isolated cars, highly viscous overcrowded situation). In the (thermodynamic) limit of infinite number of vehicles on an infinite long road the analytical solution for the fundamental diagram is in agreement with stationary solutions of the Master equation for finite systems of different sizes.

15.2 Dynamics of a Spontaneous Traffic Jam

The possible states of a highway traffic at varying densities are well known; let us briefly discuss them in review. In contrast to the phase diagram of a van der Waals gas we have a *gas phase* at low densities: the particles are considered as non–interacting (ideal gas flow); an initially homogeneous flow will remain stable; the flow–density curve in the fundamental diagram is linear, its slope is the mean velocity. The relationship between mean velocity and density can be described either by a *security distance* law, or by an *optimum velocity* law.

On the other hand at high densities we find the *liquid phase*: the particles interact strongly (condensed flow) and an initially homogeneous flow will also remain stable. The mean velocity and the flow reduce as the density gets larger, and reach zero at a finite maximal value for density (i. e., at a minimal value for the inter–vehicle distance).

At intermediate densities there is a region of phase separation between both stable phases. Since the particles interact episodically, an initially homogeneous flow become unstable. The birth of one or several aggregates as clustering by collisions takes place. The car clusters can either disappear by concurrence or rapidly reach very high local densities and very low speeds for the trapped vehicles. The clusters as binding states move backwards with a speed that is directly dependent on the rate of incoming cars, i. e. the upstream flow. The general behaviour is an alternation of free and congested areas, often called *stop–and–go waves*, or *stop–start waves*. There are two critical densities, in the vicinity of which metastabilities and bifurcations may occur.

Keeping in mind our goal of building a Master equation which could describe the stochastic evolution of the aggregation phenomenon, we have now to find a simplified but still realistic definition for a traffic jam. Many results, for example [250, 251], tell us that, in a new–born jam, the convergence towards very high density and corresponding very low individual speed is significantly higher than that of the *size* of the jam towards its stationary size. By this reason, we will give the following *definition of a car cluster:*

A cluster of size n is an aggregate of n vehicles whose individual speed is zero, and whose front–to–rear (bumper–to–bumper) distance is zero.

That we have set the minimal allowed distance between two vehicles equal to zero, does not reduce generality, since we can consider the *effective length* of a car as a minimal distance added to the real length. However, the zero–speed hypothesis is a rough simplification; a useful generalization of this definition would then be to allow density as an additional variable (besides the size) for the characterization of the cluster. A further simplification performed in this model is that we will allow the formation of only one cluster at some given time. Other work [17] have shown that the total number of cars blocked on the road is practically independent of the number of coexisting car clusters, so that our results should not be too strongly affected by this restriction.

15.3 Follow–the–Leader Behaviour

In order to allow us a description of the homogeneous flow behaviour, we have to choose a relationship between the speed of a given vehicle and the distance to its leader. As this particular problem is not the essential point of this paper, and should be further discussed and justified in confrontation with experimental data, our choice was that of a sufficiently simple, but still realistic model, namely a *sigmoidal optimum velocity* curve. Sigmoidal functions have a characteristic shape that starts at one value and rises smoothly to another value with a single inflection point. There are many algebraic expressions to represent such curves, but in applications like biological growth laws, only a few of them are encountered frequently [233]. We mention the hyperbolic tangent, which has been used by Bando *et al.* [17, 18, 19] as optimum velocity curve. We make another choice, sometimes known as Hill function of the second order, analytically written as

$$v_{opt}(h) = v_{max}\,\frac{h^2}{d^2 + h^2}\,, \tag{15.1}$$

where h is the headway (bumper–to–bumper with effective length), v_{max} is the maximal speed allowed, and d is a positive control parameter, that can be seen as a characteristic headway for the transition between non–interacting and interacting phases.

This particular form (Eq.(15.1)) of the optimum velocity curve (Fig. 15.1) has the advantage of allowing an analytical treatment for the following developments, and of being consistent with the limit $v_{opt}(0) = 0$; the fact that the upper limit $h \to \infty$ is not consistent with the periodic boundaries, does not appear to be problematic.

Besides the movement with an optimum velocity behaviour the car–car interaction has to be considered. In comparison to a granular particle flow in narrow pipes [442] the clustering of cars is driven by inelastic collisions. When two particles collide inelastically (two cars reach each other without accident), their velocities change so that the faster car adopts the velocity of

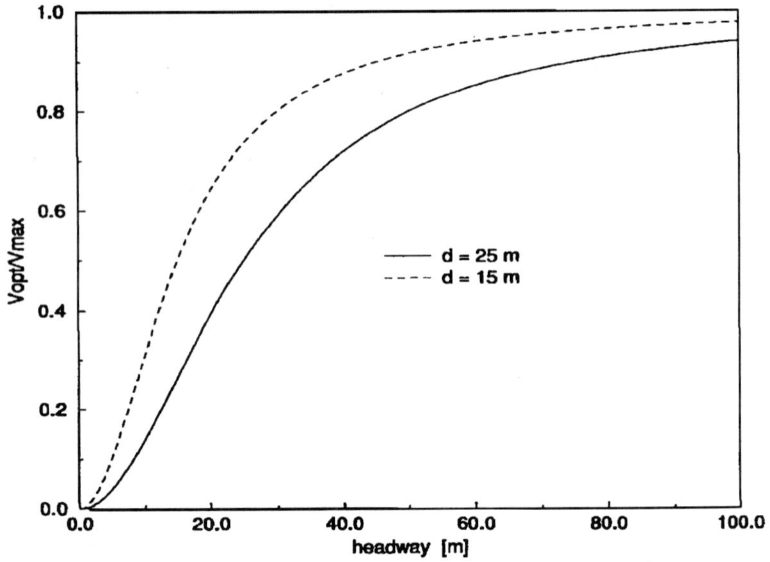

Fig. 15.1 Sigmoidal curves for optimum velocity showing the desired speed [dimensionless quantity (v_{opt}/v_{max})] in dependence of the headway h to the next car.

the slower one, and hence they remain close to one another. The proposed mechanism for the formation of aggregates allows us to derive a stochastic equation describing the creation and evolution of a car cluster of size n.

15.4 The Master Equation

The Master equation is a differential equation describing the time evolution of the probability distribution of a set of random variables for a stochastic dynamical system [211] that follows a Markov process, i. e. for which the transition probabilities depend on the actual state but not on the past states. It is expressed in terms of transition rates between the accessible states of the system. In our particular case, the only random variable is the size n of the cluster. We will furthermore accept that only one car at a time can go into or come out of the jam; this means that we do not consider the merging or splitting of aggregates, which is a consequence of our choice of a one–cluster system. As a consequence, only two kinds of transitions are allowed:

a growth transition $n \to (n+1)$ and a decay transition $n \to (n-1)$, to which we associate a growth transition rate $w^{(+)}$ and a decay transition rate $w^{(-)}$, respectively. These rates depend in general on the variable n and by conservation of cars therefore on the total number of cars on the road; they could also be time–dependent, if we would consider the possibility of an external control on them (e. g. through the action of traffic lights or a temporary adaptation of legal speed). Such cases will not be treated here. Their incorporation should be a necessary step towards practical applications, such as technical or legal measures that can be taken to prevent jamming.

We can now write our Master equation as

$$\frac{\partial}{\partial t}P(n,t) = w^{(+)}(n-1)\ P(n-1,t) + w^{(-)}(n+1)\ P(n+1,t) \\ - [w^{(+)}(n) + w^{(-)}(n)]\ P(n,t)\ . \qquad (15.2)$$

This equation can be used to determine numerically the temporal evolution of the stochastic variable $n(t)$, the probability $P(n,t)$ to find a size–n–jam at time t, the mean value as well the variance for the random variable n, and hence the desired evolution of the flow. Nevertheless, in order to deal with this equation, we still need the transition rates, which have to emerge from phenomenological considerations.

15.5 Relaxation Times and Transition Rates

A very important parameter in many highway traffic models is the relaxation time, which describes the driver's technical–psychological fastness of adaptation to the actual downstream state of the flow. Associated with an optimum velocity model, it is the time constant for the exponentially asymptotic adaptation from his actual speed to the optimum speed required for his actual headway. However, in these models it is generally assumed, implicitly, that time constants for speeding up and for slowing down are identical, an assumption that has obviously no reasonable foundation, neither technically nor psychologically. We shall on the contrary assume that *the braking time constant is much greater than the accelerating one.* Indeed, as a driver approaches the rear of the jam, he does not realize immediately that he will have to come to a real stop, therefore he will use his brakes very late and very strongly. In the opposite case, namely when the driver sees again the free flow in front of him, he will have adapted his speed, which is zero at the time, to the mean speed of the free flow, after a time which can be approximated by a *constant.*

As a consequence of these statements, we are able to define the transition rates in a very simple way. Let us discuss both cases separately. Considering the free flow upstream the jam, we remember that we assumed it as homogeneous, with an average headway h and speed $v_{opt}(h)$. At time t, the car directly behind the jam has to drive a distance h, and, if we take the extreme assumption that the decelerating time constant is zero, it needs an average time $h/v_{opt}(h)$ to get into the jam. Hence, the probability, at time $t+dt$, that the car has collided ('jumped') into the jam, is $dt\, v_{opt}(h)/h$. The transition rate, which is defined per time unit, will therefore be

$$w^{(+)}(h) = \frac{v_{opt}(h)}{h} \, . \tag{15.3}$$

On the other hand, as the head car of the jam can come free again, it will need the average time τ to leave the aggregate. Hence the corresponding transition rate is simply

$$w^{(-)}(h) = \frac{1}{\tau} \, . \tag{15.4}$$

Now, based on the conservation law of finite systems that all cars are moving on a street with periodic boundary conditions, applying the following simple relation between the headway h and the size of the cluster n

$$h(n) = \frac{L - N\, l_0}{N - n} \, , \tag{15.5}$$

where L is the total length of the road, N the total number of cars and l_0 the effective length of a single car, we finally find (see Fig. 15.2) the different forms of the curve $w^{(+)}(n)$ for different densities $\varrho = N/L$. For $w^{(-)}(n)$ we still have a straight line $w^{(-)}(n) = w^{(-)} = 1/\tau = \text{const}$. The parameters for the curves shown in the figures are chosen to be realistic enough. However, they are not empirically justified. Nevertheless, we will see that these parameters will lead to a realistic flow–density behaviour.

The curves shown in Fig. 15.2 allow us to describe the different patterns connected with different relative values of the control parameters. The first case to be examined should be the case where $(1/\tau)$ is larger than the

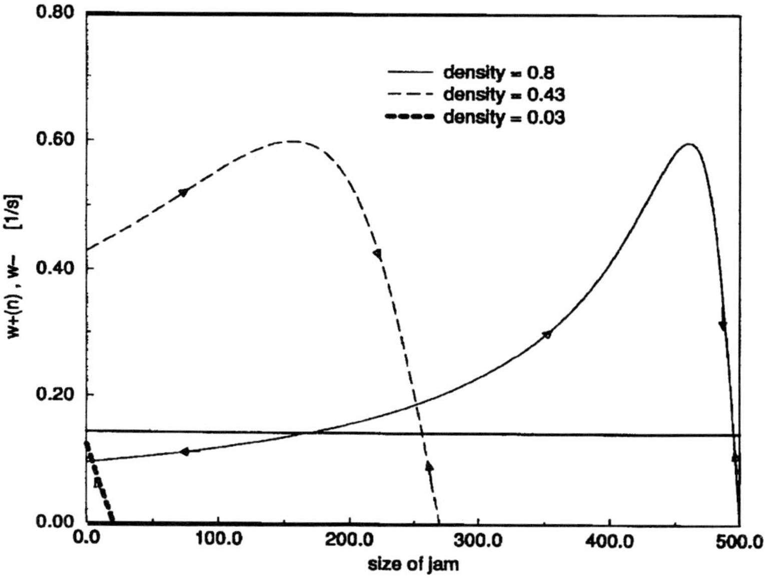

Fig. 15.2 Analytical curves for the transition rates (decay rate: horizontal line, growth rates: curves for subcritical, supercritical and very high densities) for three different regimes. The parameters are $d = 20$ m, $v_{max} = 40$ m/s, $l_0 = 8$ m, $\tau = 7$ s, $L = 5000$ m.

maximum value for $w^{(+)}$, with the condition $\tau < 2d/v_{max}$. This case leads to a stable homogeneous flow for the whole range of densities. However, for realistic values of the parameters, for example $d = 20$ m, $v_{max} = 40$ ms^{-1}, this situation will not occur, as the time constant should be less than a second. For the opposite case, the density range appears to be divided into three parts by two critical values. Below the lower critical value, the growing transition will always be less probable than the decay transition, and hence any initial perturbation will disappear. Here we find again that for low densities the homogeneous flow is stable. For intermediate densities, we find one equilibrium size, corresponding to the intersection point $w^{(+)}(n) = w^{(-)}(n)$. The respective value of n is a *stable* stationary size of the cluster. This means that, after a long time, the system approaches a stationary two–phase system, as described by several authors, where the cluster is moving backwards through a homogeneous flow. Now, for high densities, there are *two* equilibrium sizes, the greater one being stable and the smaller one being unstable; thus, for initially homogeneous or slightly perturbed flows, in other words, if the initial size of the perturbation is less than the unstable critical size, the perturbation will disappear. We find then the homogeneous congested flow described in general highway measurements. On the other hand, if the initial cluster is larger than this critical size, it will increase even more its size, until it reaches the stable stationary value. This behaviour at high densities can be seen as a metastability of the homogeneous flow, since it needs a minimal perturbation to become unstable.

From the knowledge of equilibrium cluster sizes in dependence on the density, we can already calculate analytically the mean value of the general flow at all densities. In other words we can find the *fundamental diagram*, as plotted in Fig. 15.3, as a first test of the validity of our model. We see that the shape is alike most analytical and empirical curves. We see also the influence of the parameter d, that can be seen as the 'range' of the interaction, and of the parameter l_0, that is the size of the 'hard core'.

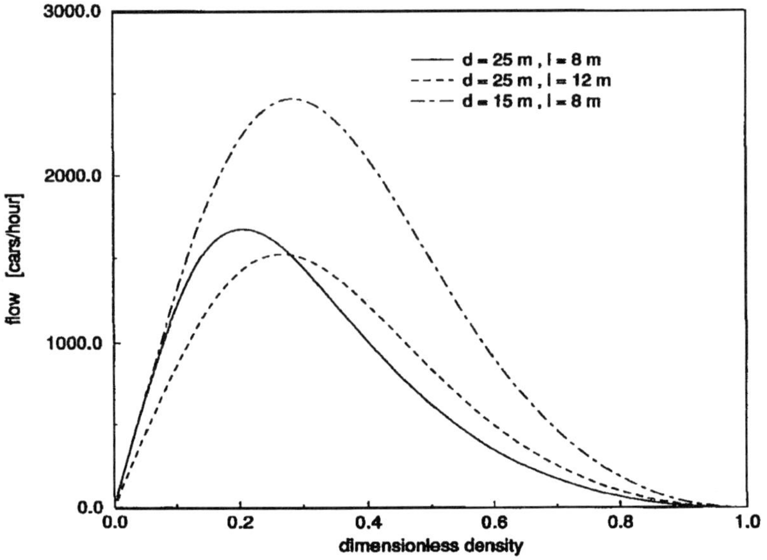

Fig. 15.3 Examples of flow–density curves obtained. The computed fundamental diagram shows the validity of our approach and agrees with measured data curves.

15.6 Jam Dynamics: Analytical and Numerical Results

The Master equation (15.2), with its associated transition rates given by Eqs. (15.3) and (15.4), with the help of Eq. (15.5), allows us to determine analytically and numerically the time evolution for the mean value of the size of the car cluster, as well as of its variance, for different sets of parameters, and in particular for different values of the density. In order to find an analytical expression for the time evolution of the mean value $\langle n\rangle(t) = \sum_{n=0}^{N} n\,P(n,t)$ from the Master equation (15.2), we have to perform the approximation $\langle w^{(+)}(n)\rangle \simeq w^{(+)}(\langle n\rangle)$, which leads to the following form

$$\frac{d}{dt}\langle n\rangle \simeq w^{(+)}(\langle n\rangle) - \frac{1}{\tau} \tag{15.6}$$

with stationary solutions of the mean car cluster size

$$\langle n\rangle_{stat} = N - \frac{L/l_0 - N}{\left(\frac{v_{max}\tau}{2\,l_0}\right) \pm \sqrt{\left(\frac{v_{max}\tau}{2\,l_0}\right)^2 - \left(\frac{d}{l_0}\right)^2}} \tag{15.7}$$

in the range of heavy traffic with supercritical densities.

Eq. (15.6) correctly describes the dynamics of the system at far–from–critical densities, and has basically the correct stationary solutions. In parallel to this, we simulate particular stochastic trajectories, which are in accordance with the results computed from the Master equation (see Figs. 15.4, 15.5).

For far–from–critical densities, the evolution is quasi–deterministic and the fluctuations are small. On the other hand, the high density trajectories show the bifurcation at a high density around the critical initial jam size. The initially inhomogeneous flow can have a long lifetime, then bifurcates either into a homogeneous congested flow ($n \to 0$), or into a stable state where many cars are blocked into a jam ($n \to n_{stable}$). For realistic values of the parameters, we see that this big cluster contains almost all the cars. As a result, if we take into account the fluctuations of the system, we can see that both final states are equivalent. Indeed, in the homogeneous flow, the cars move at a very small speed, and could be considered — with a less

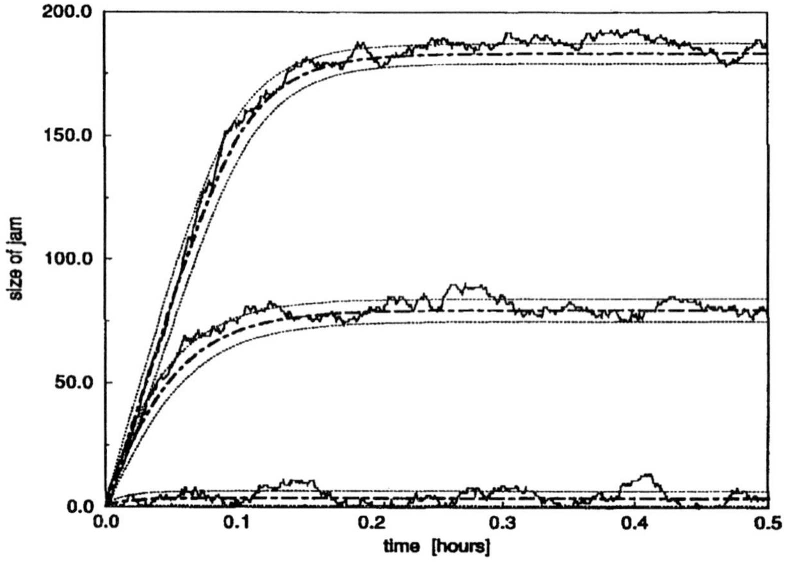

Fig. 15.4 Examples of stochastic trajectories as solutions of the Master Equation for intermediate densities (from bottom to top the total number of cars increases $N = 24, 100, 200$) compared with numerical solutions for the mean cluster size and its variance.

restrictive definition than ours for the cluster (cf. section 15.2), allowing non–zero particle velocity — being inside a jam that covers the road completely, with fluctuations around homogeneity. These fluctuations could be analyzed, for example, as the propagation of 'holes' (free spaces) in an homogeneous medium (see also [348]). However, these holes do not describe a discrete system anymore — having no finite fixed size —, and therefore do not allow for symmetry between very high and very low densities.

15.7 Stationary Solution and Flux Calculations

According to the above discussed model, the probability $P(n,t) \equiv P(n)$ to find the system at time t in a state n (cluster size n as a stochastic variable) is given by the well–known one–dimensional stochastic Master equation (see

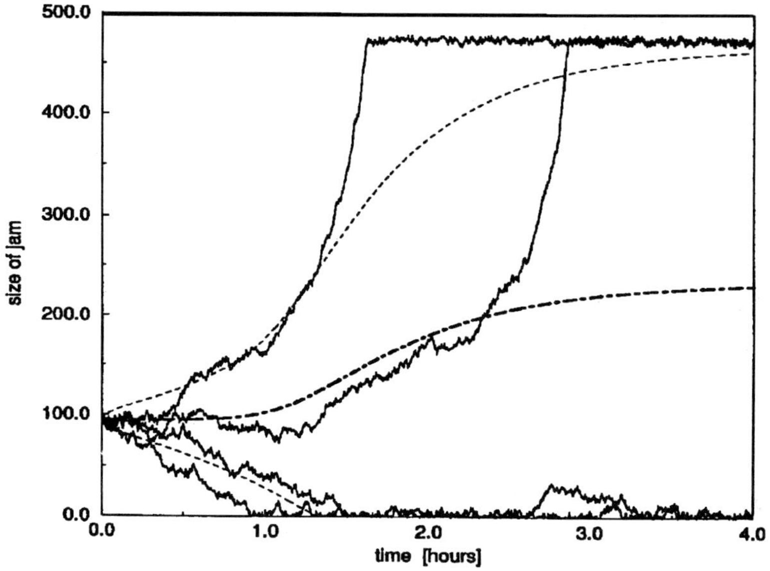

Fig. 15.5 Stochastic trajectories starting with critical initial size (unstable cluster size) at a very high density 0.68 (total number of cars $N = 480$) together with the mean value.

e. g. [166]) given by Eq. (15.2). This equation can be rewritten as follows

$$\frac{dP(n)}{dt} = j(n-1 \to n) + j(n+1 \to n) \\ -j(n \to n+1) - j(n \to n-1) \; , \tag{15.8}$$

where $j(n-1 \to n)$, $j(n+1 \to n)$, $j(n \to n+1)$ and $j(n \to n-1)$ are the probability fluxes which are equal to the corresponding terms in Eq. (15.2). The boundary conditions for Eq. (15.8) are

$$\begin{aligned} j(1 \to 0) &= j(0 \to 1) &&= 0 \\ j(N \to N+1) &= j(N+1 \to N) &&= 0 \end{aligned} \tag{15.9}$$

Remember that N is the total number of cars in the system.

Now let us consider the stationary solution of the Master equation (15.2) corresponding to the condition

$$\frac{dP(n)}{dt} = 0 \, . \tag{15.10}$$

The general solution of Eq. (15.10) is well–known and reads [166]

$$P(n+1) = P(n)\,Q(n) \qquad \text{with} \qquad Q(n) = \frac{w^{(+)}(n)}{w^{(-)}(n+1)} \tag{15.11}$$

or

$$P(n) = \frac{\prod_{m=0}^{n-1} Q(m)}{\sum_{n=1}^{N} \prod_{m=0}^{n-1} Q(m)} \qquad \text{with} \qquad Q(0) = 1 \, . \tag{15.12}$$

Here we propose a short explanation how to get this result proving also that it is the only stationary solution and detailed balance holds. From Eqs. (15.8) and (15.10) we conclude that the relation

$$j(m \to m+1) = j(m+1 \to m) \tag{15.13}$$

holds for $m = n$ if it is satisfied at $m = n - 1$. According to Eqs. (15.8) – (15.10), Eq. (15.13) holds for $m = 1$, therefore it holds for any m in the range $1 \leq m \leq N-1$. This means that the detailed balance Eq. (15.13) is true and leads unambiguously to the final solution Eq. (15.12). A series of snapshots of $P(n)$ for increasing densities, calculated on the basis of Eq. (15.12) showing the formation of a cluster on the road, is depicted in Fig. 15.6.

One of the most important characteristics of traffic flow is the fundamental diagram showing the flux j of cars as function of the total density ρ on the road. We define j as an averaged over infinite time interval local flux $\rho(x,t)v(x,t)$ where $\rho(x,t)$ is the local density and $v(x,t)$ is the local velocity of cars at a time moment t and space coordinate x, i. e.

$$j = \lim_{T \to \infty} \frac{1}{T} \int_{t=0}^{T} \rho(x,t)v(x,t)\,dt \, . \tag{15.14}$$

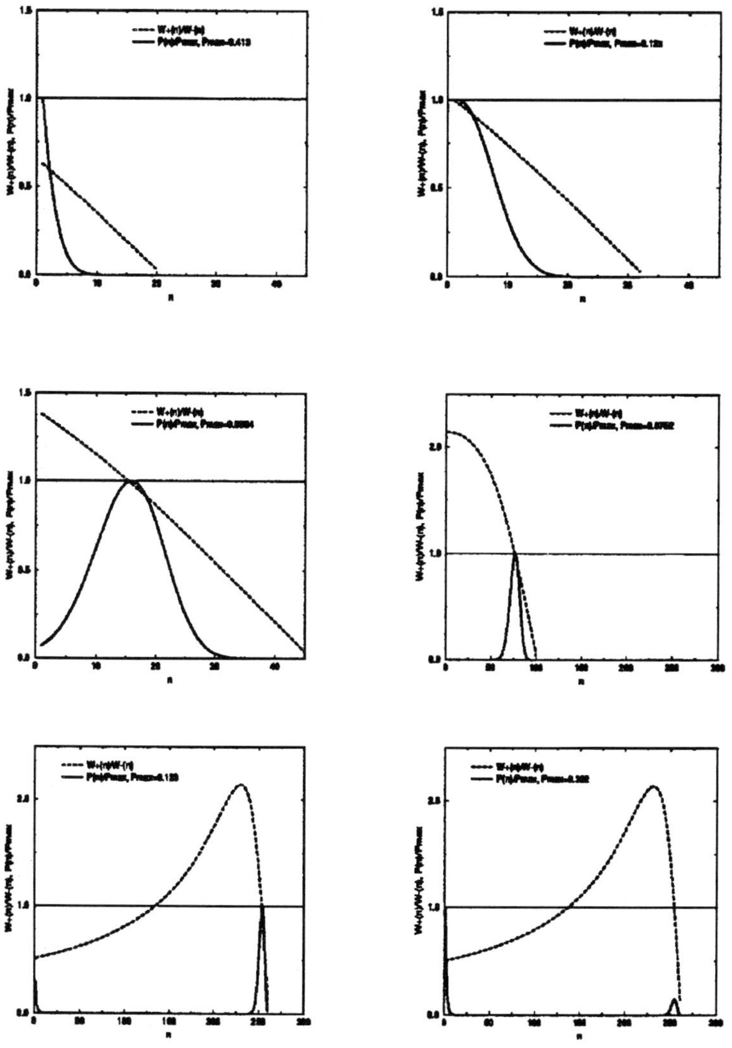

Fig. 15.6 Series of different stationary probability distributions $P(n)$ and ratios of transition rates $w^{(+)}(n)/w^{(-)}$ showing the formation of a jam of size n depending on the total number of cars N on the road (the values of N are 20, 32, 45, 100, 260, and 261). The parameters of the system are: $L = 5000$ m, $l_0 = 15$ m, $v_{max} = 30$ m/s, $\tau = 5$ s, and $d = 35$ m. The maximum of the probability distribution corresponds to the stable cluster size of congested cars.

In our model the local velocity as well as the density of cars are defined by the size of cluster n and the distance $x - x'$ between the considered local coordinate x and the coordinate x' of the first car in the jam. Thus, we have $\rho(x,t) = \rho(x - x'(t), n(t))$ and $v(x,t) = v(x - x'(t), n(t))$, and averaging over time yields

$$j = \sum_n \int dx'\, P(n,x')\, \rho(x - x', n) v(x - x', n) \,, \qquad (15.15)$$

where $P(n,x')dx'$ denotes the part of the total time during which the size of the cluster is n and the coordinate of the first car of the jam is between x' and $x' + dx'$. The cluster can be found with equal probability at any coordinate x' along the circle if an averaging over an infinite time interval T is considered. Thus we have $P(n,x') = P(n)/L$, and with our assumption that $v = 0$ holds inside the jam [338] and $v = v_{opt}(h(n))$ (see Eq. (15.1)) holds outside the jam, where $h(n) = (L - l_0 N)/(N - n + 1)$ is the distance between two cars in the free dilute region, it yields finally

$$j' = \sum_n P(n) \left(\rho' - \frac{n l_0}{L} \right) \frac{v_{opt}(h(n))}{v_{max}} \,. \qquad (15.16)$$

Here $j' = j l_0 / v_{max}$ is the dimensionless flux and $\rho' = N l_0 / L$ is the dimensionless total density of cars.

15.8 Analytical Solution of the Fundamental Diagram

Now we consider the behaviour of the system under the condition $v_{max}\,\tau/d > 2$ where always a cluster with $n/N \neq 0$ (at $N \to \infty$) emerges, i. e., a phase transition takes place at some value of ρ'. In the opposite case there is no phase transition (cluster formation) at all. An analysis of the solution of Eq. (15.12) shows that $P(y) = N^{-1}\delta(y - y_0)$ holds in the thermodynamic limit $N \to \infty$, where y is defined as $y = n/N$ with the value y_0 corresponding to the absolute maximum of $P(y)$. $y_0 = 0$ holds, if $\rho' \leq \rho_1$ or $\rho' > \rho_2$, and $y_0 = y_0' = 1 - l_0(1 - \rho')/(\rho' \tau v_{max} A)$ holds if $\rho_1 \leq \rho' < \rho_2$, where $A = 1/2 + \sqrt{1/4 - d^2/(v_{max}\tau)^2}$. The critical densities ρ_1 and ρ_2 are defined by $y_0' = 0$ or $\rho_1 = l_0/(l_0 + A\tau v_{max})$, and $\ln(P(y=0)) = \ln(P(y=y_0'))$, respectively. In conclusion the corresponding flux is given by

$$j'(\rho') = \begin{cases} \dfrac{\rho'(1-\rho')^2}{(\rho' d/l_0)^2 + (1-\rho')^2} & : \quad \rho' \in [0;\rho_1] \cup]\rho_2;1] \\[2ex] \dfrac{l_0(1-\rho')}{\tau v_{max}} & : \quad \rho' \in [\rho_1;\rho_2[\end{cases} \tag{15.17}$$

The regions $\rho' \in [0;\rho_1]$, $\rho' \in [\rho_1;\rho_2[$, and $\rho' \in]\rho_2;1]$ correspond to three different regimes of traffic flow. Besides, there is a breakpoint in j at $\rho' = \rho_1$ and a jump at $\rho' = \rho_2$, as it may be seen from Fig. 15.7 where the analytical solution (15.17) for some values of parameters is shown by dashed lines (corresponding to the first and the second formulae in Eq.(15.17)). The region $\rho' \in [0;\rho_1]$ corresponds to the free traffic flow where the relative part of cars involved in the jam tends to zero if $N \to \infty$. The region $\rho' \in [\rho_1;\rho_2[$ corresponds to the traffic flow on a partly congested road where one cluster of cars coexists with a region of free traffic. The region $\rho' \in]\rho_2;1]$ corresponds to a highly viscous overcrowded situation, where the density of cars is high and their velocity is small over the whole road.

15.9 Discussion

The fundamental diagram of traffic flow for various lengths L of the road, calculated on the basis of Eqs. (15.17) and (15.12) is shown in Fig. 15.7. It is seen from the figure that the plot of $j = j(\rho)$ converges to the analytical solution (15.17), represented by dashed lines, with increasing road lenght L. In comparison, in Fig. 15.6 the probability distribution function $P(n)$ for various densities ρ' or various total numbers N of cars on the road of length $L = 5000$ m is shown. We may see from both figures how the average size of the cluster, which is near the value of n corresponding to the maximum of $P(n)$, increases with increasing N (cases 2 to 5) within some region of densities (between ρ_1 and ρ_2) corresponding to the coexisting phase. In this case we have $w^{(+)}(n) \sim w^{(-)}(n)$ at the size n corresponding to the maximum, as it can be seen from Fig. 15.6 and the solution (15.12), whereas for other densities the maximum is at $n = 1$ indicating the absence of any macroscopic car cluster.

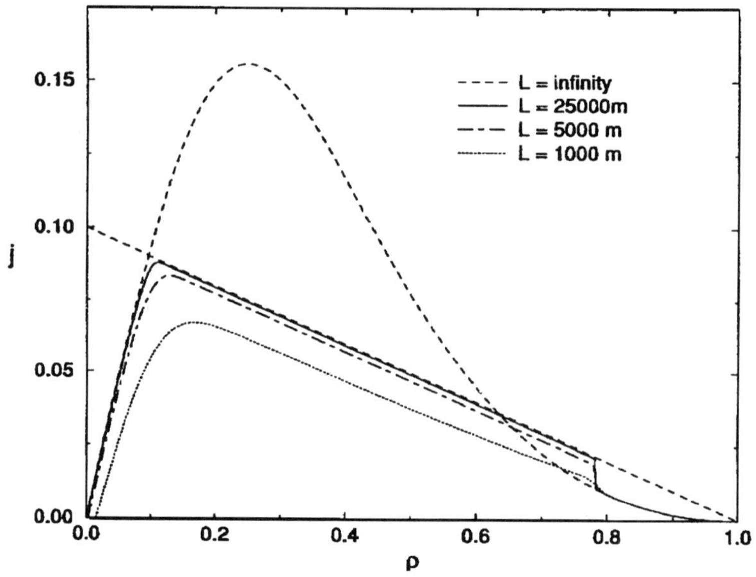

Fig. 15.7 Based on the stationary solution of the stochastic Master equation the fundamental diagram (dimensionless flow rate (flux) j vs dimensionless car density ρ) is calculated. Fixing all other parameters $l_0 = 15$ m, $v_{max} = 30$ m/s, $\tau = 5$ s, and $d = 35$ m only the length of road L varies. For finite roads ($L < \infty$) as well as for infinitely long roads ($L \to \infty$) the flow j can be divided into two homogeneous regimes (left: free flow as gaseous phase, right: heavy traffic as liquid phase) and a transition regime with free and congested vehicles (formation of a cluster).

In conclusion we emphasize that the model described here is a completely new approach to traffic problems based on a one–dimensional Master equation. By focusing on the jamming phenomenon and on its essentially stochastic properties, we have been able to describe the general behaviour of the system very realistically under very simple hypothesis and with very small computing power. In particular, our fundamental diagram (Figs. 15.3 and 15.7) shows its habitual shape. Of course, many improvements and generalizations are possible and necessary. In particular, a more subtle definition of the car cluster (cf. section 15.2) would be helpful in describing the behaviour of the system at very high densities (cf. discussion at end of section 15.6), but also at low densities — since in that case we do not expect the cars to come to a real stop by agglomerating. The decay transition (Eq. (15.4)) could be defined in a more precise way, and in particular should be density–dependent. The parametrization should be investigated and quantitative results compared with real life. Allowing coexistence of several clusters and overtaking (which could be done using an analogy with tunnel effect), examining the influence of traffic lights — that create traffic jams artificially — and of crossroads — that create jams by external stochastic processes — should be the first steps to generalization and applications.

16 Concluding Remarks

Coming to the final pages of the monograph, the question arises whether the ideas one had in mind in starting to write the book are realized or not. Looking back, the satisfaction that a variety of results could be outlined in a – as we hope – compact and clearly understandable form, becomes mixed with the feeling that an even larger part of the possible problems, one is involved in considering nucleation, growth and, more generally, aggregation phenomena, could not be even touched. However, a number of references is given, so that further details and methods can be found easily in the current literature.

Nevertheless, we hope to have clearly demonstrated that the general ideas and methods of description are quite general and may be of use thus for a large group of people working in different areas of science and technology. We would be glad if the reader of this book will share this our opinion.

Bibliography

[1] Abraham, F. F.: *Homogeneous Nucleation Theory*, Academic Press, New York, 1974

[2] Acharyya, M., Chakrabarti, B. K.: *Ising Systems in Oscillating Field: Hysteretic Response*; In: D. Stauffer (Ed.), Annual Review of Computational Physics, World Scientific, Singapore, 1994, p. 107; Rev. Mod. Phys., (1998), in press

[3] Aichelin, J., G. Peilert, A. Bohnet, A. Rosenhauer, H. Stöcker, W. Greiner, Phys. Rev. **C 37**, 2451 (1988)

[4] Aichelin, J., Phys. Rep. **202**, 233 (1991)

[5] Aitcheson, J., Brown, J. A. C.: *The Lognormal Distribution*, Cambridge University Press, Cambridge, 1973

[6] Allen S. W., Cahn, J. W., Acta Metall. **27**, 1085 (1979)

[7] Alm, Th., Röpke, G., Bauer, W., Daffin, F., Schmidt, M., Nucl. Phys. **A 587**, 815 (1995)

[8] Alt, E. O., Grassberger, P., Sandhas, W., Nucl. Phys. **B 2**, 167 (1967)

[9] Anagnostatos, G. S., von Oertzen, W. (Eds.): *Atomic and Nuclear Clusters*, Springer, Berlin, 1995

[10] Arnett, W. D., Astrophys. Journal **218**, 815 (1977)

[11] Baillie, C. F., Int. J. Mod. Phys. **C 1**, 91 (1990)

[12] Bak, P., Chen, K., Tang, C., Phys. Lett. **A 147**, 297 (1990)

[13] Bak, P., Tang, C., Wiesenfeld, K., Phys.Rev. Lett. **59**, 381 (1987)

[14] Bak, P., Tang, C., J. Geophys. Res. **94**, 15635 (1989)

[15] Bak, P., Sneppen, K., Phys. Rev. Lett. **71**, 4083 (1993)

[16] Balescu, R.: *Equilibrium and Non–Equilibrium Statistical Mechanics*, Wiley, New York, 1975

[17] Bando, M., Hasebe, K., Nakayama, A., Shibata, A., Sugiyama, Y., Japan J. Indust. and Appl. Math. **11**, 203 (1994)

[18] Bando, M., Hasebe, K., Nakayama, A., Shibata, A., Sugiyama, Y., Phys. Rev. **E 51**, 1035 (1995)

[19] M. Bando, K. Hasebe, K. Nakanishi, A. Nakayama, A. Shibata, Y. Sugiyama, J. Phys. I France **5**, 1389 (1995)

[20] Bartels, J., Schweitzer, F. Schmelzer, J., J. Non–Cryst. Solids **125**, 129 (1990)

[21] Bartels, J., *Zur Theorie von Keimbildung und Keimwachstum in kondensierten Systemen*, Dissertation, Rostock, 1991

[22] Bartels, J., Lembke, U., Pascova, R., Schmelzer, J., Gutzow, I., J. Non–Cryst. Solids **136**, 181 (1991)

[23] Bartels, J., Schmelzer, J., phys. stat. solidi **a132**, 361 (1992)

[24] Barz, H. W., Bondorf, J. P., Donangelo, R., Mishnstin, I. N., Schulz, H., Nucl. Phys. **A 448**, 753 (1986)

[25] Baxter, R. J., J. Phys. **C 6**, L445 (1973)

[26] Becker, R., Döring, W., Ann. Phys. **24**, 719 (1935)

[27] Becker, R., Annalen der Physik **32**, 128 (1938)

[28] Becker, R.: *Theorie der Wärme*, Springer, Berlin – Göttingen – Heidelberg, 1964

[29] Benedek, G., Martin, T. P., Pacchioni, G. (Eds.): *Elemental and Molecular Clusters*, Springer, Berlin, 1988

[30] Benson, S. W.: *The Foundations of Chemical Kinetics*, Wiley, New York, 1960

[31] Berg, B. A., Neuhaus, T., Phys. Lett. **B 267**, 249 (1991)

[32] Bertsch, G., das Gupta, S., Phys. Rep. **160**, 189 (1988)

[33] Beyer, M., Röpke, G., Sedrakian, A., Phys. Lett. **B 376**, 7 (1996)

[34] Beyer, M., Röpke, Phys. Rev. **C 56**, 2636 (1997)

[35] Bhattacharya, T., Lacaze, R., Morel, A., Nucl. Phys. **B 435**, 526 (1995)

[36] Bhattacharya, T., Lacaze, R., Morel, A., Nucl. Phys. (Proc. Suppl.), **B 42**, 743 (1995)

[37] Bhattacharya, T., Lacaze, R., Morel, A., Europhys. Lett. **23**, 547 (1993)

[38] Biham, O., Middleton, A. A., Levine, D.: Phys. Rev. E **46**, R6124 (1992)

[39] Binder, K., Stauffer, D., Adv. Phys. **25**, 343 (1976)

[40] Binder, K., Phys. Rev. **B 15**, 4425 (1977)

[41] Binder, K. (Ed.): *Monte Carlo Methods in Statistical Physics*, Springer, Berlin, 1979

[42] Binder, K., J. Physique (France) **41**, C4–51 (1980)

[43] Binder, K., Kalos, M. H., J. Stat. Phys. **22**, 363 (1980)

[44] Binder, K., J. Chem. Phys. **79**, 6387 (1983)

[45] Binder, K., Frisch, H. L., Jäckle, J., J. Chem. Phys. **85**, 1505 (1986)

[46] Binder, K., Ann. Rev. Phys. Chem. **43**, 33 (1992)

[47] Binder, K., Phys. Rev. Lett. **47**, 693 (1981)

[48] Binder, K., Phys. Rev. **A 25**, 1699 (1982)

[49] Binder, K. (Ed.): *Application of Monte Carlo Methods in Statistical Physics*, Springer, Berlin, 1984.

[50] Binder, K., Phys. Rev. **A 29**, 341 (1984)

[51] Binder, K., Rep. Prog. Phys. **50**, 783 (1987)

[52] Binder, K., Heermann, D. W.: *Monte Carlo Simulation in Statistical Physics. An Introduction*, Springer, Berlin, 1988 (3. Aufl. 1997)

[53] Binder, K. (Ed.): *The Monte Carlo Method in Condensed Matter Physics*, Springer, Berlin, 1992

[54] Binder, K.: *Spinodal Decomposition*, In: Materials Science and Technology, VCH Verlagsgesellschaft, Weinheim, 1992

[55] Binder, K., Vollmayr, K., Deutsch, H. P., Reger, J., Scheucher, M., Landau, D. P., Int. Journ. Mod. Phys., **3**, 253 (1992)

[56] Blander, M., Adv. Colloid Interface Science **10**, 1 (1979)

[57] Blankschtein, D., Shapir, Y., Aharoni, A., Phys. Rev. **B 29**, 1263 (1984)

[58] Blaschke, D., Röpke, R., Wiss. Z. Univ. Rostock **32**, 34 (1983)

[59] de Boer, J., Derrida, B., Flyvbjerg, H., Jackson, A. D., Wetting, T., Phys. Rev. Lett. **73**, 906 (1994)

[60] Bohr, A., Mottelson, B.: *Nuclear Structure*, Benjamin, New York, 1963

[61] Bonacic–Koutecky, V., Fantucci, P., Koutecky, J., Phys. Rev. **B 37**, 4369 (1988)

[62] Bonacic–Koutecky, V., Fantucci, P., Koutecky, J., Chem. Rev. **91**, 1035 (1991)

[63] Bondorf, J. P., Garpman, S. I. A., Zimanyi, J., Nucl. Phys. **A 296**, 320 (1978)

[64] Bonasera, A., Gulminelli, F., Molitoris, J., Phys. Rep. **243**, 2 (1994)

[65] Bondorf, J. et al., Nucl. Phys. **A 443**, 321 (1985); **A 444**, 460 (1985); **448**, 753 (1986)

[66] Boon, J. P., Dab, D., Kapral, R., Lawniczak, A., Phys. Rep. **273**, 55 (1996)

[67] Botet, R., Jullien, R., Ann. Phys. (France) **13**, 153 (1988)

[68] Boustani, I., Pewestorf, W., Fantucci, P., Bonacic–Koutecky, V., Koutecky, J., Phys. Rev. **B 35**, 9437 (1987)

[69] Brack, M., Rev. Mod. Phys. **65**, 677 (1993)

[70] Brown, G. E., Bethe, H. A., Baym, G., Nucl. Phys. **A 375**, 481 (1982)

[71] Brunisma, J. van Wageningen, R., Nucl. Phys. **A 282**, 1 (1977)

[72] Budde, A.: *Computersimulation zur Clusterbildung in Gasen und Makromelekülen*, Dissertation, Universität Rostock, 1989

[73] Buff, F. P., Lovett, R. A., Stillinger, F. H., Phys. Rev. Lett. **15**, 621 (1965)

[74] Burbidge, E. M., Burdidge, C. R., Fowler, W. A., Hoyle, F., Rev. Mod. Phys. **29**, 547 (1957)

[75] Burlatsky, S. F., Sov. Theor. Exp. Chem. **14**, 483 (1978)

[76] Burlatsky, S. F., Ovchinnikov, A. A., Sov. Phys. Chem. **52**, 2847 (1978)

[77] Burov, V. V., Eldyshev, Yu. N., Lukyanov, V. K., Pol, Yu. S., Dubna–preprint E4–8029, Joint Institute of Nuclear Research, Dubna, Russia

[78] Cahn, J. W., Hilliard, J. E., J. Chem. Phys. **28**, 258 (1958); **31** 688 (1959)

[79] Cahn, J. W., Acta Met. **8**, 554 (1960)

[80] Cameron, A. G. W., Space Sci. Rev. **15**, 121 (1973)

[81] Car, R., Parrinello, M., Phys. Rev. Lett. **55**, 2471 (1985)

[82] Cardy, J. L. (Ed.): *Finite Size Scaling,* Amsterdam, North Holland, 1988

[83] Carlson, J. M., Langer, J. S., Phys. Rev. Lett. **62**, 2632 (1989)

[84] Carmesin, H. O., Heermann, D. W., Binder, K., Z. Phys. **B 65**, 89 (1986)

[85] Chakrabarthy, S., J. Phys. G **20**, 469 (1994)

[86] Challa, M. S. S., Landau, D. P., Binder, K., Phys. Rev. B **34**, 1841 (1986)

[87] Chen, K., Bak, P., Obukhov, S. P., Phys. Rev. **A 43**, 625 (1990)

[88] Chernov, A. A., Z. phys. Chemie (Leipzig) **269**, 941 (1988)

[89] Chowdhury, D., Wolf, D. E., Schreckenberg, M., Physica **A 235**, 417 (1997)

[90] Christensen, K., Fogedby, H. C., Jensen, H. J., J. Stat. Phys. **63**, 653 (1991)

[91] Christensen, K., Olami, Z., Phys. Rev. **E 48**, 3361 (1993)

[92] Christian, J. W.: *The Theory of Transformations in Metals and Alloys*, Oxford University Press, Oxford, 1975

[93] Cohen, R. D., J. Chem. Soc. Faraday Trans. **86**(12), 2133 (1990); Proc. Royal Society (London) **A 435**, 483 (1991); Powder Technology **63**, 261 (1990)

[94] Collins, F. C., Z. Elektrochemie **59**, 404 (1955)

[95] Colonna, M., Chomaz, Ph., Randrup, J. P., Nucl. Phys. **A 567**, 637 (1994)

[96] Colonna, M., di Toro, M., Guarnera, A., Nucl. Phys. **A 580**, 512 (1994)

[97] Colton, J. S., Suh, N. P., Polymer Engineering and Science **27**, 485 (1987)

[98] Coniglio A., Klein, W., J. Phys. **A 13**, 2775 (1980)

[99] Cook, H. E., Acta Met. **18**, 297 (1970)

[100] Craievich, A. F., Olivieri, J. P., J. Appl. Cryst. **14**, 444 (1981)

[101] Crank, J.: *The Mathematics of Diffusion*, Clarendon Press, Oxford, 1975

[102] Cremer, M.:*Der Verkehrsfluß auf Schnellstraßen*, Springer, Berlin, 1979

[103] Creutz, M., Computers in Physics, March/April 198 (1991)

[104] Csahok, Z., Vicsek, T., J. Phys. **A 27**, L591 (1994)

[105] Csernai, L. P., Kapusta, J. I., Phys. Rep. **131**, 223 (1986)

[106] Cugnon, J.: *Review of Multifragmentation Theories*, in: [145], p. 244 ff

[107] Danielewicz, P., Nucl. Phys. **A 314**, 465 (1979)

[108] Danielewicz, P., Bertsch, G. F., Nucl. Phys. **A 533**, 712 (1991)

[109] Danielewicz, P., Pan, Q., Phys. Rev **C 46**, 2002 (1992)

[110] Debye, P., Electrochem. **32**, 265 (1942)

[111] Demo, P., Kozisek, Z., Phys. Rev. **B 48**, 3620 (1993)

[112] Demo, P., Kosizek, Z., J. Phys. G **23**, 971 (1997)

[113] Desbois, J., Boisgard, R., Ngo, C., Nemeth, J., Z. Phys. **A 328**, 101 (1987)

[114] Dhar, D., Phys. Rev. Lett. **64**, 1613 (1990)

[115] Dhar, D., Manna, S. S., Majumdar, S. N., Phys. Rev. **A 46**, R4471 (1992)

[116] Dhar, D., Manna, S. S., Phys. Rev. **E 49**, 2684 (1994)

[117] Doktorov, A. B., Kipryanov, A. A., Burshtein, A. I., Sov. Opt. Spectr. **45**, 497 (1978)

[118] Domb, C., Green, M. S., Lebowitz, J. L. (Eds.): *Phase Transitions and Critical Phenomena*, **Vol. 1 – 12**, Academic Press, London, 1972 – 1988

[119] Domb, C.: *The Critical Point*, Taylor & Francis, London, 1996

[120] Dreizler, R. M., Gross, E. K. U.: *Density Functional Theory*, Springer, Berlin, 1990

[121] Drossel, B., Schwabl, F., Phys. Rev. Lett. **69**, 1629 (1992)

[122] Drossel, B., Clar, S., Schwabl, F., Phys. Rev. Lett. **71**, 3739 (1993)

[123] Dukelsky, J., Röpke, G., Schuck, P., Nucl. Phys. **A 628**, 17 (1998)

[124] Dutta, P., Horn, P. M., Rev. Mod. Phys. **53**, 497 (1981)

[125] Ebeling, W.: *Strukturbildung bei irreversiblen Prozessen*, Teubner, Leipzig, 1976

[126] Ebeling, W.: *Thermodynamisch–statistische Fluktuationen und Kinetik der Keimbildung*, Berichte der Akademie der Wissenshaften der DDR, Berlin, 1981

[127] Ebeling, W., Feistel, R.: *Physik der Selbstorganisation und Evolution*, Akademie – Verlag, Berlin, 1982

[128] Ebeling, W., Malchow, H.: *Räumliche Phasenseparation im Nichtgleichgewicht*, Berichte des Dresdener Seminars für Theoretische Physik **19**, 68 (1982)

[129] Ebeling, W., Ulbricht, H.: *Selforganization by Nonlinear Irreversible Processes*, Springer – Series in Synergetics **Vol. 33**, Springer, Berlin, 1986

[130] Ebeling, W.: *Chaos, Ordnung und Information*, Urania – Verlag, Leipzig, 1989

[131] Eigen, M., Winkler, R.: *Das Spiel – Naturgesetze steuern den Zufall*, Piper – Verlag, München, 1975

[132] Ekardt, W., Phys. Rev. **B 32**, 1961 (1985)

[133] Elliot, J. B. et al., Phys. Rev. **C 49**, 3185 (1994)

[134] Emmerich, H., Rank, E., Physica A **216**, 435 (1995)

[135] Englman, R., J. Phys.: *Condens. Matter* **3**, 1019 (1991)

[136] Eschrig, H.: *The Fundamentals of Density Functional Theory*, Teubner–Texte zur Physik, B. G. Teubner, Leipzig, 1996

[137] Evesque, P., Rachenbach, J., Phys. Rev. Lett. **62**, 44 (1989)

[138] Eyring, H., Lin, S. H., Lin, S. M.: *Basic Chemical Kinetics*, Wiley, New York, 1980

[139] Farkas, L., Z. Phys. Chemie **125**, 236 (1927)

[140] Farmer, D., Toffoli, T., Wolfram, St. (Eds.): *Cellular Automata*, Proceedings of an Interdisciplinary Workshop 1983, Physica **D 10** (1984)

[141] Feder, J., Russell, K. C., Lothe, J., Pound, G. M., Adv. Phys. **15**, 117 (1966)

[142] Feder, J.: *Fractals*, Plenum Press, New York–London, 1988

[143] Feder, H. J. S., Feder, J., Phys. Rev. Lett. **66**, 2669 (1991)

[144] Feistel, R., Ebeling, W.: *Evolution of Complex Systems*, Deutscher Verlag der Wissenschaften, Berlin, 1989

[145] Feldmeier, H., Nörenberg, W. (Eds.): *Multifragmentation*; Proceedings International Workshop on Gross Properties of Nuclei and Nuclear Excitations; Hirschegg, Austria, January 17 – 22, 1994

[146] Fermi, E., Pasta, J., Ulam, S.: *Studies of Non–Linear Problems*, Los Alamos Scientific Laboratory Report LA – 1940 (1955)

[147] Fetter, A. L., Walecka, J. D.: *Quantum Theory of Many–Paricle Systems*, Mc Graw Hill, New York, 1971

[148] Filipovich, V. N., Izv. Akad. Nauk, Neorgan. Materialy, **3**, 1192 (1967)

[149] Finn, J. E., Agarwal, S., Bujak, A., Chuang, J., Gutay, L. J., Hirsch, A. S., Minich, R. W., Porile, N. T., Scharenberg, R. P., Stringfellow, B. C., Phys. Rev. Lett. **49**, 1321 (1982)

[150] Fisher, M. E., Physics **3**, 255 (1967)

[151] Fisher, M. E., In: *Critical Phenomena* **a**, Proc. 1970 Enrico Fermi International School on Physics, Ed. M. S. Green, Academic Press, New York, 1971

[152] Fisher, M. E., Rep. Prog. Phys. **30**, 615 (1967); Rev. Mod. Phys. **46**, 597 (1974)

[153] Flyvbjerg, H., Sneppen, K., Bak, P., Phys. Rev. Lett. **71**, 4087 (1993)

[154] Frank, W.: *Stationäre und Instationäre Kondensationsvorgänge bei einer Prandtl – Meyer – Expansion*, In: Strömungsmechanik und Strömungsmaschinen, Mitteilungen des Institutes für Strömungslehre und Strömungsmaschinen der Universität Karlsruhe (TH) **25**, 61 (1978)

[155] Frank–Kamenetski, D. A.: *Diffusion and Thermal Conduction in Chemical Kinetics*, Nauka, Moscow, 1967

[156] Franks, F.: *Water. A Comprehensive Treatise*, Plenum Press, New York, 1972

[157] Frenkel, Ya. I., Kontorova, T., Zh. Eksp. Teor. Fiz. **8**, 89 (1938)

[158] Frenkel, Ya. I.: *The Kinetic Theory of Liquids*, Oxford University Press, Oxford, 1946

[159] Frenkel, Ya. I., Uspekhi Fiz. Nauk **32**, 294 (1947)

[160] Freud, J., Pöschel, Th., Physica A **219**, 95 (1995)

[161] Freudental, A. M.: *Inelastic Behaviour of Engineering Materials and Structures*, Wiley, New York, 1956

[162] Fröbrich, P., Gontchar, I. I., Phys. Rep. **292**, 131 (1998)

[163] Fujimoto, M.: *The Physics of Structural Phase Transitions*, Springer, New York, 1997

[164] Furukawa, H., Binder, K., Phys. Rev. **A 26**, 556 (1982)

[165] Galli, G., Parrinello, M., In: *Computer Simulations in Materials Science*, Nato ASI Series, Vol. 205, p. 283 ff, Kluwer Academic, Dordrecht, 1991

[166] Gardiner, C. W.: *Handbook of Stochastic Methods for Physics, Chemistry and Natural Sciences*, Springer, Berlin, 1983

[167] Gebhardt, W., Krey, U.: *Phasenübergänge und kritische Phänomene*, Vieweg, Braunschweig, 1980

[168] Geguzin, Ya. E.: *Physik des Sinterns*, Dt. Verlag für Grundstoffindustie, Leipzig, 1973

[169] Gerlough, D. L.: *Simulation of Freeway Traffic by an Electronic Computer*, In: Proc. 35th Annual Meeting of Highway Research Board, Washington, D. C., 1956

[170] Gibbs, J. W.: *On the Equilibrium of Heterogeneous Substances*, In: *The Collected Works*, Vol. 1., Thermodynamics, Academic Press, New York – London – Toronto, 1928

[171] Glöckle, W.: *The Quantum Mechanical Few–Body Problem*, Springer, Berlin, 1983; Afnan, I. R., Thomas, A. W., in: *Modern Three Hadron Physics* ed. A.W. Thomas, p.1, Springer, Berlin, 1977; Schmidt, E. W., Ziegelmann, H.: *The Quantum Mechanical Three–Body Problem*, Pergamon Press, Oxford, 1974); Sandhas, W., Acta Physica Austriaca, Suppl. **IX**, 57 (1972); Belyaev, V. B.: *Lectures on the Theory of Few–Body Systems*, Spinger, Berlin 1990

[172] Goles, E., Martinez, S. (Eds.): *Cellular Automata, Dynamical Systems and Neural Networks*, Kluwer Acad. Publ., Dordrecht, 1994

[173] Gomatam, J., Amdjadi, F., Phys. Rev. E **56**, 3913 (1997)

[174] Goodman, J., Sokal, A. D., Phys. Rev. Lett. **56**, 1015 (1986); Phys. Rev. **D 40**, 2035 (1989)

[175] Gösele, U., Hantley F. A., Phys. Lett. **A 55**, 291 (1975)

[176] Grassberger, P., Manna, S. S., J. Phys. (France) **51**, 1077 (1990)

[177] Grassberger, P., Kantz, H., J. Stat. Phys. **63**, 685 (1991)

[178] Grassberger, P., J. Phys. **A 26**, 2081 (1993)

[179] Gross, D. H. E., Rep. Progr. Phys. **53**, 605 (1990); Phys. Rep. **279**, 120 (1997) and references cited therein

[180] Gunton, J. D., San Miguel, M., Sahni, P. S.: *The Dynamics of First–Order Phase Transitions*, In: *Phase Transitions and Critical Phenomena*, **Vol 8**, Eds. C. Domb, J. L. Lebowitz, Academic Press, London – New York, 1983

[181] Gutowitz, H. (Ed.): *Cellular Automata: Theory and Experiment*, Proceedings of a Workshop 1989, Physica **D 45** (1990)

[182] Gutzow, I.: *Crystallization Processes in Viscous Glass – Forming Melts.* Doctoral Thesis, Bulgarian Academy of Sciences, Sofia, 1970

[183] Gutzow, I., Glast. Berichte **46**, 219 (1973)

[184] Gutzow, I., Contemporary Phys. **21**, 121, 243 (1980)

[185] Gutzow, I., Dobreva, A., Schmelzer, J., J. Materials Science **28**, 890 (1993)

[186] Gutzow, I., Dobreva, A., Schmelzer, J., J. Materials Science **28**, 901 (1993)

[187] Gutzow, I., Schmelzer, J.: *The Vitreous State: Thermodynamics, Structure, Rheology, and Crystallization*, Springer, Berlin, 1995

[188] Haberland, H., Schindler, H.–G., Worsnop, D. R., Ber. Bunsenges. Phys. Chem. **88**, 270 (1984)

[189] Haberland, H.: *Cluster*, In: Bergmann-Schaefer, Lehrbuch der Experimentalphysik, Bd. 5, Hrsg. W. Raith, de Gruyter, Berlin, 1992, Chapter 8

[190] Haberland, H. (Ed.): *Clusters of Atoms and Molecules – Theory and Experiment* Springer, Berlin, 1994

[191] Haken, H.: *Synergetics. An Introduction*, Springer, Berlin, 1978 (1st ed.), 1983

[192] Halpern, I., Ann. Rev. Nucl. Sci. **21**, 245 (1971)

[193] Häkkinen, H., Kolehmainen, J., Koskinen, M., Lipas, P. O., Manninen, M., Phys. Rev. Lett. **78**, 1034 (1997)

[194] Hänggi, P., Talkner, P., Borkovec, M., Rev. Mod. Phys. **62**, 251 (1980)

[195] Harary, F.: *Graph Theory*, Reading, MA, Addison Wesley, 1990

[196] de Heer, W. A., Rev. Mod. Phys. **65**, 611 (1993)

[197] Heermann, D. W., Klein, W., Stauffer, D., Phys. Rev. Lett. **49**, 1262 (1982)

[198] Heermann, D. W., Coniglio, A., Klein, W., Stauffer, D., J. Stat. Phys. **36**, 447 (1984)

[199] Heermann, D. W., Phys. Rev. Lett. **52**, 1126 (1984); Z. Phys. **B 61**, 311 (1985)

[200] Heermann, D. W.: *Computersimulation Methods in Theoretical Physics*, Springer, Berlin, 1990

[201] Helbing, D.: *Verkehrsdynamik. Neue physikalische Modellierungskonzepte*, Springer, Berlin, 1997

[202] Held, G. A., Solina, D. H., Keaton, D. T., Haag, W. J., Horn, P. M., Grinstein, G., Phys. Rev. Lett. **65**, 1120 (1990)

[203] Heiselberg, H., Pethick, C. J., Ravenhall, D. G., Phys. Rev. Lett. **61**, 818 (1988)

[204] Herrmann, H. J., Janke, W., Karsch, F. (Eds.): *Dynamics of First Order Phase Transitions*, World Scientific, Singapore, 1992

[205] Herz, A. V. M., Hopfield, J. J., Phys. Rev. Lett. **75**, 1222 (1995)

[206] Hillert, M.: *A Theory of Nucleation for Solid Metallic Particles*, PhD – Thesis, MIT, Cambridge, 1956

[207] Hilliges, M.: *Ein phänomenologisches Modell des dynamischen Verkehrsflusses in Schnellstraßennetzen*, Dissertation, Universität Stuttgart, 1995

[208] Hilliges, M., Weidlich, W., Transportation Research **B 29**, 407 (1995)

[209] Hirth, J. P., Pound, G. M.: *Condensation and Evaporation*, Pergamon Press, London, 1963

[210] Hodgson, A. W., Adv. Colloid Interface Science **21**, 303 (1984)

[211] Honerkamp, J.: *Stochastische Dynamische Systeme*, VCH Verlagsgesellschaft, Weinheim, 1990; *Stochastic Dynamical Systems*, VCH, New York, 1994

[212] Honerkamp, J., Römer, H.: *Klassische Theoretische Physik*, Springer, Berlin, 1986

[213] Houston, E. L., Cahn, J. W., Hilliard, J. E., Acta Metall. **14**, 1685 (1966)

[214] Hüfner, J., Phys. Rep. **125**, 130 (1985)

[215] Huse, D. A., Phys. Rev. **B 34**, 7845 (1986)

[216] Hwa, T., Kardar, M., Phys. Rev. Lett. **62**, 1813 (1989)

[217] Itzykson, C., Drouffe, J. M.: *Statistical Field Theory*, Vol. I, Cambridge University Press, Cambridge, 1989

[218] Ivashkevich, E. V., Ktitarev, D. V., Priezzhev, V. B., Physica **A 209**, 347 (1994)

[219] Ivashkevich, E. V., Ktitarev, D. V., Priezzhev, V. B., J. Phys. **A 27**, L 585 (1994)

[220] Jaeger, H. M., Lin, C.-H., Nagel, S. R., Phys. Rev. Lett. **62**, 40 (1989)

[221] Jaeger, H. M., Nagel, S. R., Science **255**, 1523 (1992) and references cited therein

[222] Janke, W., In: *Computer Simulation Studies in Condensed Matter Physics*, Springer Proceedings in Physics, Vol. **78**, Eds. D. P. Landau, K. K. Mon, Berlin, 1994

[223] Janke, W., Sauer, T., J. Stat. Phys. **78**, 759 (1995)

[224] Jayanth, C. S., Nash, P., J. Mater. Science **24**, 3041 (1989)

[225] Johnson, W. C., Howe, J. M., Laughlin, D. E., Soffa, W. A. (Eds.): *Solid to Solid Phase Transformations*, Proceedings of the International Conference on Solid to Solid Phase Transformations in Inorganic Materials PTM94; Nemacolin Woodlands, Farmington, July 1994

[226] Jones, R. O., Gunnarson, O., Rev. Mod. Phys. **61**, 689 (1989)

[227] Kadanoff, L. P., Baym, G.: *Quantum Theory of Many-Particle Systems*, Mc Graw-Hill, New York, 1962

[228] Kaischew, R., Stranski, I. N., Z. Phys. Chem. **A 170**, 295 (1934); **B26**, 317 (1934)

[229] Kaischew, R., Stoyanov, S., Commun. Department of Chemistry, Bulgarian Academy of Sciences **2**, 127 (1969)

[230] Kämpfer, B., Lukacs, B., Paal, G.: *Cosmic Phase Transitions*, Teubner Texte zur Physik, Bd. 29, Teubner, Leipzig, 1994

[231] Kang, K., Redner, S., Phys. Rev. Lett. **52**, 955 (1984)

[232] Kang, K., Redner, S., Phys. Rev. **A 32**, 4355 (1985)

[233] Kaplan, D., Glass, L.: *Understanding Nonlinear Dynamics*, Springer, New York, 1995

[234] Kashchiev, D., Surface Science **14**, 209 (1969); **18**, 389 (1969)

[235] Kashchiev, D., Surface Science **18**, 293 (1969)

[236] Kashchiev, D., J. Chem. Phys. **76**, 5098 (1982)

[237] Kashchiev, D., Cryst. Res. Technol. **19**, 1413 (1984)

[238] Kashchiev, D., Firoozabadi, A., J. Chem. Phys. **98**, 4690 (1993)

[239] Kastaleyn, P. W., Fortuin, C. M., J. Phys. Soc. Japan (Suppl.) **26**, 11 (1969)

[240] Katz, J. L., Wiedersich, H., J. Colloid Interface Science **61**, 351 (1977)

[241] Katz, J. L., Spaepen, F., Phil. Mag. **B37**, 137 (1978)

[242] Katz, J. L., Donohue, M. D., Adv. Chem. Phys. **40**, 137 (1979)

[243] Katz, J. L., Pure Appl. Chem. **64**, 1661 (1992)

[244] Katz, J. L., Fisk, J. A., Rudek, M. M.: *Nucleation of Single Component Supersaturated Vapors*, In: Proc. 14-th International Conference on Nucleation and Atmospheric Aerosols, Helsinki, Finland, 26.–30. 8. 1996, p. 1; Eds. P. Wagner and M. Kulmala

[245] Kawasaki, K., Enomoto, Y., Physica **A 150**, 463 (1988)

[246] Keizer, J.: *Statistical Thermodynamics of Nonequilibrium Processes*, Springer, Berlin, 1983

[247] Kelton, K. F., Greer, A. L., Thompson, C. V., J. Chem. Phys. **72**, 6261 (1983)

[248] Kelton, K. F., Greer, A. L., J. Non–Cryst. Solids **79**, 295 (1986)

[249] Kelton, K. F., Greer, A. L., Phys. Rev. **B38**, 10089 (1988)

[250] Kerner, B. S., Konhäuser, P., Phys. Rev. E **48**, R2335 (1993)

[251] Kerner, B. S., Konhäuser, P., Phys. Rev. E **50**, 54 (1994)

[252] Kerner, B. S., Rehborn, H., Phys. Rev. E **53**, R1297 (1996)

[253] Kerner, B. S., Rehborn, H., Phys. Rev. E **53**, R4275 (1996)

[254] Kessler, D. A., Koplik, J., Levine, H., Adv. Phys. **37**, 255 (1988)

[255] Kiang, C. S., Phys. Rev. Lett. **24**, 47 (1970)

[256] Kirkaldy, J. S., Young, D. J.: *Diffusion in the Condensed State*, The Institute of Metals, London, 1987

[257] Kirkwood, J. G., J. Chem. Phys. **76**, 479 (1935)

[258] Kittel, C.: *Thermal Physics*, Wiley, New York, 1969

[259] Klimontovich, J. L.: *Statistical Physics*, Nauka, Moskau, 1982

[260] Kohn, W., Sham, L. J., Phys. Rev **A 140**, 1139 (1965)

[261] Kohyama, T.: J. Stat Phys. **63** 637 (1991)

[262] Kotomin, E. A., Kuzovkov, V. N., Chem. Phys. **76**, 479 (1983)

[263] Kotomin, E. A., Kuzovkov, V. N., Rep. Progr. Phys. **55**, 2079 (1992)

[264] Kotomin, E. A., Kuzovkov, V. N., Frank, W., Seeger, A., J. Phys. A: Math. Gen. **27**, 1453 (1994)

[265] Kozisek, Z. In: *Kinetic Phase Diagrams*, Ed. Z. Chvoj, J. Sestak, A. Triska, Elsevier Science Publishers, New York, 1991, p. 277

[266] Kozisek, Z., Demo, P., J. Crystal Growth **132**, 491 (1993)

[267] Kraeft, W.-D., Kremp, D., Ebeling, W., Röpke, G.: *Quantum Statistics of Charged Particle Systems*, Akademie–Verlag, Berlin and Plenum, New York, 1986

[268] Kramers, H. A., Physica **7**, 284 (1940)

[269] Kubo, R.: *Thermodynamics*, North–Holland–Publishing Company, Amsterdam, 1968

[270] Kühne, R.: *Fernstraßenverkehrsbeeinflußung und Physik der Phasenübergänge*, Phys. in unserer Zeit **15**, 84 (1984)

[271] Kühne, R.: *Verkehrsablauf auf Fernstraßen. Autofahren als Beispiel für nichtlineare Dynamik*, Phys. Bl. **47** (1991) 201

[272] Küpper, W. A., Wegmann, G., Hilf, E. R., Ann. Phys. **88**, 454 (1974)

[273] Kulmala, M., Wagner, P. E. (Eds.): *Nucleation and Atmospheric Aerosols 1996*, Proceedings of the Fourteenth International Conference on Nucleation and Atmospheric Aerosols, Helsinki, August 26 – 30, 1996

[274] Kuzovkov, V. N., Kotomin, E. A., J. Phys. **C 13**, L499 (1980)

[275] Kuzovkov, V. N., Kotomin, E. A., Phys. Stat. Sol. **b 105**, 789 (1981)

[276] Kotomin, E. A., Kuzovkov, V. N., Phys. Stat. Sol. **b 108**, 37 (1981)

[277] Kuzovkov, V. N., Kotomin, E. A., Chem. Phys. Lett. **87**, 575 (1982)

[278] Kuzovkov, V. N., Kotomin, E. A., Chem. Phys. **81**, 335 (1983)

[279] Kuzovkov, V. N., Kotomin, E. A., J. Phys. **C 17**, 2283 (1984)

[280] Kuzovkov, V. N., Kotomin, E. A., Chem. Phys. Lett. **117**, 266 (1985)

[281] Kuzovkov, V. N., Kotomin, E. A., Chem. Phys. **98**, 351 (1985)

[282] Kuzovkov, V. N., Kotomin, E. A., Rep. Progr. Phys. **51**, 1479 (1988)

[283] Kuzovkov, V. N., Kotomin, E. A., J. Stat. Phys. **72**, 127 (1993)

[284] Kuzovkov, V. N., Kotomin, E. A., J. Chem. Phys. **98**, 9107 (1993)

[285] Kuzovkov, V. N., Kotomin, E. A., Physica Scripta **50**, 720 (1994)

[286] Kuzovkov, V. N., Kotomin, E. A., Physica Scripta **47**, 585 (1993)

[287] Kuzovkov, V. N., Kotomin, E. A., von Niessen, W., (to be published)

[288] Labudde, D.: *Zur Bindungsenergie von Clustern – Thermodynamische Beschreibung und Simulationen*, Diplomarbeit, Universität Rostock, 1993

[289] Labudde, D.: *Clusterbildung in finiten und expandierenden Systemen*, Dissertation, Universität Rostock, 1997

[290] Labudde, D., Mahnke, R., Frischfeld, V., Computer Phys. Comm. **106**, 181 (1997)

[291] Lamb, D. O., Lattimer, J. M., Pethick, C. J., Ravenhall, D. G., Phys. Rev. Lett. **41**, 1623 (1978)

[292] Landau, L. D., Lifschitz, E. M.: *Mechanik* (Lehrbuch der theoretischen Physik, Bd. 1), Akademie–Verlag, Berlin, 10. Auflage, 1981

[293] Landau, L. D., Lifshitz, E. M.: *Mechanics of Continuous Media*, Nauka, Moscow, 1953 (in Russian)

[294] Landau, L. D., Lifschitz, E. M.: *Elektrodynamik der Kontinua*, Akademie–Verlag, Berlin 1967

[295] Landau, L. D., Lifschitz, E. M.: *Statistische Physik*, Akademie – Verlag, Berlin 1969

[296] Landau, L. D., Lifschitz, E. M.: *Elastizitätstheorie*, Akademie–Verlag, Berlin, 1976

[297] Landau, L.D., Lifschitz, E.M.: *Lehrbuch der theoretischen Physik, Band 10: Physikalische Kinetik*, Akademie – Verlag, Berlin, 1983

[298] Landolt–Börnstein: Numerical and Functional Relationships in Science and Technology, Comprehensive Index 1988; Ed.: Mandelung, O.; Springer, Berlin, 1988

[299] Langer, J. S., Ann. Phys. (NY) **41**, 108 (1967); **54**, 258 (1969)

[300] Langer, J. S., Bar–on, M., Miller, H.–D., Phys. Rev. **A 11**, 1417 (1975)

[301] Latora, V., Belkacem, M., Bonasera, A., Phys. Rev. Letters **73**, 1765 (1994)

[302] Lee, K. C., Phys. Rev. Lett., **73**, 2801 (1994)

[303] Leibfried, G.: In *Bestrahlungseffekte in Festkörpern*, Teubner, Stuttgart, 1965, p. 266

[304] Leutzbach, W.: *Einführung in die Theorie des Verkehrsflusses*, Springer, Berlin, 1972; *Introduction to the Theory of Traffic Flow*, Springer, Berlin, 1988

[305] Levich, V. G.: *Physico–Chemical Hydrodynamics*, Academy of Science Publishers, Moscow, 1952 (in Russian)

[306] Li, T. et al., Phys. Rev. Lett. **70**, 1924 (1993); Phys. Rev **C 49**, 1630 (1994)

[307] Lichtenberg, A. J., Lieberman, M. A.: *Regular and Stochastic Motion*, Springer, New York, First Edition, 1983; *Regular and Chaotic Dynamics*, Springer, New York, Second Edition, 1992

[308] Lifshitz, I. M., Slezov, V. V., J. Exper. Theor. Phys. (USSR) **35**, 479 (1958)

[309] Lifshitz, I. M., Slezov, V. V., Fiz. Tverd. Tela (USSR), **1**, 1401 (1959)

[310] Lifshitz, I. M., Slezov, V. V., J. Phys. Chem. Solids **19**, 35 (1961)

[311] Lindenberg, K., West, B.J., Kopelman, R., Phys. Rev. Lett. **60**, 1777 (1988)

[312] Lindner, A.: *Grundkurs Theoretische Physik*, Teubner Studienbücher Physik, B. G. Teubner, Stuttgart, 1994

[313] Lock, G. S. H.: *Latent Heat Transfer*, Oxford University Press, Oxford, 1996

[314] Lothe, J., Pound, G. M., J. Chem. Phys. **36**, 2080 (1962)

[315] Lothe, J., Pound, G. M., J. Chem. Phys. **45**, 630 (1966)

[316] Lubetkin, S. D.: *Bubble Nucleation and Growth*, In: *Controlled Particle, Droplet and Bubble Formation*, Ed. D. J. Wedlock, Butterworth – Heinemann Publ., Oxford, 1994

[317] Lubetkin, S. D.: *The Fundamentals of Bubble Evolution*, Chem. Soc. Reviews 243 (1995)

[318] Luding, S., Schnörer, H., Kuzovkov, V. N., Blumen, A., J. Stat. Phys. **65**, 1261 (1991)

[319] Ludwig, F.–P.: *Über die Kinetik der Phasenumwandlungen in kondensierten Systemen*, Diplomarbeit, Universität Rostock, 1993

[320] Ludwig, F.–P., Schmelzer, J., Bartels, J., J. Materials Science **29**, 4852 (1994)

[321] Ludwig, F.–P., Schmelzer, J., Milchev, A., Phase Transitions **48**, 237 (1994)

[322] Ludwig, F.–P., Schmelzer, J., J. Colloid Interface Science **181**, 503 (1996)

[323] Ma, Y. G., Shen, W. Q., Phys. Rev. **C 51**, 710 (1995)

[324] Mahnke, M.: PC Simulation Program VERKEHR, Rostock, 1997

[325] Mahnke, R., Schmelzer, J., Z. Phys. Chemie (Leipzig) **266**, 1028 (1985)

[326] Mahnke, R., Wiss. Z. Uni. Rostock **N 35**, 13 (1986)

[327] Mahnke, R., Budde, A.: Complexity of Patterns Generated by One–Dimensional Cellular Automata. In: *Selforganisation by Nonlinear Irreversible Processes* (Eds.: W. Ebeling, H. Ulbricht), Springer, Berlin, 1986

[328] Mahnke, R., Budde, A., Wiss. Z. Uni Rostock, **N36**, 50 (1987)

[329] Mahnke, R., Budde A., Wiss. Z. Uni. Rostock **N38** 71 (1989) and **N39** 111 (1990)

[330] Mahnke, R.: *Zur Evolution in nichtlinearen dynamischen Systemen*, Habilarbeit, Universität Rostock, 1990

[331] Mahnke, R., Budde, A.: Z. Phys. Chemie (Leipzig) **271**, 857 (1990)

[332] Mahnke, R., Urbschat, H., Budde, A., Z. Phys. **D 20**, 399 (1991)

[333] Mahnke, R., Budde, A.: Pattern Formation by Cellular Automata, Syst. Anal. Model. Simul. **10**, 133 (1992)

[334] Mahnke, R., Schmelzer, J., Röpke, G.: *Nichtlineare Phänomene und Selbstorganisation*, Teubner Studienbücher Physik, B. G. Teubner, Stuttgart, 1992

[335] Mahnke, R.: *Nichtlineare Physik in Aufgaben*, Teubner Studienbücher Physik, B. G. Teubner, Stuttgart, 1994

[336] Mahnke, R., Labudde, D., acta physica slovaca **44**, 275 (1994)

[337] Mahnke, R., Seemann, M.: Aggregation Phenomena in a Flow Channel. In: *Traffic and Granular Flow* (Eds.: D. E. Wolf, M. Schreckenberg, A. Bachem), World Scientific Publ., Singapore, 1996, see [570]

[338] Mahnke, R., Pieret, N., Phys. Rev. **E 56**, 2666 (1997)

[339] Mahnke, R., Kaupužs, J.: *One more Fundamental Diagram of Traffic Flow*, Presentation at Second Workshop on *Traffic and Granular Flow*, Univ. Duisburg, October 1997, see [496]

[340] Mahnke, R., Z. Phys. Chemie **204**, 85 (1998)

[341] Mahnke, R., Kaupužs, J., : *Stochastic Theory of Freeway Traffic*, submitted for publication, 1998

[342] Malchow, H., Schimansky–Geier, L.:*Noise and Diffusion in Bistable Nonequilibrium Systems*, Teubner–Texte zur Physik, Teubner, Leipzig, 1985

[343] Mandelbrot, B. B.: *Die fraktale Geometrie der Natur*, Akademie–Verlag, Berlin, 1987

[344] Mather, J. C. et al., Astrophys. J. **420**, 439 (1994)

[345] Mathews, G. J., Ward, R. A., Rep. Prog. Phys. **48**, 1371 (1985)

[346] Mazurek, T. J., Lattimer, J. M., Brown, G. E., Astrophys. J. **229**, 713 (1979)

[347] Meinel, R., Neugebauer, G., Steudel, H.: *Solitonen*, Akademie–Verlag, Berlin, 1991

[348] Migowsky, St., Wanschura, T., Rujan, P., Z. Phys. B **95**, 407 (1994)

[349] Mirold, P., Binder, K., Acta Met. **25**, 1435 (1977)

[350] Majumdar, S. N., Dhar, D., J. Phys. **A 24**, L357 (1991)

[351] Majumdar, S. N., Dhar, D., Physica **A 185**, 129 (1992)

[352] Manna, S. S., J. Stat. Phys. **59**, 509 (1990)

[353] Marder, M., Phys. Rev. **A36**, 858 (1987)

[354] Marro, J., Bortz, A. B., Kalos, M. H., Lebowitz, J. L., Phys. Rev. **B 12**, 2000 (1975)

[355] Marqusee, J., Ross, J., J. Chem. Phys. **79**, 373 (1983); **80**, 536 (1984)

[356] Matsuzaki, M., Takayasu, H., J. Geophys. Res. **96**, 19925 (1991)

[357] Mayer–Kuckuk, T.: *Kernphysik*, Teubner, Stuttgart, 1984

[358] Mazenko, G. F., Valls, O. T., Zannetti, M., Phys. Rev. **B 38**, 520 (1988)

[359] Mc Coy, B. M., Wu, T. T.: *The Two-Dimensional Ising Model*, Harvard University Press, Cambridge, Mass. 1973

[360] Mc Hardy, I., Czerny, B., Nature **325**, 696 (1987)

[361] Meakin, P.: *Fractal Aggregates*, Adv. Colloid Interface Science **28**, 249 (1988)

[362] Mehlig, B., Forrest, B. M., Z. Phys. **89**, 89 (1992)

[363] Meier, H., Strobl, G. R., Macromolecules **20**, 649 (1988)

[364] Mekijan, A. Z., Phys. Rev. Lett. **64**, 2125 (1990); Phys. Rev. **C 41**, 2103 (1990)

[365] Metropolis, N. Rosenbluth, A. W., Rosenbluth, M. N., Teller, A. H., Teller, E., J. Chem. Phys. **21**, 1087 (1953)

[366] Milchev, A., Binder, K., Surf. Sci. **164**, 1 (1985)

[367] Milchev, A., Heermann, D. W., Binder, K., Z. Phys. **B 63**, 521 (1986)

[368] Milchev, A., Heermann, D. W., Binder, K., J. Stat. Phys. **44**, 749 (1986)

[369] Milchev, A., Heermann, D. W., Binder, K., Acta Metall. **36**, 377 (1988)

[370] Milchev, A., Fraggis, Th., Pnevmatikos, St., Phys. Rev. **B 45**, 18 (1992)

[371] Milchev, A., Gerroff, I., Schmelzer, J., Z. Phys. **B 94**, 101 (1994)

[372] Möller, J., Schmelzer, J., Gutzow, I., Pascova, R., phys. stat. solidi **b180**, 315 (1993)

[373] Möller, J., Schmelzer, J., phys. stat. solidi **b180**, 331 (1993)

[374] Möller, J.: *Zur theoretischen Beschreibung von Phasenumwandlungen in elastischen und viskoelastischen Medien*, Dissertation, Universität Rostock, 1994

[375] Möller, J., Schmelzer, J., Avramov, I., phys. status solidi **b 196**, 49 (1996)

[376] Morawetz, K., Danielewicz, P., Ayik, S.: *Cluster Formation in Heavy Ion Reactions*, Preprint, 1995

[377] Mosner, W. K., Drossel, B., Schwabl, F., Physica **A 190**,205 (1992)

[378] Müller, W. F. J. et al.: *Boiling Nuclei*, GSI–Nachrichten 03–95, Eds. U. Grundinger, V. Metag, Darmstadt, 1995, see also: Möhlenkamp, T.: In: GSI Scientific Report, 1994, p. 46; U. Grundinger (Ed.), Darmstadt, 1994 and references cited therein

[379] Mutaftschiev, B.: *Nucleation Theory*. In: *Handbook of Crystal Growth*, T. D. J. Hurle (Ed.), Elsevier Science Publishers B. V., Amsterdam 1993, p. 187

[380] Nabarro, F. R. N., Proc. Royal Society (London) **A 175**, 519 (1940)

[381] Nagatani, T., Physica A **218**, 145 (1995)

[382] Nagatani, T. Physica A **223**, 137 (1996)

[383] Nagel, K., Phys. Rev. E **53**, 4655 (1996)

[384] Nagel, K., Paczuski, M., Phys. Rev. E **51** 2909 (1995)

[385] Nagel, K., Schreckenberg, M., J. Phys. I France **2**, 2221 (1992)

[386] Nagel, K., Herrmann, H. J., Physica **A 199**, 254 (1993)

[387] Nakanishi, H., Phys. Rev. **A 41**, 7086 (1990)

[388] Narsimhan, G., Ruckenstein, E., J. Colloid Interface Science **128**, 549 (1989)

[389] Nesterenko, V. O., Sov. J. Part. Nucl. **23**, 726 (1992)

[390] Nesterenko, V. O., Kleinig, W., Gudkov, V. V., Iudice, N. L., Kvasil, J., Phys. Rev. **A 56**, 607 (1997)

[391] von Neumann, J.: *Theory of Self–Reproducing Automata* (Ed.: A. W. Burks), University of Illinois Press, Urbana, 1966

[392] Newton, J. D.: *Nuclear Fission Induced by Heavy Ions*, Physics of Elementary Particles and Atomic Nuclei **21**, 821 (1990) (in Russian)

[393] Nicolis, G., Prigogine, I.: *Self–Organization in Non–Equilibrium Systems*, Wiley, New York, 1977

[394] Nowakowski, B., Ruckenstein, E., J. Chem. Phys. **94**, 1397 (1991)

[395] Ogilvie, C. A. et al., Phys. Rev. Letters **67**, 1214 (1991)

[396] O'Lami, Z., Feder, H. J. S., Christensen, K., Phys. Rev. Lett. **68**, 1244 (1992)

[397] Olemski, A. I., Koplyk, I. V., Uspekhi Fiz. Nauk **165**, 1105 (1995)

[398] Olson, T., Hamill, P., J. Chem. Phys. **104**, 210 (1996)

[399] Onsager, L., Phys. Rev. **65**, 117 (1944)

[400] Oono, Y., Puri, S., Phys. Rev. **A 38**, 434 (1989)

[401] Oshanin, G. S., Burlatsky, S. F., Ovchinnikov, A. A., Phys. Lett. A **139**, 245 (1989)

[402] Ostwald, W., Z. Phys. Chemie **22**, 282 (1897)

[403] Ostwald, W.: *Lehrbuch der Allgemeinen Chemie*, Engelmann, Leipzig, 1896 – 1901

[404] Ostwald, W.: *Analytische Chemie*, Leipzig, 1901

[405] Ovchinnikov, A. A., Zeldovich, Ya. B. Chem. Phys. **28**, 215 (1978)

[406] Ovchinnikov, A. A., Burlatzky, S. F., JETP Lett. **43**, 494 (1986)

[407] Ovchinnikov, A. A., Timashev, S. F., Belyi, A. A.: *Kinetics of Diffusion–Controlled Chemical Processes in Chemistry*, Nova Science Publishers, New York, 1988

[408] Oxtoby, D. W. et al., J. Chem. Phys. **89**, 7521 (1988); **94**, 4472 (1991); **103**, 3686 (1995)

[409] Oxtoby, D. W., Kashchiev, D., J. Chem. Phys. **100**, 7665 (1994)

[410] Packard, N. H., Wolfram, St., J. Stat. Phys. **38**, 901 (1985)

[411] Pan, J., das Gupta, S., Phys. Lett. **B 344**, 30 (1995); Phys. Rev. **C 51**, 1384 (1995)

[412] Panagiotou, A. D. et al., Phys. Rev. Lett. **52**, 496 (1984); Phys. Rev. **C 31**, 55 (1985)

[413] Papp, G., Nörenberg, W.: *The Path of Hot Nuclei Towards Multifragmentation*, GSI–Preprint 95–30

[414] Pascova, R., Gutzow, I., Glastechnische Berichte **56**, 324 (1983)

[415] Pascova, R., Gutzow, I., Tomov, I., J. Materials Sci. **25**, 913 (1990)

[416] Pascova, R., Gutzow, I., Schmelzer, J., J. Materials Sci. **25**, 921 (1990)

[417] Peitgen, H. O., Richter, P. H.: *The Beauty of Fractals*, Springer, Berlin, 1986

[418] Penrose, O., Lebowitz, L., Marro, J., Kalos, M. H., Sur, A., J. Stat. Phys. **19**, 243 (1978)

[419] Penrose, O., J. Stat. Phys. **89**, 305 (1997)

[420] Pethick, C. J., Ravenhall, D. G., Nucl. Phys. **A 471**, 19c (1987)

[421] Petrov, Ju. I.: *Physics of Small Particles*, Moscow, Nauka Publishers, Moscow, 1982 (in Russian)

[422] Phillips, A. C., Nucl. Phys. **A 107**, 209 (1968)

[423] Plessas, W. *et al.*, Few–Body Systems Suppl. **7**, 251 (1994); Mathelitsch, L., Plessas, W., Schweiger, W., Phys. Rev. **C 26**, 65 (1982); Haidenbauer, J., Plessas, W., Phys. Rev. **C 30**, 1822 (1984)

[424] Pochodzalla, J. et al., Phys. Rev. Lett. **75**, 1040 (1995)

[425] Pocsik, I., Z. Phys. **D 20**, 392 (1991)

[426] Poggi, G. et al., Nucl. Phys. **A 586**, 755 (1995)

[427] Polak, L. S.: *Non–Equilibrium Chemical Kinetics and its Applications*, Nauka, Moscow, 1982

[428] Porile, N. T. et al., Phys. Rev. **C 39**, 1914 (1989)

[429] Potts, R. B., Proc. Camb. Philos. Soc. **48**, 106 (1952)

[430] Priezzhev, V. B., J. Stat. Phys. **74**, 955 (1994)

[431] Priezzhev, V. B., Ktitarev, D. V., Ivashkevich, E. V., Phys. Rev. Lett. **76**, 2093 (1996)

[432] Prigogine, I., Herman, R.: *Kinematic Theory of Vehicular Traffic*, Elsevier, New York, 1971

[433] Privman, V. (Ed.): *Finite Size Scaling and Numerical Simulation of Statistical Systems*, World Scientific, Singapore, 1990

[434] Rademann, K., Kaiser, B., Ewen, U., Hensel, F., Phys. Rev. Lett. **59**, 2319 (1987)

[435] Rajewsky, N., Schreckenberg, M.: Physica **A 245** 139 (1997)

[436] Reinecke, T. L., Ying, S. C., Phys. Rev. Lett. **35**, 311 (1975)

[437] Reif, F.: *Fundamentals of Statistical and Thermal Physics*; Mc Graw Hill Book Company, New York, 1965

[438] Reiss, H., J. Chem. Phys. **18**, 840 (1950)

[439] Reiss, H. et al., J. Chem. Phys. **95**, 9209 (1991); **97**, 5766 (1992); **99**, 5374 (1993)

[440] Reynolds, P. J. (Ed.): *On Clusters and Clustering*, North–Holland, Amsterdam, 1993

[441] Rickert, M., Nagel, K., Schreckenberg, M., Latour, A., Physica **A 231**, 534 (1996)

[442] Riethmüller, T., Schimansky-Geier, L.,. Rosenkranz, D., Pöschel, Th., J. Stat. Phys. **86**, 421 (1997)

[443] Risken, H.: *The Fokker-Planck-Equation*, Springer, Berlin, 1984 (2. Aufl. 1989)

[444] Röpke, G., Wiss. Z. Univ. Rostock **N 31**, 61 (1982); **N 32**, 30 (1983); **N 33**, 33 (1984)

[445] Röpke, G., phys. stat. sol. **(b) 57**, 571 (1973); El-Mamoun, E., Röpke, G., phys. stat. sol. **(b) 82**, 617 (1977)

[446] Röpke, G., Seifert, T., Stolz, H., Zimmermann, R., phys. stat. sol. **(b) 100**, 215 (1980)

[447] Röpke, G., Blaschke, D., Schulz, H., Astrophys. Space Sci. **95**, 417 (1983)

[448] Röpke, G., Münchow, L., Schulz, H., Nucl. Phys. **A 379**, 536 (1982)

[449] Röpke, G., Münchow, L., Schulz, H., Phys. Lett. **B 110**, 21 (1982)

[450] Röpke, G., Blaschke, D., Schulz, H., Astrophys. Space Sci. **95**, 417 (1983)

[451] Röpke, G., Schmidt, M., Münchow, L., Schulz, H., Nucl. Phys. **A 399**, 587 (1983)

[452] Röpke, G.:*Statistische Mechanik des Nichtgleichgewichts*, Deutscher Verlag der Wissenschaften, Berlin, 1987

[453] Röpke, G., Phys. Lett. **B 185**, 281 (1987)

[454] Röpke, G., Phys. Rev. **A 38**, 3001 (1988)

[455] Röpke, G., Schulz, H., Nucl. Phys. **A 477**, 472 (1988)

[456] Röpke, G., Ann. Physik (Leipzig) **3**, 145 (1994)

[457] Rothman, D. H., Zaleski, St., Rev. Mod. Phys. **66**, 1417 (1994)

[458] Rowlinson, J. S.: Translation of J. D. van der Waals' *The Thermodynamic Theory of Capillarity under the Hypothesis of Continuous Variation of Density*, J. Stat. Phys. **20**, 197 (1979)

[459] Rowlinson, J. S., Widom, B.: *Molecular Theory of Capillarity*, Clarendon Press, Oxford 1982

[460] Ruckenstein, E., Nowakowski, B., J. Chem. Phys. **94**, 1397 (1991)

[461] Rundle, J. B., Jackson, D. D., Bull. Seism. Soc. Am. **67**, 1363 (1977)

[462] Sadiq, A., Binder, K., J. Stat. Phys. **35**, 617 (1984)

[463] Sandhas, W., Acta Physica Austriaca, Suppl. **IX**, 57 (1972)

[464] Sasvari, M., Kertesz, J., Phys. Rev. **E 56**, 4104 (1997)

[465] Scheck, F.: *Mechanik. Von den Newtonschen Gesetzen zum deterministischen Chaos*, Springer, Berlin, 1988

[466] Schindler, T., Berg, C., Niedner-Schatteburg, G., Bonybey, V.E., Chem. Phys. Lett. **229**, 57 (1994)

[467] Schmelzer, J., Z. Phys. Chemie (Leipzig) **266**, 1057 (1985)

[468] Schmelzer, J.: *Thermodynamik finiter Systeme und die Kinetik von thermodynamischen Phasenübergängen 1. Art*, Dissertation B, Universität Rostock, 1985

[469] Schmelzer, J., Mahnke, R., J. Chem. Soc., Faraday Trans. I, **82**, 1413 (1986)

[470] Schmelzer, J., J. Chem. Soc., Faraday Trans. I, **82**, 1421 (1986)

[471] Schmelzer, J., Ulbricht, H., J. Colloid Interface Sci. **117**, 325 (1987); **128**, 104 (1989)

[472] Schmelzer, J., Gutzow, I., Z. phys. Chemie (Leipzig) **269**, 753 (1988)

[473] Schmelzer, J., phys. stat. sol. **b 161**, 173 (1990)

[474] Schmelzer, J., Gutzow, I., Pascova, R., J. Crystal Growth **104**, 505 (1990)

[475] Schmelzer, J., Pascova, R., Gutzow, I., phys. stat. sol. **a 117**, 363 (1990)

[476] Schmelzer, J., Milchev, A., Phys. Letters **A 158**, 307 (1991)

[477] Schmelzer, J., Möller, J., J. Phase Transitions **38**, 261 (1992)

[478] Schmelzer, J., Ulbricht, H., Mahnke, R.: *Aufgabensammlung zur klassischen theoretischen Physik*, Aula–Verlag, Wiesbaden, 1994

[479] Schmelzer, J., Möller, J., Slezov, V. V., J. Phys. Chem. Solids **56**, 1013 (1995)

[480] Schmelzer, J., Slezov, V. V., Milchev, A., Phase Transitions **54**, 193 (1995)

[481] Schmelzer, J., Gutzow, I., Schmelzer, J. (jn.), J. Colloid Interface Science **178**, 657 (1996)

[482] Schmelzer, J., Röpke, G., Ludwig, F.–P., Phys. Rev. **C 55**, 1917 (1997)

[483] Schmelzer, J., Röpke, G., Schmelzer, J. (jn.), Slezov, V. V.: *Shapes of Cluster Size Distributions Evolving in Nucleation – Growth Processes*, submitted for publication

[484] Schmelzer, J., Röpke, G., Labudde, D.: *Nucleation and Cluster Growth in Freely Expanding Gases*, Physica (1998), in press

[485] Schmelzer, J., Röpke, G.: *Nuclear Multifragmentation and Nucleation Theory*, In: Proceedings of the VIth International Conference on Heavy Ion Physics, Dubna, Russia, 22. – 27. 9. 1997, World Scientific Publ., Singapore, 1998

[486] Schmelzer, J., Vasilevskaya, T. N., Andreev, N. S.: *On the Initial Stages of Spinodal Decomposition*, submitted for publication

[487] Schmelzer, J. (jn.), Röpke, F.: *Nuclear Fragmentation and Dissipation: A Qualitative Model*, Joint Institute of Nuclear Research, Dubna, Russia, September, 1995 (unpublished)

[488] Schmidt, M, Röpke, G., Schulz, H., Ann. Phys. (NY) **202**, 57 (1990)

[489] Schmidt, R., Lutz, H. O., Dreizler, R. (Eds.): *Nuclear Physics Concepts in the Study of Atomic Clusters*, Lecture Notes in Physics **404**, Springer, Berlin, 1991

[490] Schnell, A., Röpke, G., Lombardo, U., Schulze, H.–J., Phys. Rev. **C 57**, 806 (1998)

[491] Schnörer, H., Kuzovkov, V. N., Blumen, A., Phys. Rev. Lett. **63**, 805 (1989)

[492] Schnörer, H., Kuzovkov, V. N., Blumen, A., J. Chem. Phys. **92**, 2310 (1990)

[493] Schnörer, H., Sokolov, I., Blumen, A., Phys. Rev. **A 42**, 7075 (1990)

[494] Schpolski, E. W.: *Atomphysik*, Deutscher Verlag der Wissenschaften, Berlin, 1985

[495] Schreckenberg, M., Schadschneider, A., Nagel, K., Ito, N., Phys. Rev. **E 51** 2939 (1995)

[496] Schreckenberg, M., Wolf, D. E. (Eds.): *Traffic and Granular Flow '97*, Springer, Singapore, 1998

[497] Schuck, P., Z. Phys. **241**, 395 (1971); Schuck, P., Villars, F., Ring, P., Nucl. Phys. **A 208**, 302 (1973); Dukelsky, J., Schuck, P., Nucl. Phys. **A 512**, 466 (1990); Schuck, P., Nucl. Phys. **A 567**, 78 (1994)

[498] Schulz, H., Münchow, L., Röpke, G., Schmidt, M., Phys. Lett. **B 119**, 12 (1982)

[499] Schulz, H., Röpke, G., Toneev, V. D., Gudima, K. K., Phys. Rev. **C 33**, 1095 (1986)

[500] Schuster, H. G.: *Deterministic Chaos. An Introduction*, VCH Verlagsgesellschaft, Weinheim, 1989

[501] Schwarz, P. *et al.*, Nucl. Phys. **A 398**, 1 (1983)

[502] Schweitzer, F., Schimansky–Geier, L., J. Colloid Interface Sci. **119**, 67 (1987)

[503] Schweitzer, F., Schimansky–Geier, L., Ebeling, W., Ulbricht, H., Physica **A 150**, 261 (1988); **A 153**, 573 (1988)

[504] Schweitzer, F. (Ed.): *Self–Organisation Of Complex Structures. From Individual to Collective Dynamics*, Gordon and Breach Sci. Publ., Amsterdam, 1997

[505] Seifert, B.: *Zur theoretischen Beschreibung von Fragmentationsprozessen*, Belegarbeit zum Forschungspraktikum am Vereinigten Institut für Kernforschung Dubna, Rußland, 1997

[506] Singhal, S. P., Herman, H., Kostorz, G., J. Appl. Crystallogr. **11**, 572 (1978)

[507] Skripov, V. P., Skripov, A. V., Uspekhi Fiz. Nauk **128**, 193 (1979)

[508] Slezov, V. V., Sagalovich, V. V., Uspekhi Fiz. Nauk **151**, 67 (1987)

[509] Slezov, V. V., Schmelzer, J., Möller, J., J. Crystal Growth **132**, 419 (1993)

[510] Slezov, V. V., Schmelzer, J., J. Phys. Chem. Solids **55**, 243 (1994)

[511] Slezov, V. V., Schmelzer, J., Tkatch, Ya. Y., J. Chem. Phys. **105**, 8340 (1996)

[512] Slezov, V. V., Fiz. Tverdogo Tela **37**, 2879 (1995) (in Russian)

[513] Slezov, V. V., J. Phys. Chem. Solids **58**, 455 (1997)

[514] Slezov, V. V., Schmelzer, J., Tkatch, Ya. Y., J. Materials Science **32**, 3739 (1997)

[515] Slezov, V. V., Phys. Rep. **288**, 389 (1997)

[516] Slezov, V. V., Schmelzer, J.: *Kinetics of First–Order Phase Transitions: The First Stages*, submitted for publication

[517] Slezov, V. V., Schmelzer, J.: *Comments on Nucleation Theory*, submitted for publication

[518] Smirnov, B. M., Uspekhi Fiz. Nauk **164**, 665 (1994)

[519] von Smoluchowski, M., Phys. Z. **17**, 557, 585 (1916)

[520] von Smoluchowski, M., Z. Phys. Chem. **92**, 129 (1917)

[521] Sokal, A. D., In: *Quantum Fields on the Computer*, Ed. M. Creutz, World Scientific, Singapore, 1992, p. 211

[522] Sokolov, I. M., JETP Lett. **44**, 53 (1986)

[523] Sokolov, I. M., JETP **94**, 199 (1988)

[524] Sokolov, I. M., Blumen, A., Europhys. Lett. **21**, 855 (1993)

[525] Sokolov, I. M., Argyrakis, P., Blumen, A., Fractals **1**, 470 (1993)

[526] Sokolov, I. M., Argyrakis, P., Blumen, A., J. Phys. Chem. **98**, 7256 (1994)

[527] Special issue of J. Stat. Phys. **65**, 5/6 (1991)

[528] Stanley, H. E.: *An Introduction to Phase Transitions and Critical Phenomena*, Oxford University Press, Oxford, 1971

[529] Stauffer, D.: *Introduction into Percolation Theory*, Taylor & Francis, London, 1985

[530] Stauffer, D., Stanlay, H. E.: *From Newton to Mandelbrot. A Primer in Theoretical Physics*, Springer, Berlin, 1990

[531] Stephenson, G. B., J. Non–Crystalline Solids **66**, 393 (1984); Scr. Metall. **20**, 465 (1986); Acta Metall. **36**, 2663 (1988)

[532] Stranski, I. N., Totomanov, D., Z. Phys. Chemie **A 163**, 399 (1933)

[533] Sugano, S.: *Microcluster Physics*, Springer, Berlin, 1991

[534] Szilard, L., Z. Phys. **32**, (1929)

[535] Thompson, J. M. T., H. B. Stewart, H. B.: *Nonlinear Dynamics and Chaos*, Wiley, New York, 1986

[536] Tiggesbäumker, J., Köller, L., Meiwes–Broer, K.–H., Liebsch, A., Phys. Rev. **A 48**, R1749 (1993)

[537] Tokuyama, M., Kawasaki, K., Physica **A 123**, 386 (1984)

[538] Toneev, V. D., Schulz, H., Gudima, K. K., Röpke, G., Sov. J. Part. Nucl. Phys. **17**, 1093 (1986); **21**, 364 (1990) (in Russian)

[539] Toral, R., Chakrabarti, A., Phys. Rev. **B 42**, 2445 (1990)

[540] Toussaint, D., Wilczek, F. J., J. Chem. Phys. **78**, 2624 (1983)

[541] Trautmann, W., Milkau, U., Lynen, U., Pochodzalla, J., Z. Phys. **A 344**, 447 (1993)

[542] Treiterer, J. et al.: *Investigation of Traffic Dynamics by Aerial Photogrammetric Techniques*, Interim Report EEs 278–3, Ohio State University, Columbus, 1970

[543] Trinkaus, H., Yoo, M. H., Philosophical Magazin **A55**, 269 (1987)

[544] Ulbricht, H., Schmelzer, J., Mahnke, R., Schweitzer, F.: *Thermodynamics of Finite Systems and the Kinetics of First–Order Phase Transitions*, Teubner Texte zur Physik, Bd. 17, Teubner, Leipzig, 1988

[545] Ulbricht, H., Schweitzer, F., Mahnke, R.: Nucleation Theory and Dynamics of First–Order Phase Transitions in Finite Systems. In: *Selforganisation by Nonlinear Irreversible Processes* (Eds.: W. Ebeling, H. Ulbricht), Springer, Berlin, 1986

[546] Unger, C., Klein, W., Phys. Rev. **B 29**, 2698 (1984)

[547] Vasilevskaya, T. N., Andreev, N. S., Glass Physics and Chemistry **22**, 510 (1996)

[548] van der Waals, J. D., Thesis, Leiden, 1873

[549] van der Waals, J. D.: *Die Kontinuität des gasförmigen und flüssigen Zustandes*, 2. Auflage, Leipzig 1899 – 1900

[550] van Kampen, N. G.: *Stochastic Processes in Physics and Chemistry*, North–Holland, Amsterdam, 1981

[551] Vautherin, D., Brink, D. M., Phys. Rev. **C 5**, 626 (1972)

[552] Vitukhnovsky, A. G., Pyttel, B. L., Sokolov, I. M., Phys. Lett. **128**, 161 (1988)

[553] Velasco, E., Toxvaerd, S., Phys. Rev. Lett. **71**, 388 (1993)

[554] Vilar, L. C. Q., de Souza, A. M. C., Physica **A 211**, 84 (1994)

[555] Vollmayr, K., Reger, J., Scheucher, M., Binder, K., Z. Phys. **B 91**, 113 (1993)

[556] Volmer, M., Weber, A., Z. Phys. Chemie **119**, 227 (1926)

[557] Volmer, M.: *Kinetik der Phasenbildung*, Th. Steinkopff, Dresden 1939

[558] Voorhees, P. W., Glicksman, M. E., Acta Met. **32**, 2001, 2013 (1984)

[559] Voorhees, P. W., J. Stat. Phys. **38**, 231 (1985)

[560] Wagner, C., Z. Elektrochemie **65**, 581 (1961)

[561] Wagoner, R. V., In: *Physical cosmology* (Eds.: R. Balian et al.), North Holland, Amsterdam, 1980, p. 397

[562] Waite, T. R., Phys. Rev. **107**, 463 (1957)

[563] Waite, T. R., Phys. Rev. **107**, 471 (1957)

[564] Waite, T. R., J. Chem. Phys. **28**, 103 (1958)

[565] Wehner, M. F., Wolfer, W. G., Phil. Magazine **A 52**, 189 (1985)

[566] von Weimarn, P. P., Chem. Rev. **2**, 217 (1926)

[567] Weinberg, S.: *Dreams of a Final Theory*, Hutchinson Radius Publishers, London, 1993

[568] Whitham, G. B.: *Linear and Nonlinear Waves*, John Wiley & Sons, New York, 1974

[569] Wissel, C.: *Theoretische Ökologie. Eine Einführung*, Springer, Berlin, 1989

[570] Wolf, D. E., Schreckenberg, M., Bachem, A. (Eds.): *Traffic and Granular Flow*, World Scientific Publ., Singapore, 1996

[571] Wolfram, St.: *Theory and Applications of Cellular Automata*, World Scientific Publ., Singapore, 1986

[572] Wörner, A.: *Kritisches Verhalten in der Kernfragmentation*, Dissertation, Johann Wolfgang Goethe Universität, Frankfurt am Main, 1995

[573] Wu, F. Y., Rev. Mod. Phys. **54**, 235(1982); **55**, 315(E) (1983)

[574] Yang, J., Turner, M. S., Steigman, G., Schramm, D. N., Olive, K. A., Astrophys. J. **281**, 493 (1984)

[575] Yannouleas, C., Broglia, R. A., Phys. Rev. **A 44**, 5793 (1991)

[576] Yi–Cheng, Z., Phys. Rev. Lett. **59**, 1725 (1987)

[577] Yourgreau, W., van der Merwe, A., Raw, G.: *Treatise on Irreversible and Statistical Thermophysics*, New York – London, 1966

[578] Yunson Du, Hao Li, Kadanoff, L. P., Phys. Rev. Lett. **74**, 8 (1995)

[579] Zeldovich, Ya. B., Sov. Phys. JETP **12**, 525 (1942); Acta Physicochim. USSR **18**, 1 (1943)

[580] Zeldovich, Ya. B., Sov. Electrochem. **13**, 677 (1977)

[581] Zeldovich, Ya. B., Ovchinnikov, A. A., JETP Lett. **26**, 588 (1977)

[582] Zeldovich, Ya. B., Ovchinnikov, A. A., JETP **74**, 1588 (1978)

[583] Zener, C., J. Appl. Phys. **20**, 950 (1949)

[584] Zettlemoyer, A. C. (Ed.): *Nucleation*, Marcel Decker, New York, 1969

[585] Zettlemoyer, A. C. (Ed.): *Nucleation Phenomena*, Adv. Colloid Interface Science **7** (1977)

[586] Zhabotinsky, A. M.: *Concentration Auto–Oscillations*, Nauka, Moscow, 1974

[587] Zhang, Y. C., Phys. Rev. Lett. **63**, 470 (1989)

[588] Ziabicki, A., J. Chem. Phys. **48**, 4368 (1968)

[589] Ziman, J. M.: *Models of Disorder*, Cambridge University Press, Cambridge, 1979

[590] Zittarz, J., Phys. Rev. **154**, 529 (1967)

[591] Zubarev, D. N.: *Nonequilibrium Statistical Thermodynamics*, Plenum, New York 1974

[592] Zubarev, D. N., Morozov, V., Röpke, G.: *Statistical Mechanics of Non–Equilibrium Processes*, Vol. 1: Basic Concepts, Kinetic Theory; Vol. 2: Relaxation and Hydrodynamic Processes, Akademie - Verlag Berlin, VCH–Wiley Publishers Group, 1996, 1997

[593] Applegate, J. H., Phys. Rep. **163**, 141 (1988); Rauscher, T. et al., Astrophys. J. **429**, 499 (1994)

[594] Barnett, R. M. et al., Phys. Rev. **D 54**, 1 (1996)

[595] Kahana, D. E. et al., Phys. Rev. **C 54**, 338 (1996)

[596] Röpke, G., Phys. Lett. **B 121**, 223 (1983)

Authors Index

Subject Index